Proceedings of the International Workshop on

Applications of Neural Networks to Telecommunications 2

Edited by

Joshua Alspector
Bell Communications Research

Rodney Goodman
California Institute of Technology

Timothy X Brown
Bell Communications Research

THE INTERNATIONAL NEURAL NETWORKS SOCIETY SERIES

Harold Szu, Editor

Alspector/Goodman/Brown • *Proceedings of the International Workshop on Applications of Neural Networks to Telecommunications* (1993)

Alspector/Goodman/Brown • *Proceedings of the International Workshop on Applications of Neural Networks to Telecommunications, Vol. 2* (1995)

Freeman • *Societies of Brains: A Study in the Neuroscience of Love and Hate* (1995)

Lendaris/Grossberg/Kosko • *World Congress on Neural Networks: Proceedings of the 1993 INNS Annual Meeting*

Levine/Elsberry • *Optimality in Biological and Artificial Networks* (in prep.)

Pribram • *Rethinking Neural Networks: Quantum Fields and Biological Data* (1993)

Pribram • *Origins: Brain and Self Organization* (1994)

Sobajic • *Neural Network Computing for the Electric Power Industry* (1993)

Proceedings of the International Workshop on

Applications of Neural Networks to Telecommunications

Edited by

Joshua Alspector
Bell Communications Research

Rodney Goodman
California Institute of Technology

Timothy X Brown
Bell Communications Research

LAWRENCE ERLBAUM ASSOCIATES, PUBLISHERS
1995 Hillsdale, New Jersey Hove, UK

Lawrence Erlbaum Associates, Inc., Publishers
365 Broadway
Hillsdale, New Jersey 07642

ISBN 0-8058-2084-1

Books published by Lawrence Erlbaum Associates are printed
on acid-free paper, and their bindings are chosen
for strength and durability.

Printed in the United States of America

10 9 8 7 6 5 4 3 2 1

Table of Contents:

PREFACE:

Welcome to the second International Workshop on the Application of Neural Network to Telecommunications (IWANNT*95). The first IWANNT was held in Princeton, NJ, USA on Oct. 21–23, 1993 close to the heart of the American telecommunications industry which includes AT&T and Bellcore. This meeting is being held about a year and a half later on May 22-24, 1995 in Stockholm, Sweden, the home of another large telecommunications company, Ericsson.

This workshop, like the last, is motivated by needs in the rapidly changing telecommunications industry. The last workshop had an electronic version of the proceedings which was accessible to interested people on the internet. It is hard to believe how unique that experiment seemed that short time ago. Since then, the enormous growth of the internet and the availability of visual network browsers like Mosaic have made remote information access seem commonplace. We have seen further evidence of the revolution in the industry from a structure where telecommunications networks, equipment, and services are controlled by large vertically integrated companies like pre-divestiture AT&T to many specialized horizontally structured firms. Traditional monopoly telecommunications is evolving into a diverse and dynamic information infrastructure. There will be firms specializing as local access network providers (cable, copper, wireless) while others will provide long distance networks (fiber, microwave, satellite). Still others will provide network equipment (switches, routers, computers) and terminal or user equipment (handsets, PDA's, screen phones, laptops, set-top boxes, multi-media terminals). Added to the above mix, which was typically the province of a large vertically integrated and regulated monopoly telephone company, will be companies not traditionally thought of as telecommunications providers. These include information providers such as entertainment companies (movie studios, television networks, music studios, video games), news sources (wire services, newspapers, magazines), and other information sources (libraries, financial data, weather). There will also be firms specializing in software to control the networks and enable information service applications such as movies-on-demand. Perhaps most significant, the convergence of computers and communications has blurred these industries till they are nearly indistinct.

In this environment, centralized control of the information infrastructure (as was done by a monopoly provider) will neither be advisable or even possible. The answer will be a form of adaptive or "market-based" management. Each of the players in this diverse marketplace will adopt an ecological niche defined by their abilities and the way they can adapt to continually changing conditions. Adaptive and intelligent systems which make use of tools such as neural networks will give the players in this fast changing world the flexibility necessary for survival.

There are approximately 50 papers in this proceedings. They represent neural network approaches to a variety of needs. There is the need to adapt to changing channel conditions (adaptive non-linear equalization) and the growing importance of wireless access. The convergence of digital services in asynchronous transfer mode (ATM) packet networks requires new methods of access control and neural networks promise a unique approach. There are also papers on applications in switching and routing. There will be a new class of applications in information filtering, data compression, and speech and pattern recognition and understanding which will make use of neural network technology. The distributed nature of networks will stimulate research in network management and decision tools, fault prediction and error recovery, and new methods of managing the large software systems that will be necessary. These proceedings show how neural network technology contributes a new tool to tackling the challenging problems of the information age.

The papers in this workshop were brought together with the generous support of IEEE Communications Society, Ericsson, Bell Communications Research, Swedish Neural Network Society, Japanese Neural Network Society, and the International Neural Network Society. We also thank the other members of the organizing committee for their time and effort: Rodney Goodman, Timothy Brown, Anthony Jayakumar, Miklos Boda, Bengt Asker, Atul Chhabra, C. Lee Giles, Harald Brandt, Tzi-Dar Chiueh, Fogelman Soulie, Michael Gell, Lawrence Jackel, Thomas John, Adam Kowalczyk, S-Y Kung, Tadashi Sone, Bernard Widrow, Steve Weinstein, and Betty Greer

Josh Alspector
General Chair
February, 1995

Analysis of Recurrent Neural Networks as Digital Communication Channel Equalizers

M.J.Bradley and P.Mars

School of Engineering, University of Durham, South Road, Durham DH1 3LE, UK.
m.j.bradley@durham.ac.uk
philip.mars@durham.ac.uk

Abstract

The method by which simple recurrent neural networks perform equalization is highlighted and compared to the operation of the Decision Feedback Equalizer. The design of multinode networks which use quantizing node nonlinearities is addressed and uses nodes to provide a soft-decision capability. The performance of the resulting structures are compared with both conventional and other neural network based equalizers and are seen to have a performance improvement only when zero decision delay is permitted.

1 Introduction

The field of adaptive equalization has seen a renewed interest in the application of nonlinear structures over the past few years. This is partially due to the availability of efficient processing hardware which has made practical implementations of such algorithms viable. The computational requirements of a Maximum Likelihood Sequence Estimator (MLSE) operating on channels with impulse responses spanning many symbols, such as H.F. channels, still present a significant burden for this optimum technique. Equalizer structures which have improved bit error rate performance over conventional Decision Feedback equalizers but allow computational savings over MLSE techniques provide an attractive compromise.

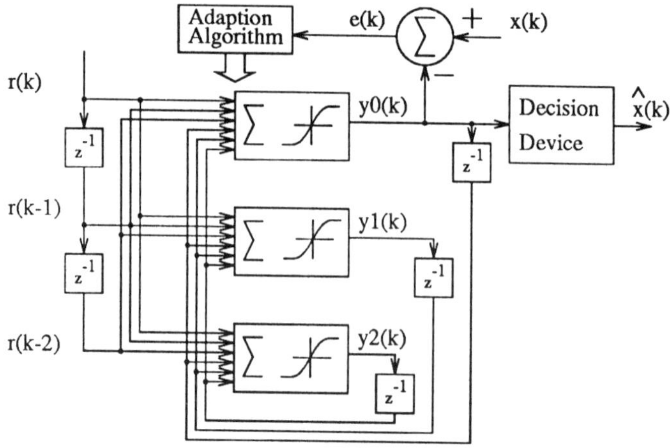

Figure 1: Structure of a typical RNN equalizer

The motivation for applying neural networks to equalization of digital communication channels lies in their pattern classification abilities. This contrasts with the more traditional method of treating equalization as an inverse filtering problem rather than a classification problem. Neural network equalizers generally operate on a symbol-by-symbol decision basis rather than on a symbol sequence basis as in the MLSE and its derivatives. The Multilayer Perceptron (MLP) [1] divides received channel states, formed from sets of received samples, with a nonlinear boundary into partitions and a decision device assigns each partition to a particular symbol. The Radial Basis Function equalizer (RBF) [2] operates similarily to the extent of partitioning the received channel states but uses a knowledge of the assumed additive channel noise probability density function to generate a maximum *a posteriori* decision procedure. The RBF equalizer is a realization of the optimal symbol-by-symbol decision equalizer but has a significant computational overhead. The Recurrent Neural Network equalizer (RNN) [3] has recently been proposed as an alternative structure with a low computational cost and the performance compared with linear equalizers.

This paper examines the mechanism by which simple RNN equalizers achieve equalization and identifies the node nonlinearity as a limiting factor. Improved RNN structures are formulated and compared with the RBF equalizer and the conventional Decision Feedback Equalizer (DFE).

The network (Fig. 1) consists of a small number of fully interconnected nodes with delayed samples of the received signal as network inputs. Each node forms a weighted sum of its inputs and passes this activation signal through a hyperbolic tangent nonlinearity. One node output is designated as the network output and is classified by a decision device which, for PAM data, is a slicer. The network estimates the symbol amplitude and the network weights are adjusted by a gradient descent based algorithm (RTRL) [4] using the error signal obtained from the desired response and the network output.

2 Single node RNN equalizers

Initial simulations performed by the authors showed that, whilst equalizing channels with two symbol spaced paths and using a binary PAM transmission scheme, the RNN network reduced to a single node structure. This network is essentially an IIR filter with a single feedback tap and a nonlinearity appearing after the summing junction. The step response of this structure can exhibit a variety of dynamics, dependent on the network weights, including rapid convergence to an equilibrium state and limit cycles.

For this simple structure a theoretical design of the network weights is feasible given a two path channel [5]. The design criteria are that the node has a stable response to a constant input (equivalent to a repeatedly transmitted symbol) and that the network outputs a symbol estimate at each iteration. The theoretical weight values depend upon the node nonlinearity, the channel model taps and the output levels of the node corresponding to the 2 symbols of the binary PAM transmission scheme. These weight settings have been observed in networks trained using the RTRL algorithm indicating an equivalence between the MSE criterion of this algorithm and the design criteria.

The node nonlinearity limits the performance of the network in a number of ways. Firstly, certain nonminimum phase channels require weight settings that result in limit cycles. Secondly, weight settings cannot be defined for channels with longer impulse responses unless variations in the desired node output levels are tolerated. Thirdly, noise present in the received samples causes variations in the node output which is fed back to the node input and adversely affects the subsequent symbol estimate. The effects of all 3 problems can be reduced by increasing the desired node output levels representing the two symbols towards the ± 1 limits of the node nonlinearity. Thus the nonlinearity is in effect a quantizer or slicer and the resulting structure is precisely that of a conventional DFE with a single feedback tap. Identifying this similarity between the single node RNN and the DFE, the theorectical weight settings of the RNN are seen to effectively cancel the post-cursors of the channel impulse response.

3 Multinode RNN equalizers

Multinode RNN equalizers have a large number of feedback paths creating complex dynamics and the design of such networks may be simplified by specifying a desired response for each node. The motivation for this is the observation that, whilst equalizing channels with longer impulse responses, some nodes are used as symbol spaced delays and the structure is again similar to that of a DFE. In particular each node can operate with a different decision delay so that at time $t = k$, node (i) estimates the symbol transmitted at time $t = (k - i)$. This allows cancellation of more than one channel post-cursor and the network has the capability of soft-decision feedback, i.e. a symbol estimated by node (i) at time $t = k$ may be updated at time $t = (k + k_1)$ by node $(i + k_1)$. Due to the dispersion of the transmitted symbol energy by the channel, delaying a hard decision on the symbol until more symbol energy has been received will generally increase the quality of the decision. This can reduce the effects of incorrect decision feedback which degrades the performance of the conventional DFE.

Two differrent RNN structures may be formulated both of which replace the node hyperbolic tangent nonlinearity with a quantizer. The first, RNN1, uses purely post-cursor cancellation of the combined impulse response of the channel and FIR filter formed by each node from the received sample delay line. This structure is shown in Figure 2 where $r(k)$ is the received sample and $\widehat{x}(k)$ is the symbol estimate at time $t = k$. Connections to a node from the received sample delay line only exist for samples containing energy from the transmitted symbol that the node is estimating. The second RNN structure, RNN2, uses both post-cursor and pre-cursor cancellation and is a more fully recurrent network. Pre-cursor cancellation is only applicable for nodes with decision delays of 2 or more symbol periods and uses the output of nodes with low decision delay to cancel the intersymbol interference (ISI) present in received samples.

The received sample input to node 0 has $(n_c - 1)$ interfering symbol terms for a channel impulse response length of n_c symbol periods. Thus the network requires $(n_c - 1)$ nodes for complete cancellation of these terms. However, this does not provide a sufficient number of symbol estimates for the cancellation of ISI present at the input of the remainder of the nodes. An additional delay line, fed by the ouput of node $(n_c - 2)$, provides these symbol estimates and results in this node having the same structure as a DFE.

The network may be adapted using the LMS algorithm in conjunction with a training signal. The algorithm is essentially similar to the LMS algorithm applied to the DFE with the error signal being formed from the training signal and the node activation rather than the node output due to the undifferentiable node nonlinearity.

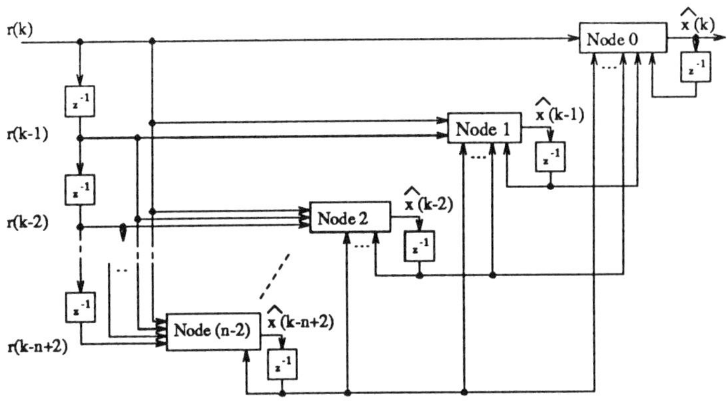

Figure 2: Structure of RNN1 equalizer

4 Performance simulation

The performance of the 2 RNN structures were compared against the DFE and RBF equalizers by simulating the bit error rates (BER) as a function of signal-to-noise ratio for 2 stationary channels, $H_1(z) = 0.4082 + 0.8164z^{-1} + 0.4082z^{-2}$ and $H_2(z) = 0.2041 + 0.4082z^{-1} + 0.6124z^{-2} + 0.6124z^{-3} + 0.2041z^{-4}$. Transmitted symbols were in a binary PAM format and the received signals were corrupted by additive white Gaussian noise. Each equalizer was trained on 2000 symbols using LMS type algorithms with a step size of $\mu = 0.05$ and, in the case of the RBF equalizer, an FIR channel estimator of the correct order was used in order to generate values for the node centres. The training period was followed by the transmission of 10^4 symbols and the number of classification errors were accumulated. This process was repeated until 10^5 or 10^6 nontraining symbols had been transmitted depending on the final BER.

For the partial response channel, $H_1(z)$, RNN2 is not appropriate as the network only requires 2 nodes with decision delays $d = 0$ and $d = 1$. RNN1 had 2 nodes, an additional delay line of 1 symbol estimate and up to 2 received sample inputs. Both the DFE and RBF had a received sample delay line of length $m = 1$, a decision delay of $d = 0$ and used $n = 2$ previous symbol estimates to either cancel ISI or to select a subset of the network nodes in the case of the RBF [2].

Figure 3 shows the BER against SNR for the 3 equalizers together with the BERs obtained using correct symbol feedback (csfb) in each of the equalizers. The difference between the BER curves for estimated and correct symbol feedback show that RNN1 suffers the least from incorrect symbol feedback. The DFE and RBF structures used in this simulation are appropriate only for zero decision delay and as soon as a delay (1 or 2 symbol periods) is used both equalizers outperform the RNN1.

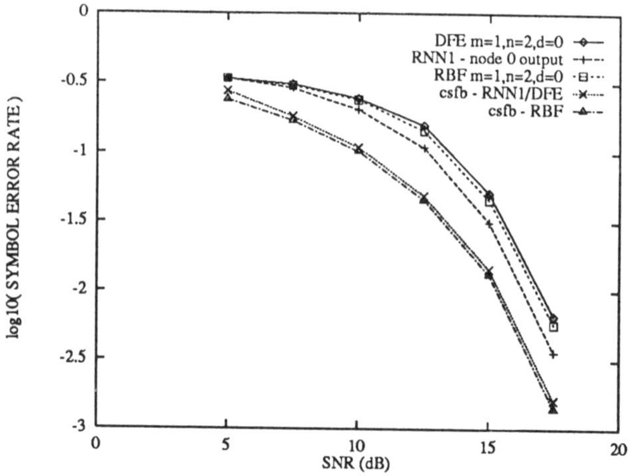

Figure 3: BER curves for channel $H_1(z)$

The nonminimum phase channel $H_2(z)$ was used to compare the performance of both RNN structures together with the DFE and RBF equalizers. The RNN structures had 4 nodes, an additional delay line of 3 symbol estimates and up to 4 received sample inputs. RNN2 had feedback paths from node 0 to nodes 2 & 3 and from node 1 to node 3 in order to cancel as much pre-cursor ISI as possible. The BER of each of the 4 nodes was compared to DFE and RBF equalizers with similar decision delays. These equalizers had a received sample delay line of $m = (d + 1)$ samples and $n = 4$ past symbol estimates as feedback signals.

Figure 4 shows the BER against SNR for the equalizers with zero decision delay together with the curves obtained using correct symbol feedback. RNN1 again suffers the least from error propagation and the RBF has the best performance under ideal conditions. It is noted that RNN2 performs worse than RNN1 except possibly in high SNR conditions. The reason for this is that the generally poor quality of symbol estimates produced by node 0 feed back and increase the pre-cursor ISI present at nodes with larger decision delays. This causes a larger number of errors at these nodes which are fed back to node 0 resulting in a performance degradation. This effect may be seen in Figure 5 which shows the BER curves for the RNNs' node 2 outputs compared to DFE and RBF equalizers with $d = 2$ decision delay. The lower curve shows the improvement in BER attained by pre-cursor cancellation given correct symbol feedback. The actual BER is worse than the equivalent DFE with purely post-cursor cancellation. The RBF performance is now the best as more of the symbol energy is available to aid classification. The performance of RNN1 node 2 output is essentially similar to that of the DFE as there is only a single soft-decision feedback

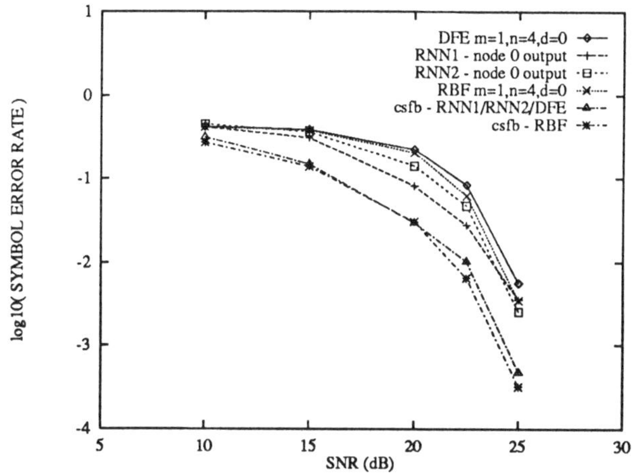

Figure 4: BER curves for channel $H_2(z)$, zero decision delay

path for this node.

Further simulations have shown that the relative performance of the RNN structures is dependent on how the symbol energy is distributed by the channel. If most of the energy occurs at the beginning of the channel impulse response then the RNN structures provide little, if any, performance gains even for zero decision delay equalizers.

5 Conclusions

The nonlinearity in the RNN is seen to limit the performance of the equalizer in a number of ways whilst operating on two symbol spaced path channels. Employing a quantizing nonlinearity removes these problems and the operation of the RNN then becomes similar to that of the DFE.

Two alternative RNN structures are formulated based on the use of soft-decision feedback. The second structure uses both post and pre-cursor cancellation to reduce the ISI present on the equalizer inputs and is close in structure to a fully recurrent network. However the actual benefits obtained are significantly poorer due to the difficulty of accurate pre-cursor cancellation.

The RNN structure which only uses post-cursor cancellation has an improved performance over the DFE and RBF only when the decision delay is minimal. The DFE and RBF equalizers are operating in sub-optimal conditions and allowing a decision delay greatly improves their performance. The relative performance of the RNN1 structure is also dependent on the channel characteristics and only improves over the DFE and RBF when a decision delay greatly improves the quality of symbol estimates allowing

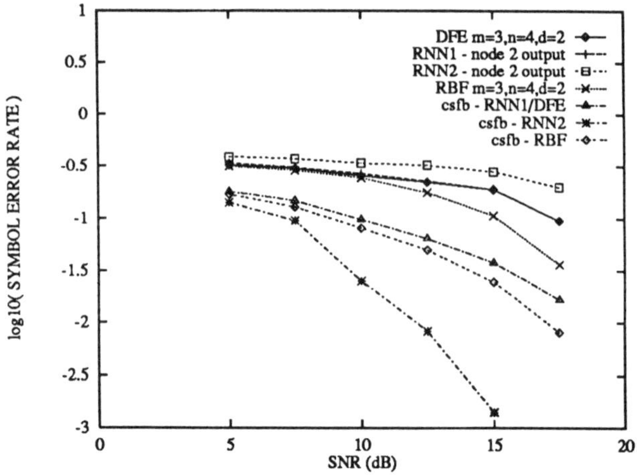

Figure 5: BER curves for channel $H_2(z)$, one decision delay

soft-decision feedback to significantly reduce error propagation.

Acknowledgements

The authors gratefully acknowledge the support of GEC-Marconi and the UK Engineering and Physical Sciences Research Council.

References

[1] G.J.Gibson, S.Siu and C.F.N.Cowan, "Multilayer perceptron structures applied to adaptive equalisers for data communications," in *Proceedings of ICASSP-89,* 1989, p. 1183-1186.

[2] S.Chen, B.Mulgrew and S.McLaughlin, "Adaptive Bayesian equalizer with decision feedback," in *IEEE Transactions on Signal Processing,* vol. 41, no. 9, 1993, pp. 2918-2927.

[3] G.Ketchriotis, E.Zervas and E.S.Manolakos, "Using recurrent neural networks for adaptive communication channel equalization," in *IEEE Transactions on Neural Networks,* vol. 5, no. 2, 1994, pp. 267-278.

[4] R.J.Williams and D.Zisper, "A learning algorithm for continually running fully recurrent neural networks," in *Neural Computation,* vol. 1, 1989, pp. 270-280.

[5] M.J.Bradley and P.Mars, "Application of recurrent neural networks to communication channel equalization," presented at ICASSP-95 , Detroit, USA, May 8-12, 1995.

Neural Network Speech Enhancement for Noise Robust Speech Recognition

Bert de Vries[1], Chi Wei Che[2], Roger Crane[1], Jim Flanagan[2], Qiguang Lin[2], and John Pearson[1]

[1]*David Sarnoff Research Center*
CN5300 Princeton, NJ 08543-5300
{bdevries, rcrane, jpearson}@sarnoff.com

[2]*CAIP Center, Rutgers University*
CoRE Building, Busch Campus
Piscataway, NJ 08855-1390
{werche, qlin, jlf}@caip.rutgers.edu

Abstract

We have developed neural net equalizers that compensate for the effects of mismatched acoustics in the training and operational environments of speech recognizers. We show that neural nets can be used to significantly boost recognition accuracy, without retraining the speech recognizer.

1 Introduction

The performance of current state-of-the-art Hidden Markov Model (HMM) based speech recognizers has improved to the point that commercial applications are feasible. These recognition systems work best for high quality, "close-talking" speech and require consistent environments for training, testing and operation. Typically, the performance of speech recognizers drops drastically if training and operational conditions are not matched.

Degradation of speech signal quality by environmental conditions is due to noise, room acoustics, non-linearities in recording equipment, and compensation by the speaker himself.

Robust speech recognition refers to the art of producing graceful performance degradation when training and testing data set conditions differ. Research on reducing the data set mismatch takes three forms ([1] Gong and Treurniet, 1993). In the first approach the search is for *noise-resistant speech features* which are used in the recognizer pre-processor. It is currently believed that cepstral features are more resistant to noise than the more common discrete Fourier transform coefficients ([2] Erell and Weintraub, 1993). In the second approach the goal is to map noisy speech into cleaner speech. These methods are collectively called *speech enhancement* techniques. The third approach does not work on the received speech signal but instead adapts the recognizer's model parameters to accommodate the changed environmental conditions. Such methods are referred to as *model compensation* techniques.

In this study we explore neural net room acoustics equalizers to expand the use of existing speech recognizers to practical environments where users need not be encumbered by hand-held or attached microphone systems. Applications include combat information centers, large group conferences (tele-conferencing) and mobile hands-busy, eyes-busy maintenance tasks.

In the next section we present our approach to neural net speech enhancement. In section 3 we report on experimental results.

2 Systems Analysis

2.1 Experimental Methodology

We recorded speech data from 50 male and 30 female speakers at a sampling frequency of 16 KHz and used 16-bit linear quantization. Each person spoke 20 isolated commands, 10 digits and 10 continuous sentences. Data were recorded in two sessions. During the first session, speech was simultaneously recorded by a high-quality head-mounted microphone (type HMD-224; these data will be referred to as the close-talking or head-mounted data) and a one-dimensional beamforming line-array microphone located 3 meters from the speaker. In the second session the data were recorded simultaneously by the same head-mounted microphone and a desk-mounted microphone (type PCC 160) at 3 meters distance from the speaker. The recording environment was a regular hard-walled laboratory room (6x6x2.7 meters). Ambient noise in the room was from several workstations, fans, and a large video display machine. The recording and data processing configuration is shown in Figure 1.

To quantify performance improvements achieved with neural networks, we used SPHINX-I ("sphinx-one"), a HMM continuous speech recognition system, developed by Carnegie Mellon University, and a simple dynamic time warping (DTW) recognizer, which we developed ourselves. Before the received speech data is fed to the speech recognizer, a feature extractor converts the speech waveform signal into a cepstral vector signal, which in our case consists of 13 coefficients. Whereas the speech waveform is sampled at 16 KHz, the feature extractor produces a cepstral feature vector at a frequency of 0.1 KHz (the "frame rate" is 10 msecs). As is shown in Figure 1, there are two locations in this configuration where a speech enhancement filter can be placed. We can enhance speech before the feature extractor (the light-shaded box in Figure 1) or after preprocessing (the dark-shaded boxes). Before presenting experimental results, we discuss the consequences of either choice.

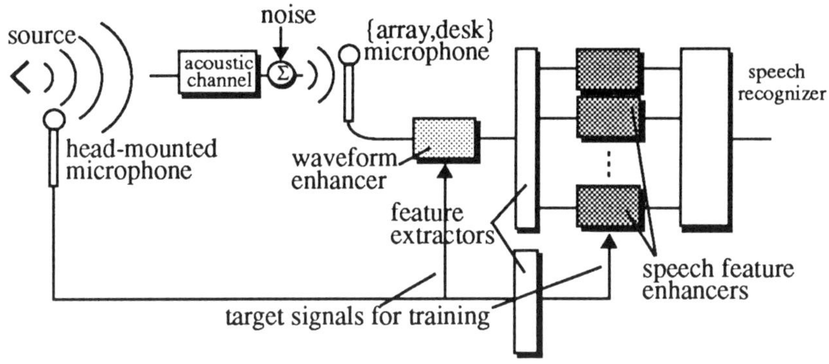

Figure 1 The recording and data processing configuration.

2.2 Speech Waveform versus Feature Enhancement

If a speech enhancement filter is placed before the feature extractor, we might anticipate that a linear equalizer would suffice. We could justifiably use a received speech signal model of the form

$$x(t) = h(t) * u(t) + n(t) \tag{1}$$

where $h(t)$ is the impulse response model for the acoustic channel, $u(t)$ the speech signal at the speaker's end, $n(t)$ a noise source and $x(t)$ the received signal at the distant microphone. Without any information about $n(t)$, the best an equalizer can do is approximate $g(t)$ such that $g(t) * h(t) = \delta(t-k)$, that is, the speech enhancement filter is the inverse channel response (modulo some delay).

The speech enhancement filter can also be positioned after the feature extractor, a configuration which we call (speech) feature enhancement. Since the feature extractor already focuses on rather noise-robust features, the task of the enhancement filter seems easier than in the case of waveform enhancement. For this reason we elected to perform our experiments on feature enhancement rather than waveform enhancement.

The feature extraction stage performs irreversible operations such as time averaging (since we go from 16Khz to a 0.1KHz sample rate), as well as non-linear operations (logarithms in the case of cepstral synthesis). Consequently we should not expect that a linear filter is optimal at this stage. In this study we have compared the benefits of non-linear and recurrent nets over linear filters for feature enhancement.

While the ultimate goal of speech enhancement is to improve recognition performance (in terms of word recognition accuracy), we cannot use the word accuracy, a piecewise constant function, as a training cost function for the speech enhancements filters. Instead we must use related continuously differentiable cost functions such as output signal mean square error (MSE). In this paper we study experimentally the relation between signal MSE and recognition performance.

Feature enhancement in comparison to waveform enhancement adds another difficulty since we are dealing now with a vector signal (13 channels in our case) instead of a scalar signal. Should we use a single net with 13 inputs and 13 outputs, or is it better to assign different networks to the individual channels? There are arguments for either approach. We do not know if a particular "number of channels per network" assignment is consistently better than other alternatives. Experimental results are reported only for the one-net-per-channel approach.

3 Experiments

3.1 Comparative Evaluation across Microphone Types

In the first set of experiments we used the SPHINX-I speech recognizer. We used one neural network per channel for a total of 13 different neural nets. The networks had similar architectures for all channels. The network configuration is shown in Figure 2. The input layer has 9 taps, covering the current frame, the four preceding and four following frames. The input layer fully connects to a 4-node hidden layer of sigmoid nodes, which connect in turn to a single linear output node.

testing dataset identified by microphone type	accuracy (%) on **pre-trained** SPHINX	accuracy (%) after **re-training** on close-talking	accuracy (%) after **re-training** on line-array
close-talking	88	95	-
line-array (LA)	16	21	82
LA+neural net	82	85	-

Table 1: Results on SPHINX for varying dataset environmental conditions.

The input data is drawn from the (cepstral coefficients of the) line array data and the target data is taken from the simultaneously recorded head-mounted microphone data set. These neural nets were individually trained to minimize the MSE between corresponding frames of the input and target data set. We trained on approximately 4 seconds of speech material (isolated words) recorded from a single speaker. The networks were trained in speaker-dependent mode. The goal of the experiments is to estimate the maximal performance gain achievable with neural network enhancement. Results for speaker-independent training would very likely be less favorable. We also want to compare results across different microphone types. The results are captured in Table 1. See also ([3], Che et al. 1994) for a more elaborate account. When we used the SPHINX recognizer, which was pre-trained by the developers, on data recorded by the head-mounted microphone, 88% of the words were correctly classified. When we used line-array data and no neural nets, the performance dropped to 16%. A substantial improvement to 82% performance is observed when the line array data were used in conjunction with feature enhancement neural nets.

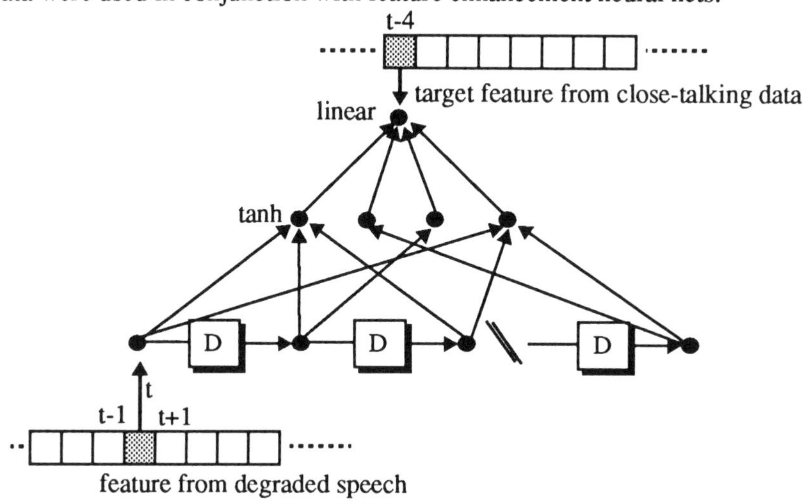

Figure 2 The basic configuration of the neural nets used in our experiments.

Next we re-trained the SPHINX system both on data from the head-mounted microphone as well as on line array data. As expected the performance is very good when the training and testing sets are matched. We found 95% and 82% word accuracy on the head-mounted and line-array data respectively. The results in Table 1 illustrate the main result of this experiment: *whereas it is possible to re-train the HMM recognizer on a matching dataset in order to recover good performance, it is faster to train a (set of) small neural nets with comparable performance results* (relative to HMM re-training). Furthermore it is remarkable that re-training on the (noisy) line array data (82%) instead of the cleaner (better SNR ratio) head-mounted data (21%) greatly improves the performance when we test on line array data. Apparently, matching training and testing data sets is preferable over training on clean data.

3.2 Comparison across Network Architectures

In a second set of experiments we used a simple dynamic time warping recognizer which we developed ourselves. The purpose of these experiments was to determine the benefits of non-linear and recurrent nets relative to linear filters. Furthermore we want to determine the relation between the MSE criterion for neural net training and the DTW word accuracy performance.We trained three different architectures, a linear feedforward net (adaline, adaptive transversal filter), a non-linear layered feedforward net, and a non-linear recurrent network. We used 13 nets per experiment, one for each channel. The transversal filter had 9 taps. The non-linear feedforward net had 9 taps in the input layer, 5 sigmoidal hidden units and 1 linear output node. The recurrent net was similarly configured but the hidden layer was fully connected to itself as well (an "Elman" network). The line array data were used as input data and the head-mounted data set as target data for neural net training. We used batch training, backpropagation with adaptive learning rate and no momentum. Twenty words from 6 different speakers (120 words total) were used both for training and testing (where, of course, the testing set made use of different words than the training set). The DTW recognizer was trained speaker-dependently on the head-mounted data set. The recognizer performance was tested after every neural net training epoch on data from the line array. Results are displayed in Figure 3, where we plotted both the sum of the MSE over the 13 nets as well as the DTW performance as a function of training epoch. These graphs provide some insight in the relation between neural net MSE and DTW performance. The graphs in Figure 3 are all normalized between 0 and 1. For all networks, the DTW performance asymptotes at approximately 0.85, i.e., 85% word recognition accuracy. When we did not use a neural net, a performance of 38% was observed when the DTW recognizer, trained on close-talking data, was tested on line array data. Hence, the feature enhancement nets were able to boost recognition performance from 38% to approximately 85%. It is surprising that the linear transversal filter works as well as the non-linear feedforward and recurrent nets. In the previous section we hypothesized that we needed non-linear processing for feature enhancement. These experiments seem to indicate that linear processing performs just as well, although we do not want to draw generally valid conclusions from one experiment.

An interesting phenomenon is the unstable performance of the DTW recognizer at early stages of learning. It is clear from these graphs that recognition accuracy is not monotonically related to neural net MSE. Yet, if we train long enough the DTW recognizer performance stabilizes for all networks. In Figure 3(d) we plotted the MSE for the individual adaline nets for channels 0, 6, and 12. This plot in combination with Figure 3(a) suggests that the recognizer performance stabilizes only after the "slowest-learning" network (for channel 12) has converged (after about 100 epochs). Apparently the higher-order channels are instrumental for reliable recognition.

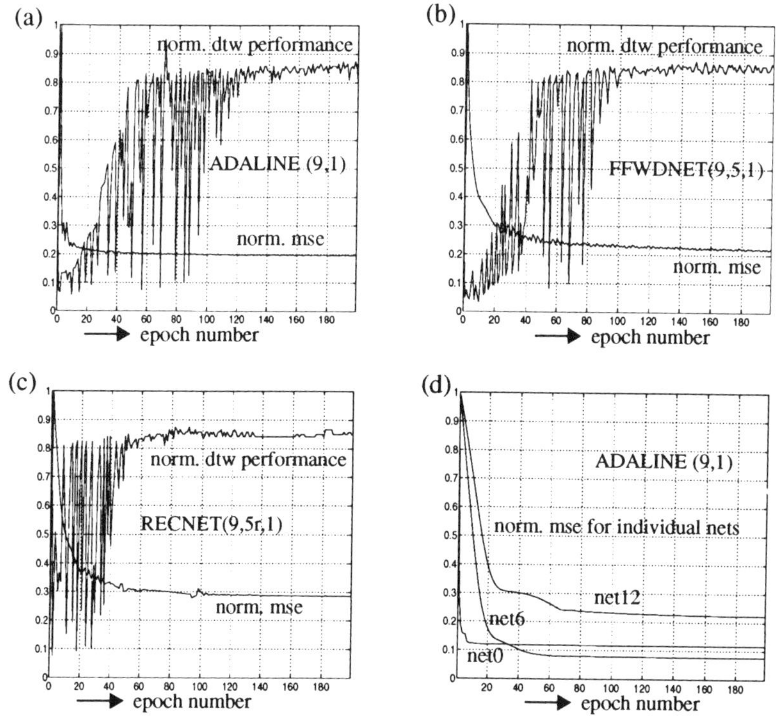

Figure 3 Results on DTW word accuracy using different neural nets as speech feature enhancers. All curves are normalized between 0 and 1. Neural net training proceeded in batch mode, with adaptive learning rates (and no momentum term). The DTW performance curves are relative to the test set. The MSE curves are with respect to the training data set.

4 Conclusions

We have used neural networks for speech feature enhancement in a speech recognizer. From a more general perspective, these experiments are about boosting performance when the training and testing sets are not well matched. This theme is unfortunately rather common in practical applications for adaptive systems, since real world environments are typically non-stationary. We do not claim general truths here, but we noticed some interesting results in our experiments. In general, the technique works very well. In our first set of experiments we were able to boost recognition performance from about 20% to about 80%. In our second set of experiments, on another recognizer, with different data sets, recognition accuracy was improved from 38% to about 85%.

An important conclusion is that recognition performance with neural net speech enhancers in mismatched training and testing data sets appears comparable to re-trained HMM recognizers (without neural nets). The neural net approach avoids re-training of the recognizers, which is a tedious and time-consuming process. We also noticed that it is not always a priority to train on the cleanest data available. It is more important to have matching conditions for the datasets. If the testing environment is noisy, it appears to be better to also train on data with low signal-to-noise ratio. We found that for the case of cepstral feature enhancement, quite surprisingly, the linear transversal adaptive filter performs as well as non-linear and recurrent networks. This result needs further investigation. Furthermore, as expected, the MSE cost criterion for neural net training appeared not to be monotonically related to word recognition accuracy. On the other hand, when training continued long enough, i.e., until the MSE for all nets converged, the DTW performance was stable and MSE training did lead to an excellent performance boost.

Acknowledgments

We are grateful to Dr. Richard Stern and the speech group at Carnegie Mellon University for providing us with a copy of the SPHINX speech recognition system. This work is supported by ARPA contract no. DABT63-93-C-0037.

References

[1] Gong Y. and Treurniet W.C., Speech recognition in noisy environments: a survey, Technical report CRC-TN 93-002, Communications Research Centre, Department of Communications, Ottawa, 1993.

[2] Erell A. and Weintraub M., Filterbank-energy estimation using mixture and Markov model recognition of noisy speech, *IEEE transactions on speech and audio processing*, vol.1, no.1, 1993.

[3] Che C., Lin Q., Pearson J., de Vries B., and Flanagan J., Microphone arrays and neural networks for robust speech recognition, *proceedings of ARPA workshop on Human Language Technology*, Plainsboro, NJ, 1994.

Non-Linear Channel Equalisation using Associative Memory Neural Networks

D.S. Reay

Department of Computing and Electrical Engineering,
Heriot-Watt University,
Edinburgh EH14 4AS, United Kingdom
dsr@cee.hw.ac.uk

Abstract

Previous studies of non-linear channel equalisation using neural networks have concentrated on multilayer perceptron and radial basis function architectures. This paper describes the application of an associative memory network to the problem. It is demonstrated that this neural network is capable of acting as an adaptive equaliser for linear, non-linear and time-varying channels and its performance is compared with that of the optimal Bayesian discriminator. The associative memory equaliser is particularly simple to implement.

1 Introduction

Equalisation of digital communication channels may be viewed as a classification problem. An equaliser is required to decide which of a finite set of possible transmitted signals, $x(k)$ is represented by a sequence of continuous-valued observations of the received signal, $y(k)$. Figure 1 illustrates the discrete-time model of a digital communication channel and equaliser considered here.

Even if the desired channel states (the finite set of possible channel outputs in the noise free case) are linearly separable, in the presence of additive noise, a linear decision boundary is not necessarily optimal. For this reason, non-linear solutions to the equalisation problem have been sought. Previous work on the use of neural networks as equalisers has concentrated on multilayer perceptron

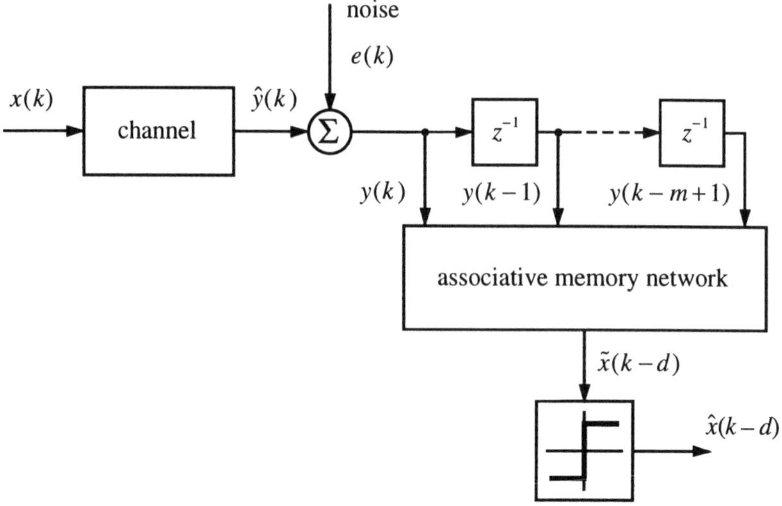

Figure 1: Discrete Time Data Transmission System and Equaliser

and radial basis function architectures [1] [2].

Associative memory networks are well suited to approximating non-linear functions [3]. Here it is demonstrated that, employed as equalisers, after very little training, their performance can approach that of the optimal Bayesian solution. Such a system makes few assumptions about the nature of the channel and is substantially simpler to implement than other neural network architectures.

2 Associative Memory Neural Networks

The form of associative memory neural network used was the cerebellar model articulation controller [4]. The simplicity of implementing this network is particularly apparent when its operation is characterised as the computation of a number of addresses in memory as a function of an input vector, and the summation of the weights stored at those addresses to form an output. The network is essentially a lookup table with local generalisation properties.

3 Training Method

Each weight, $\theta(k)$, accessed in response to an input, $y(k)$, is adjusted according to the equation

$$\theta(k+1) = \theta(k) + \beta \,\mathrm{sgn}\big(x(k\text{-}d) - \hat{x}(k\text{-}d)\big) \tag{1}$$

where $x(k-d)$ is the desired output value, $\hat{x}(k-d)$ is the equaliser output and β is an increment by which the weights are modified.

During training, the associative memory network constructs a representation of the *a posteriori* statistics of the training data, simply by storing it in memory. The local generalisation properties of the associative memory enable its output $\tilde{x}(k-d)$ to approximate the form of the Bayesian discriminant function

$$g(\mathbf{y}) = p(\mathbf{y}|w_1)P(w_1) - p(\mathbf{y}|w_2)P(w_2) \tag{2}$$

given only a small quantity of training data.

Here $\mathbf{y}(k) = [y(k), y(k-1), \cdots, y(k-m+1)]^{\mathrm{T}}$, $p(\mathbf{y}|w_i)$ represents a state conditional probability density function for $\mathbf{y}$ and $P(w_i)$ is the *a priori* probability of class w_i. In this application, classes w_1 and w_2 are represented by the values +1 and -1. Neither the form of $p(\mathbf{y}|w_i)$ nor the value of $P(w_i)$ is assumed by the equaliser.

The training mechanism of equation (1) is the zero-forcing version of the least mean squares (LMS) algorithm. It will adjust the network weights so as to minimise the mean squared error between $\tilde{x}(k-d)$ and $x(k-d)$. However, in this application, where the desired output value $x(k-d)$ only ever has the value ± 1 and, due to noise is not a deterministic function of $\mathbf{y}(k)$, the approximation to the discriminant function $g(\mathbf{y})$ is of more interest. Training information is contained not in the instantaneous values of the desired outputs, $x(k-d)$ but in the relative frequency of their occurrence for different values of input, $\mathbf{y}(k)$.

If the equaliser is required to track the characteristics of a time-varying channel then it is necessary to discard or forget older training examples.

4 Simulation Results

Presented here are the results of the application of an associative memory based equaliser to three different channels. The first is linear, described by

$$\hat{y}(k) = 0.5x(k) + x(k-1) \tag{3}$$

In the case of a second order equaliser, i.e. $m = 2$, and delay $d = 0$, the required decision boundary is highly non-linear. Figure 2 shows the decision boundary formed by an associative memory equaliser that has been trained on 100 symbols transmitted through the channel (3). The variance of the additive white Gaussian noise was $\sigma_n^2 = 0.2$. Shown dotted is the optimal decision boundary.

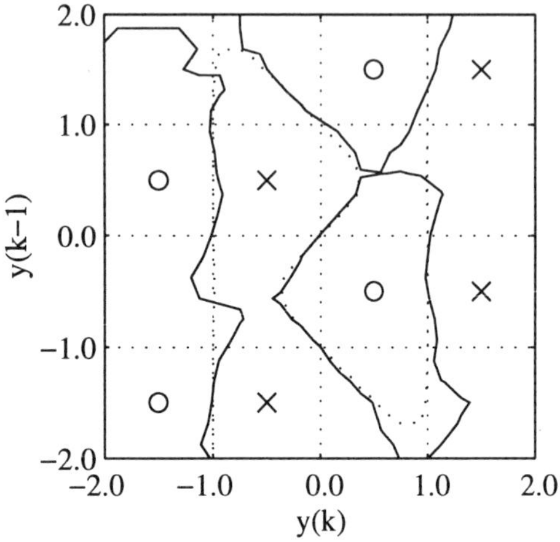

Figure 2: Decision Boundaries for Channel (3), Gaussian Noise $\sigma_n^2 = 0.2$,

Equaliser Order $m = 2$, Delay $d = 0$.

--- associative memory equaliser

· · · optimal Bayesian discriminator

The second channel is non-linear, described by

$$\hat{y}(k) = x(k) + 0.5x(k-1) - 0.9(x(k) + 0.5x(k-1))^3 \qquad (4)$$

Figure 3 shows the decision boundaries formed by an associative memory equaliser that has been trained on 100 samples of a pseudo-random sequence of transmitted symbols corrupted by additive Gaussian noise of variance $\sigma_n^2 = 0.2$, and by the optimal Bayesian discriminator. Of more importance than these boundaries are the results of an experimental comparison of bit error rates using the two different equalisers. These are shown in figure 4. Also shown is the performance of the second order Wiener filter. In this experiment, the associative memory equaliser decision boundary formed during training at a signal to noise ratio of 8dB was fixed.

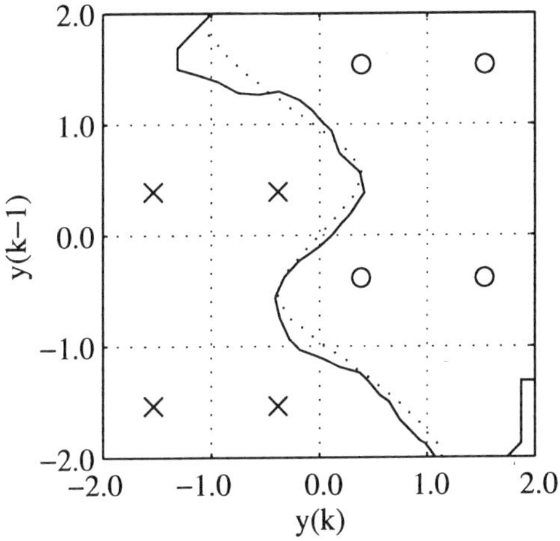

Figure 3: Decision Boundaries for Channel (4), Gaussian Noise $\sigma_n^2 = 0.2$, Equaliser Order $m = 2$, Delay $d = 1$
--- associative memory equaliser
· · · optimal Bayesian discriminator

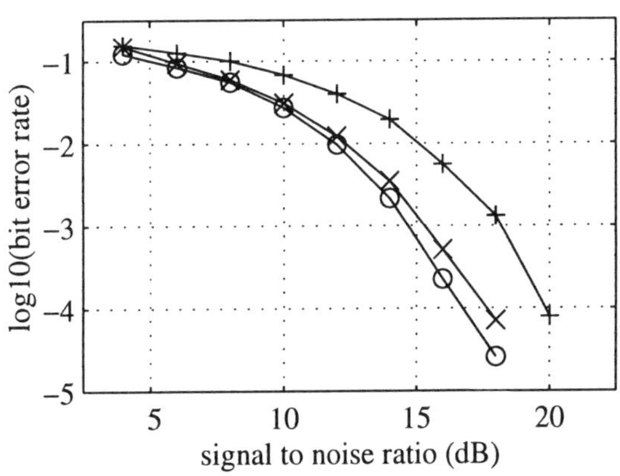

Figure 4: Comparison of Bit Error Rates
-x- associative memory equaliser
-o- optimal Bayesian discriminator
-+- Wiener filter

The performance of the associative memory equaliser is closest to the optimum at that signal to noise ratio. At a bit error rate of 10^{-4} the associative memory equaliser shows an improvement in signal to noise ratio of 2.2dB over the Wiener filter compared with the 2.9dB improvement given by the optimal solution.

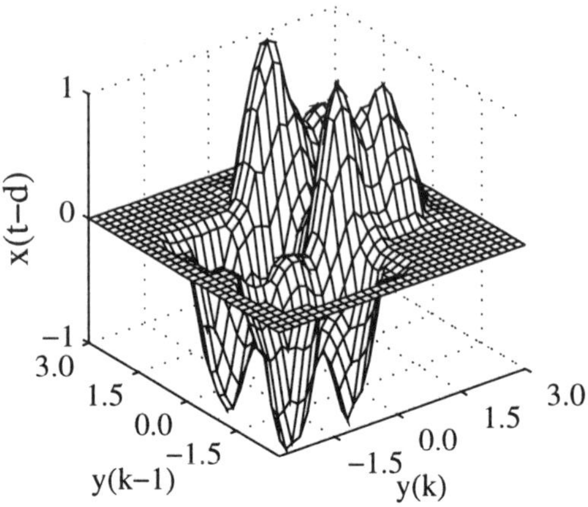

Figure 5: Associative Memory Equaliser Output $\tilde{x}(k-d)$, after 100 Training Examples from Channel (4)

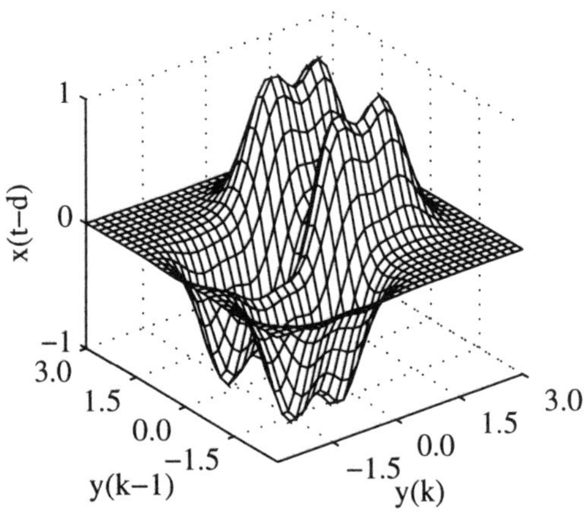

Figure 6: Optimal Bayesian Discriminator Output for Channel (4)

The network used had a total of 2048 weights and the experiment was carried out using 5×10^5 symbols transmitted at each different noise level.

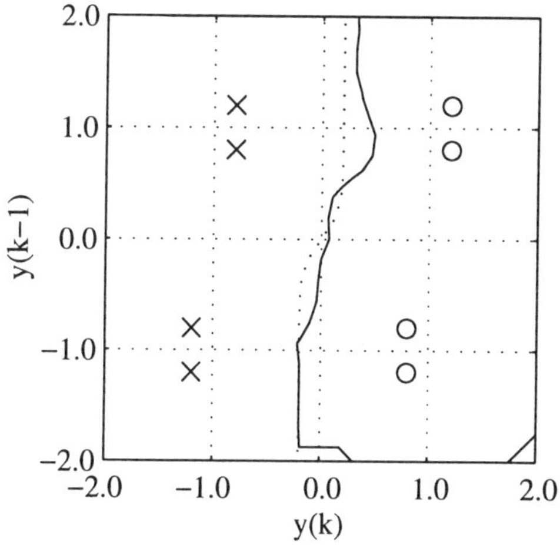

Figure 7: Decision Boundaries for Channel (5) after 100 Training Examples, Gaussian Noise $\sigma_n^2 = 0.125$, Equaliser Order $m = 2$, delay $d = 0$
--- associative memory equaliser · · · optimal Bayesian discriminator

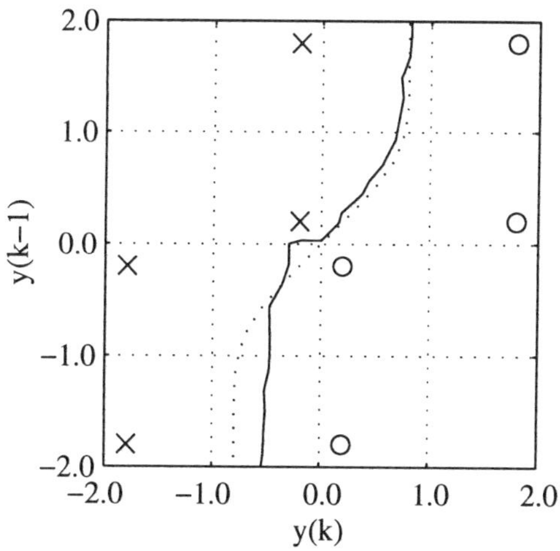

Figure 8: Decision Boundaries for Channel (5) after 300 Further Training Examples, Gaussian Noise $\sigma_n^2 = 0.125$, Equaliser Order $m = 2$, delay $d = 0$
--- associative memory equaliser · · · optimal Bayesian discriminator

Chen et al. [3] demonstrated for a case of non-stationary channel statistics, that a Gaussian radial basis function network in which the basis function centres were adjusted was capable of tracking the statistics of that channel. This demonstration emphasised the parametric nature of Chen's approach to estimating the *a posteriori* statistics of the channel output. An associative memory equaliser is capable of adapting on-line also. Figures 7 and 8 show the results of applying the equaliser to a channel described by

$$\hat{y}(k) = x(k) + (0.2 + 0.002k)x(k-1) \tag{5}$$

An initial decision boundary was formed by training the equaliser on 100 symbols transmitted through the channel

$$\hat{y}(k) = x(k) + 0.2x(k-1) \tag{6}$$

Training then continued (k assumed reset to 0) as the channel characteristic was altered, resulting in the decision boundary shown at instance $k = 300$. Shown dotted in figures 7 and 8 are the corresponding optimal decision boundaries.

5 Conclusions

The generalisation properties of associative memory neural networks make them well suited to the estimation of the *a posteriori* statistics of training data and hence of acting as a non-linear channel equaliser. They are capable of learning quickly, on-line and may adapt to time-varying channel characteristics. Compared to other neural network equalisers, associative memory has a very simple implementation.

References

[1] S. Chen, G.J. Gibson, C.F.N. Cowan, P.M. Grant, "Adaptive Equalisation of Finite Non-Linear Channels using Multilayer Perceptrons," *Signal Processing*, vol. 20, pp. 107-119, 1990

[2] S. Chen, B. Mulgrew, P.M. Grant, "A Clustering Technique for Digital Communications Channel Equalization using Radial Basis Function Networks," in *Trans. IEEE on Neural Networks*, vol.4, no.4, pp. 570-579, July 1993

[3] P.E. An, M. Brown, C.J. Harris, A.J. Lawrence, C.G. Moore, "Comparative Aspects of Associative Memory Networks for Modelling," in *Proc. ECC '93*, pp. 454-459, June 1993, Groningen, NL

[4] J.S. Albus, "A New Approach to Manipulator Control: The Cerebellar Model Articulation Controller (CMAC)," *Trans. ASME, J. Dynamic Systems, Measurement and Control*, vol. 97, no.3, pp. 312-324, 1975

Compensation of Bandpass Nonlinearities by Look-Up-Tables and CMAC

A. Artés-Rodríguez†, F.J. González-Serrano‡, A.R. Figueiras-Vidal†, Luis Weruaga-Prieto†

†*E.T.S.I. Telecomunicación - U.P.M.*
Ciudad Universitaria, 28040 Madrid - Spain
`antonio@gtts.ssr.upm.es`
‡*E.T.S.I. Telecomunicación - Univ. de Vigo - Spain*
`frank@dtc.uvigo.es`

Abstract

This paper presents two new data predistortion schemes to compensate bandpass nonlinearities in digital transmission systems employing QAM modulation schemes. The first one consist on a full size adaptive LUT that considers the actual and pasts symbols, achieving a total compensation of finite length nonlinearities. In the second one, the LUT is replaced by a modified CMAC scheme which reduces the complexity of the full size LUT at the expense of getting only a partial compensation.

1 Introduction

Nonlinear effects that limit the use of bandwidth-efficient digital modulation techniques appear in a great number of communication channels. In some cases, like microwave radio, this nonlinearity is due to the High Power Amplifier (HPA) used in the transmitter. The effect of HPA nonlinearity can be reduced by backing off the amplifier from its saturation point, but this reduces the transmitted signal power.

To make better use of the available HPA power, some compensation techniques can be used either at the transmitter (predistortion) or at the receiver. The optimal solution among the compensation techniques at the receiver is the maximum likelihood sequence estimation [1], and other sub-optimal approaches are nonlinear equalization [2] and nonlinear Intersymbol Interference (ISI) cancellation [3]. The main limitation of these techniques is their complexity.

2 Predistortion Techniques

Basically, there are two types of predistortion techniques: data predistortion and continuous-time analog signal predistortion, which is usually performed at IF. Before discussing the types of data predistorters, let us analyze the effect of the nonlinearity in the signal constellation. As we can see in Figure 1 for a 16-QAM modulation, each constellation point becomes a cluster due to linear and nonlinear ISI, and the respective centers of gravity of the clusters do not correspond to the original constellation points (constellation warping).

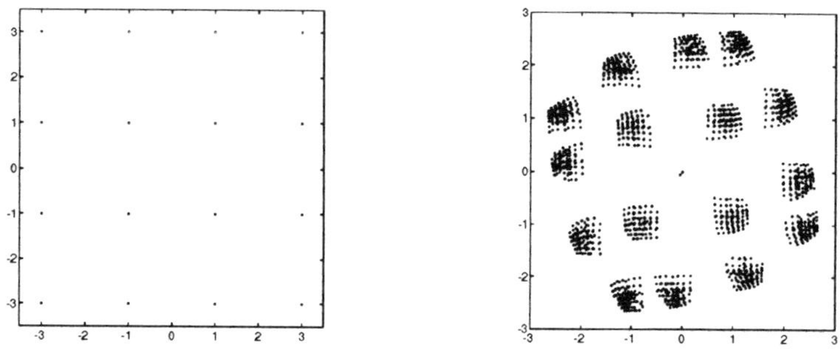

Figure 1: Original and distorted constellations

Data predistorters can be divided in two types: with memory and memoryless. Memoryless data predistorters can only compensate the constellation warping and consist on a Look-Up-Table (LUT) that stores the predistorted constellation points, usually in a magnitude-phase format. An algorithm to adapt this LUT was proposed by Saleh and Salz [4] using the scheme that Figure 2 shows, in which the error measurement is performed by implementing a local receiver.

Predistorters with memory process the data sequence using a nonlinear filter. The developed techniques are based on discrete-time Volterra filters [5], nonlinear data interpolation [6] (in order to ensure no distortion in more than one point during a symbol interval), constellation partitioning [7] (in order to simplify a full-size LUT), and (a preliminary work using)

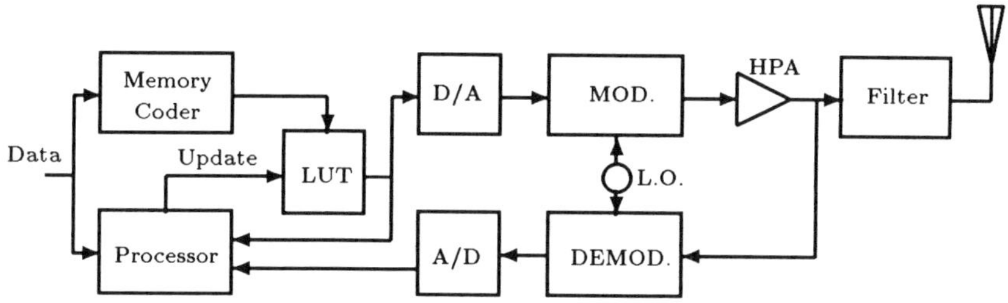

Figure 2: Scheme of the data predistorter.

multilayer perceptrons [8]. Only Volterra filters and multilayer perceptrons admit adaptive estimation procedures.

3 Proposed Schemes

In this paper we present two new adaptive predistortion schemes with memory. Both of them start from the scheme proposed by Saleh and Salz, the reasons of that choice being: 1) the use of a LUT does not assumes any model for the nonlinearity; 2) the nonlinearities that the HPA's produce are, in practice, of short duration and, so, the output of the nonlinearity is a function of a finite number of digital inputs; and, 3) it is easy to obtain adaptive procedures for this type of predistorters.

3.1 Predistorter based on LUT's

The first scheme is an extension of that proposed by Saleh and Salz in two ways: by considering not only the actual symbol, but also the past n symbols (predistortion with memory), and by using a rectangular instead of a polar representation of the symbols. So, the predistorter consists on a LUT of size $N^{(n+1)}$, where N is the size (number of points) of the original constellation, directioned by the actual and the n past symbols, and adapted according to:

$$S[k+1] = S[k] + \beta e[k]A[k] \tag{1}$$

where S is the content of the LUT, β is the adaptation step, $e[k]$ is the error between the output of the LUT and the desired response, and $A[k]$ is the director vector of the LUT composed by $N^{(n+1)} - 1$ "0s" and aa "1".

Figure 3 represents the evolution of the rms errors using the proposed algorithm for n=0,1 and 2. The modulation (as in Figure 1) is a 16-QAM, and the nonlinearity is composed by a root raised cosine filter with a roll-off factor equal to 0.75, followed by an AM/AM and AM/PM distorters that

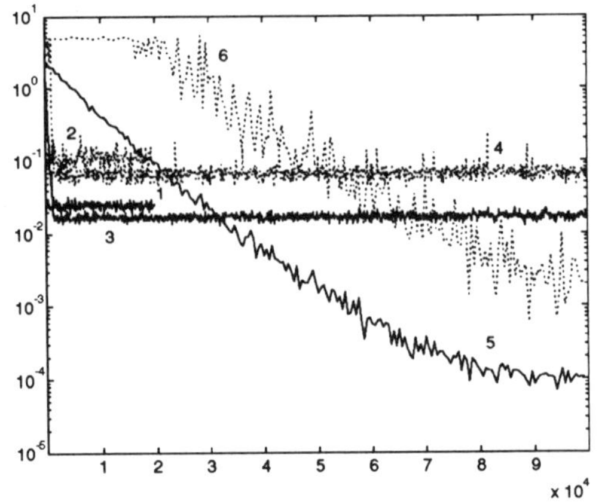

Figure 3: Rms error using LUT's of size 16 (curves 1 and 2), 256 (curves 3 and 4) and 4096 (curves 5 and 6). Solid lines represent the mean value of rms during 100 iterations, and dotted lines represent the maximun value of rms during the same period.

obeys the formulas

$$a_o = \frac{2a_i}{1 + a_i^2} \tag{2}$$

$$\phi_o = \phi_i + \frac{a_i}{1 + a_i^2} \tag{3}$$

(a_i and a_o being the input and output amplitudes, and ϕ_i and ϕ_o the input and output phases, respectively), and other root raised cosine filter that guarantees, in absence of the AM/AM and AM/PM distortion, an ISI-free channel. The value of parameter β is 1.

We can observe how, as n increases, the compensation is more effective, but the costs to pay for this improvement is increasing the size of the LUT, that can make the scheme unpractical, and decreasing the speed of convergence, too slow. The difference between the mean and maximum value of the rms error indicates how noisy the algorithm is, and it is necessary to guarantee that its maximum value does not produce errors in absence of noise in the digital link. Obviously, in that sense, the performance of the algorithm can be improved using a variable adaptation step strategy.

There are not differences between the performances of the original algorithm of Saleh and Salz and our algorithm for $n = 0$, and, so, there is not necessary to perform the two polar to rectangular conversions needed in the original algorithm.

3.2 Predistorter based on the CMAC

In the second proposed scheme, the LUT is substituted by the Cerebellar Model Articulation Controller (CMAC) [9], which, in the limit (generalization parameter ρ equal to 1), gives a LUT. The inputs to the CMAC are the actual and the past n symbols, and the ordering of the points of the constellation is made according to the philosophy of the CMAC, *i.e.*: the closest points to a given one (in an Euclidean sense) are stored in the closest positions to that of this point. In order to analyze the applicability of the CMAC to our problem, Figure 4 shows the contents of the LUT of size 256 after the adaptation: we can observe how the surface is sufficiently smooth; so, the CMAC could obtain good results even with high values of the regularization parameter.

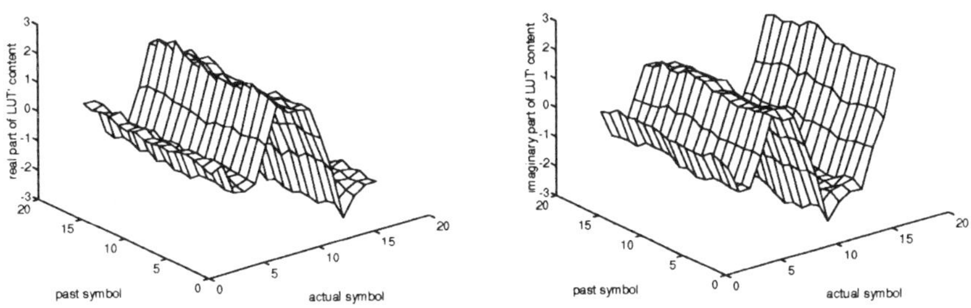

Figure 4: Contents of the LUT after the adaptation

Figure 5 shows the evolution of the rms error with $n = 2$ using the same nonlinearity, and generalization parameters 2 and 15 (the maximum), that need LUTs of sizes 1241 and 120, respectively. Comparing it with Figure 3, the compensation of the nonlinearity is worse than that provided by the adaptive LUT for $n = 2$, but better than that got with the adaptive LUT for $n = 1$, even using smaller LUTs.

However, the behavior of the CMAC algorithm is not satisfactory in two aspects: 1) the adaptation parameter must be small (0.1 and 0.05 for sizes of 1241 and 120, respectively, in the above example) in order to ensure the convergence of the algorithm; then, the convergence is very slow; and 2) when the generalization parameter is high, the high difference of the hypervolumes associated to each position of memory causes "jumps" in the convergence, and that "jumps" (that can be observed in curve 4 of Figure 5) may cause errors in the digital link even in absence of additive noise.

To avoid this problem, the error can be assigned to each position of

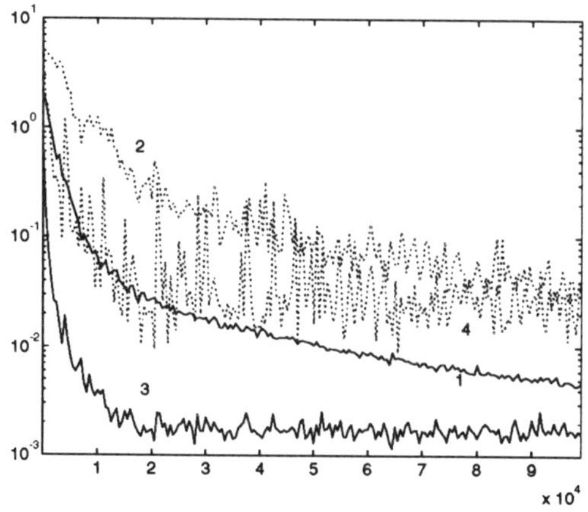

Figure 5: Rms error using the CMAC algorithm with $n = 2$ and ρ equal to 2 (curves 1 and 2) and 15 (curves 3 and 4). Same interpretation as above.

memory in an inverse proportio to its hypervolume [10], according to:

$$S[k+1] = S[k] + e[k]B[k]A[k] \tag{4}$$

where $A[k]$ is the director vector of the memory composed by $N^{(n+1)} - \rho$ zeroes and ρ ones, and

$$B[k] = \beta \frac{1}{A[k]^T C A[k]} C \tag{5}$$

C being a diagonal matrix that contains the inverse of the hypervolumes.

Figure 6 shows the results using this algorithm with the adaptation step equal to 1, where we can appreciate how the convergence is faster and the peaks are reduced. The peak observed in the curve that represents the maximum value of the rms error when the regularization parameter is equal to 15 and the convergence is achieved indicates that, for this problem and this adaptation rule, the adaptation step or the regularization parameter must be lowered. Other possibilities are indicated in the closing section of this paper.

4 Conclusions and further lines

Two new data predistortion schemes have been presented to compensate bandpass nonlinearities in digital communication systems. The schemes extend previous data predistortion methods based on LUT's, offering better performances whith different implementation complexity.

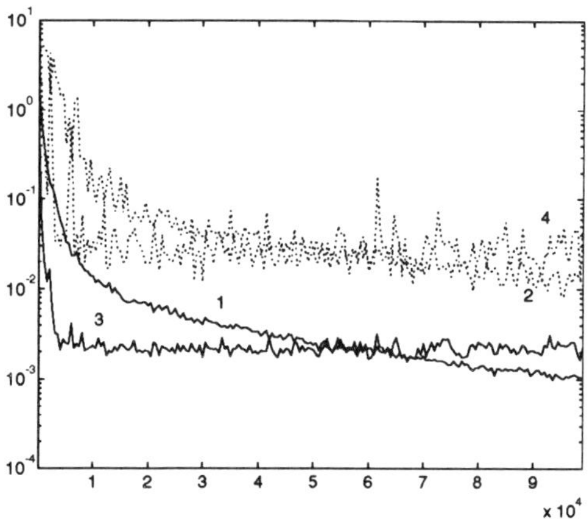

Figure 6: Rms error using the proposed modification of the Albus rule with $n = 2$ and ρ equal to 2 (curves 1 and 2) and 15 (curves 3 and 4).

An additional research effort must go along the lines that are opened in this paper:

- different weighting strategies to assignate the error to each position of the CMAC;
- other space partitionig schemes different of the CMAC according to the behaviour of the nonlinearity; and
- initialization of the CMAC different of zero.

References

[1] W. Van Etten, F. Van Vugt. "Maximum Likelihood Receivers for Data Sequences Transmitted over Nonlinear Channels", *AET*, vol. 34, pp. 216-223, 1980.

[2] S. Benedetto, E. Biglieri. "Nonlinear Equalization of Digital Satellite Channels", *IEEE J. on Sel. Areas on Communications*, vol. SAC-1, no. 1, pp. 57-62, Jan. 1983.

[3] E. Biglieri, A. Gersho, R.D. Gitlin, T.L. Lim. "Adaptive Cancellation of Nonlinear Interference for Voiceband Data Transmission", *IEEE J. on Sel. Areas on Communications*, vol. SAC-2, pp. 765-777, Sept. 1984.

[4] A.A.M. Saleh, J. Salz. "Adaptive Linearization of Power Amplifiers in Digital Radio Systems", *BSTJ*, vol. 62, no. 4, pp. 1019-1033, April 1983.

[5] G. Lazzarin, S. Pupolin, A. Sarti. "Nonlinear Compensation in Digital Radio Systems", *IEEE Trans. on Communications*, vol. 42, no. 2/3/4, pp. 988-999, Feb. 1994.

[6] G. Karam, H. Sari. "Data Predistortion Techniques Using Intersymbol Interpolation", *IEEE Trans. on Communications*, vol. 38, no. 10, pp. 1716-1723, Oct. 1990.

[7] G. Karam, H. Sari. "A Data Predistortion Technique with Memory for QAM Radio Systems". *IEEE Trans. on Communications*, vol. 39, no. 2, pp. 336-344, Feb. 1991.

[8] A. Bernardini, S. De Fina. "A New Predistortion Technique Using Neural Nets", *Signal Processing*, vol. 34, pp. 231-243, 1993.

[9] J.S. Albus. "A New Approach to Manipulator Control: the Cerebellar Model Articulation Controller", *J. Dyn. Systems, Measurements and Control*, pp. 220-227, Sept. 1975.

[10] L. Weruaga-Prieto. "Cancelación no Lineal de Ecos en Transmisión Digital", PhD dissertation, ETSI Telecomunicación – Universidad Politécnica de Madrid, 1994.

Experimental Analog Neural Network based Decision Feedback Equalizer for Digital Mobile Radio

Anthony Jayakumar and Joshua Alspector

Bellcore
445, South Street
Morristown, NJ 07960-6438, USA
Tel: (201) 829 4929; Fax: (201) 829 5888
tony@faline.bellcore.com

Abstract

The emerging digital cellular and Personal Communication Services (PCS) require stringent performance characteristics such as speed, power and complexity for channel acquisition. Conventional channel equalization methods that calculate gradients might be too complex, slow and power hungry to operate in such systems where the channel characteristics vary rapidly and which are also constrained by limited battery power. We propose a novel method where the error is minimized by learning in a neural network, aided by simulated annealing to settle the network at global minima. We experimentally show the feasibility of such an approach with an experimental prototype custom analog VLSI chip-set and demonstrate large reduction in power consumption with optimized architectures. A decision feedback equalizer (DFE) structure has also been modelled and we show improved performance in a channel subject to frequency selective multipath fading.

1. Introduction

Wireless data communications require high bandwidth through rapidly changing channels (the GSM Pan-European digital mobile radio standard specifies a bit rate of 270.8 kb/s). This is usually accomplished by coding and spectrally efficient modulation techniques. In addition, mobile radio systems must consume very little power due to the limited battery energy storage capacity in the portable. This precludes the use of conventional adaptive equalizers to increase channel capacity and to combat frequency selective fading due to the power-cost associated with fast digital signal processors and peripherals such as Analog-to-Digital (A/D) converters.

In this paper, we describe a novel approach to the equalization of mobile radio channels subject to multi-path and noise interference using a Boltzmann learning chip. This chip has been described in earlier publications [1,2]. Due to its fast on-chip learning capability, the chip can acquire rapidly varying channels. Since it uses low power analog computation, it can be implemented with a very low power-cost in the portable. This technique is compatible with the Time Division Multiple Access (TDMA) technology which is the most common of the digital technologies (GSM and North American Digital Cellular (IS-54)) and with *M*-ary Quadrature Amplitude Modulation (QAM) schemes.

2. The Learning System

The equalization experiments were carried out on a general purpose learning system that is built around an analog neural network chip-set. The experimental research prototype chips use the Boltzmann Machine (BM) and Mean Field (MF) learning algorithms for on-chip weight updates. This learning system has been described in detail elsewhere [3]. The system consists of a VME board containing both these devices along with glue logic to interface with a fast digital workstation (Sparc/2™) [3]. X-windows™ based graphical user interface enables easy access to the learning system. This makes possible fast prototyping of neural network hardware based embedded applications.

3. The Channel Equalization Problem

A typical communications channel can be characterized, in the baseband by

$$r_k = h_k s_0 + \sum_{j \neq k,\, j = 1}^{m} h_j s_{(k-j)} + n_k \tag{1}$$

where r_k is the received signal at the sampling instant, $\{s_j\}$ is the sequence of transmitted symbols, $\{h_j\}$ is the impulse response of the channel, n_k the additive noise at the sampling period (assumed to be white Gaussian with variance of σ^2) and m is the channel memory. The second term on the right hand side of equation (1) is the Inter-Symbol Interference (ISI) caused by the dispersive effects of the channel. In a Pulse Amplitude (PAM) M-ary modulation scheme, the transmitted symbol $s_k \in \{\pm 1, \pm 3, \ldots, \pm(M-1)\}$ (where M is an integral power of 2) modulates a unit amplitude rectangular pulse of duration T seconds.

A conventional adaptive linear equalizer consists of a tapped delay line FIR filter whose output $\hat{s}_{k-d}$ is an estimate of the symbol d sampling periods ago.

$$\hat{s}_{k-d} = W_k^T \cdot \{r_k\} \tag{2}$$

An adaptive algorithm tries to minimize an error measure by adjusting the tap weight vector:

$$W_k = \{w_0, w_1, ..., w_{m-1}\}_k \tag{3}$$

Inspection of equation (2) reveals a structure that is similar to a one neuron perceptron with the tapped delay line at the input layer and the summation and thresholding performed at the neuron. This observation leads to the question whether the performance of a linear equalizer can be improved by additional neurons and layers. Various authors [4] have demonstrated this to be true and have shown that the optimum decision boundary for a linear channel is non-linear. In addition, mobile communications channels exhibit deep spectral amplitude nulls (due to multi-path) that can easily defeat a linear equalizer. Hence, intuitively, the non-linear mapping capability of neural networks could offer performance superior to conventional linear techniques.

4. Neural Network based Adaptive Equalization

Conventional non-linear equalization techniques such as the Decision Feedback Equalizer (DFE) and probabilistic algorithms like the Maximum *a posteriori* probability (MAP) criterion and Maximum-likelihood Sequence Estimation (MLSE) are computationally very expensive. Although fast digital signal processing (DSP) chips can accommodate these computational requirements, they and other necessary peripheral chips such as fast A/D converters consume too much power for portable use.

The popular Back Propagation (BP) learning algorithm when applied to the equalization task, is computationally $O(m^2)$ expensive. The cost of computation is prohibitively expensive for the BM algorithm when run by software on a serial computer due to the need for stochastic settling of many variables over the range of the simulated annealing schedule. The computational advantage is realized only when the BM learning algorithm is implemented in parallel analog hardware. Since this algorithm uses local correlations, it could be implemented without much interconnection complexity. The gradient descent is aided through simulated annealing, both stochastic and deterministic, and this can also be implemented with low-power analog circuits. The weight adaptation rule is a difference operation and hence by using the network as a measuring 'scale', on-line learning is possible at each synapse using low resolution (6 ~ 8 bits) analog circuits.

5. Experimental results

The equalizer structure that is mapped to the hardware is shown in Figure 1. It consists of an analog tapped delay line that feeds the input layer. The output is fed to a high gain thresholding comparator that acts as a hard decision decoder. Note the direct connection between the input and the output layer. This speeds up convergence [4], as this arrangement can be considered as a linear equalizer in parallel with a non-linear structure.

Four different channel models were chosen for the experiment. They are:

1) $H_{mp}(z) = 0.894 + 0.447z^{-1}$

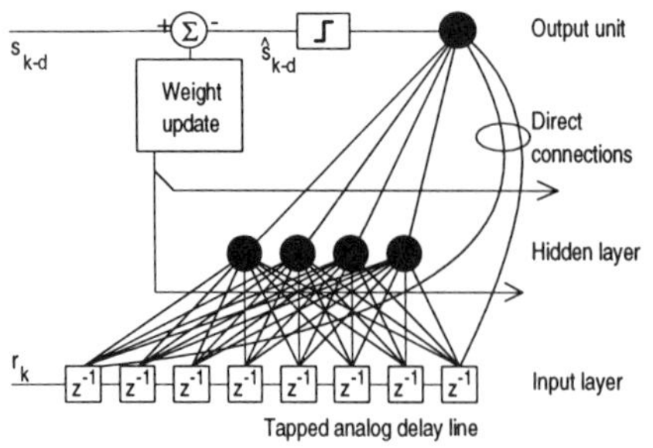

Fig. 1. Structure of the neural network equalizer mapped in the hardware

2) $H_{nmp}(z) = 0.447 + 0.894\, z^{-1}$
3) $H_{fade}(z) = 0.174 + 0.60z^{-1} + 0.60z^{-2} + 0.174z^{-3}$
4) $H_{4qam}(z) = 0.6627 + 0.5928z^{-1} + 0.4126z^{-2} + 0.1983z^{-3}$

The first is a fairly simple channel while the second is a non-minimum phase channel. The latter is frequently encountered in mobile radio systems when the direct path is more attenuated than the indirect paths. The third channel has a deep spectral amplitude null that is characteristic of frequency selective multi-path fading. A 4-QAM modulation scheme is chosen for the fourth channel which has a memory depth of 4. For the sake of simplicity, baseband equalization and perfect timing recovery are assumed. All of the experiments had the minimum number of input taps that is required for training. They are 2, 2, 6 and 6 for the four channels respectively. Co-channel interference is not considered and additive noise is assumed to be white Gaussian (AWGN).

Synthesized analog data that represents the signal output of the wireless receiver demodulator was fed to the equalizer. Note that no A/D conversion is done on the demodulator output and the equalizer is trained on the raw analog signal. The network was seeded with random initial weights before training. The equalizer was trained using the Mean Field mode of the learning chip, although the Boltzmann Machine mode would be expected to give similar results.

Figure 2 shows the evolution of mean Bit Error Rate (BER) vs. training time at a received Signal to Noise ratio (SNR) of 15 dB (a) and 20 dB (b) for the 4 channels. The plotted results are the average of 35 consecutive trials. We can see that the channels $H_{mp}(z)$ and $H_{nmp}(z)$ are learned within 10^{-2} BER in less than 150 patterns. This implies that a new channel can be acquired within 1.5 msec in our general purpose hardware which is well within the coherence time of 5.4 msec for a vehicle velocity of 100 km/hr at 2 GHz carrier frequency. Channel $H_{fade}(z)$ is considerably harder as some portions of the classification regions overlap each other. Learning occurs at a slower rate and converges at an irreducible BER for a given SNR.

An extension of the bi-orthogonal signalling scheme is the QAM which consists of 2 PAM channels on the same carrier phase shifted with respect to one another by 90° (in quadrature). In baseband they reduce to two independent PAM channels. We have mapped an equalizer for the 4-QAM modulation scheme on

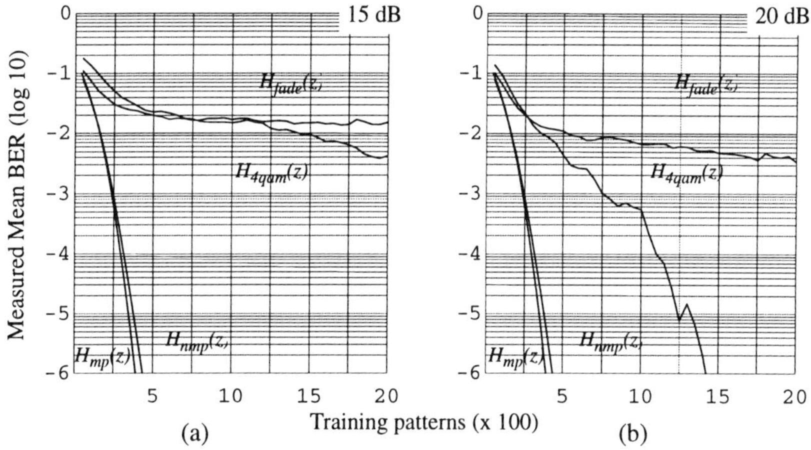

Fig. 2. Evolution of mean BER vs. training for the 4 channels under different SNRs.

the chip for the channel $H_{4qam}(z)$. This consists of two equalizers - one for the in phase signal and another for the quadrature phase component with three input taps for each. At 15 dB of SNR, the training time is longer than the simpler channels and training converges to a BER of about 7.0 10^{-2} (Fig. 3). Above 20 dB, much better performance is observed.

Improvements in the convergence rate and BER can be achieved by incorporating more input taps than the minimum necessary as used in these experiments. In a practical system, where some kind of source coding, interleaving and/or diversity is employed, the BER can be further reduced by about four orders of magnitude [5].

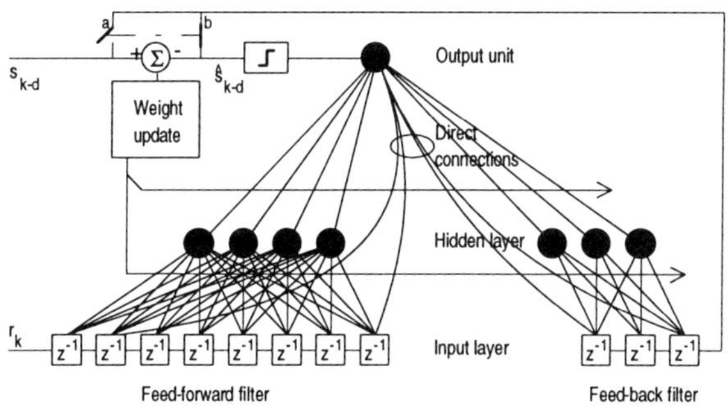

Fig. 3. Structure of the neural network based decision feedback equalizer (DFE) mapped in the hardware.

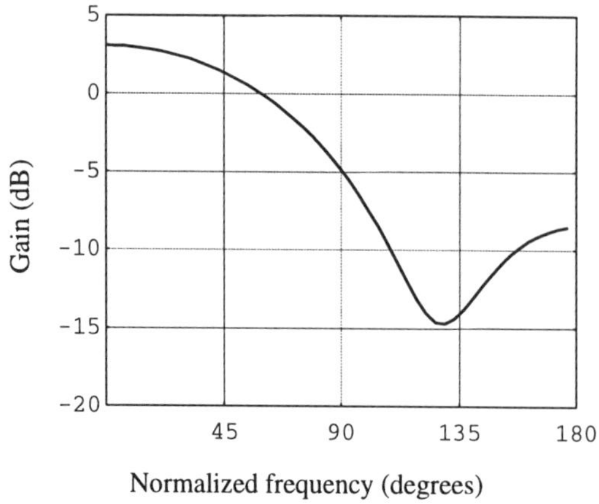

Fig. 4. Frequency response of the experimental channel H_{dfade} exhibiting deep fade.

6. Decision Feedback Equalizer structure

The decision feedback equalizer consists of two filters: a conventional feed-forward filter and a feed-back filter [5] as shown in Figure 3. The feedback filter is used to remove the ISI caused by previously detected symbols from the present estimate (post-cursor ISI) which results in better performance over the conventional equalizer. Switches *a* and *b* in the feedback path determine whether the correct or the detected symbol is fed back. Typically, during training the correct symbols are fed back and then the feedback path is switched over to the detected symbols. We have mapped the DFE to the hardware for a channel H_{dfade} with a

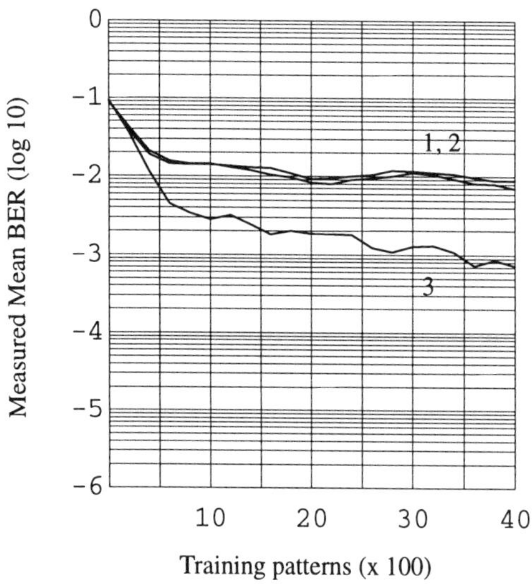

Fig. 5. Evolution of mean BER vs. training Channel model: H_{dfade} SNR: 20 dB

severe frequency selective distortion where

$$H_{dfade}(z) = 0.1887 + 0.468z^{-1} + 0.701z^{-2} + 0.468z^{-3} + 0.1887z^{-4}$$

The frequency response of H_{dfade} exhibits a notch (> 15 dB) within its bandwidth as shown in Figure 4. We use a DFE equalizer with 3 taps in the feed-forward filter and 4 taps in the feedback filter in this experiment.

Figure 5 shows the evolution of training in the hardware for the three cases:1) conventional feed-forward equalizer, 2) DFE with the detected symbols fed back and 3) DFE with the correct symbols fed back. The plots are an average of 50 consecutive runs on the hardware. We can see no appreciable difference between cases 1 and 2 while we get an order of magnitude reduction in BER in case 3. This implies that during training, the error due to incorrect symbols fed back (case 2) offsets any gain achieved by post-cursor ISI cancellation. However, when the correct symbols are fed back (case 3) the gain in performance due to post-cursor ISI cancellation is apparent. This demonstrates that fast training with acceptable BER is achievable with the DFE structure.

7. Implementation Issues

In the GSM transmission system, each time slot consists of 148 bits with a 26 bit 'midamble' for training. Training is performed using the midamble sequence and then the entire time slot is classified. This necessitates the storage of the entire burst for a period equal to the decoding time. The entire 148 analog sequence can be stored efficiently in a circulating Charge Coupled Device (CCD) ring as electronic charges. The equalizer input taps can be integrated with this CCD ring to provide a compact and high speed equalizer structure in about the same chip area as the A/D converter and digital memory that is required in conventional systems.

The equalizer can be integrated right at the RF front end of the mobile receiver. This portion of the receiver has all the high speed analog electronics fabricated with advanced semiconductor processes such as BiCMOS and HEMT. This enables the analog components such as the CCD, capacitors, resistors, low threshold transistors etc., to be fabricated with a lower marginal cost and thereby reducing the cost of the entire system. With such optimized architectures, the performance of the equalizer can easily be increased tenfold over that of the general purpose prototype CLNN32 chip.

8. Discussion and Conclusions

The low power consumption of the analog hardware is of critical importance in portable and mobile applications. Consider the equalization of the $H_{mp}(z)$ channel mentioned earlier using the conventional linear equalizer running on the TMS320C40 processor and a fast A/D converter using the LMS algorithm and on the CLNN32 chip. A simple energy calculation shows that the DSP and the A/D converter consumes approximately 700 μJ to acquire the channel while only 20 μJ is dissipated in the Boltzmann chip for the same task. When optimized for embedded applications, the energy consumption can be further reduced making it an attractive candidate for mobile radio applications.

We have demonstrated, in low-power analog hardware, the equalization of linear, minimum and non-minimum phase, noisy and frequency selective channels. Performance figures indicate that this technique can be used in the stringent mobile digital cellular radio and PCS applications for providing high bandwidth and fast channel acquisition, concomitant with low power. This approach could also be an enabling method in other highly demanding channels.

Acknowledgments

This research was supported by the Advanced Research Projects Agency of the Department of Defense and was monitored by the Air Force Office of Scientific Research under contracts F49620-90-C-0042 and F49620-92-C-0075. The United States Government is authorized to reproduce and distribute reprints for governmental purposes notwithstanding any copyright notation hereon.

References

[1] A. Jayakumar, J. Alspector, "A Cascadable Neural Network Chip Set with On-Chip Learning using Noise and Gain Annealing", presented at the *1992 IEEE Custom Integrated Circuits Conference*, May 1992, Boston, Massachusetts and in the *Proceedings of the 1992 IEEE Custom Integrated Circuits Conference* pp. 19.5.1-19.5.4.

[2] J. Alspector, A. Jayakumar, S. Luna, "Experimental Evaluation of Learning in a Neural Microsystem," in *Neural Information Processing Systems 4*, ed. J.E. Moody, S.J. Hanson, R.P. Lippmann, Morgan Kauffman, San Mateo, CA, 1992, pp. 871–878.

[3] A. Jayakumar, J. Alspector, "An Analog Neural Network Co-Processor System for Rapid Prototyping of Telecommunications Applications", at the *IEEE/INNS International Workshop on Application of Neural Networks to Telecommunications*, October 18-20, 1993, Princeton, New Jersey and in the *Proceedings of the International Workshop on Application of Neural Networks to Telecommunications*, Lawrence Erlbaum Associates, Hillsdale, New Jersey, 1993, pp. 13-19.

[4] T. X. Brown, "Neural Networks for Adaptive Equalization", at the *IEEE/INNS International Workshop on Application of Neural Networks to Telecommunications*, October 18-20, 1993, Princeton, New Jersey and in the *Proceedings of the International Workshop on Application of Neural Networks to Telecommunications*, Lawrence Erlbaum Associates, Hillsdale, New Jersey, 1993, pp. 27-33.

[5] J. G. Proakis, *Digital Communications*, McGraw-Hill, New York, 1989.

Intelligent Techniques for the GSM Handover Process

Martin Junius and Olrik Kennemann

Communication Networks, Aachen University of Technology
D-52056 Aachen, Germany
EMail: `mj@dfv.rwth-aachen.de` *and* `okm@dfv.rwth-aachen.de`

Abstract

The *Global System for Mobile Communications* (GSM) provides measurement data about the radio propagation situation that is used for power control and handover decisions. This paper presents different methods to improve the performance of the handover procedure, using fuzzy control for the handover process and pattern recognition techniques for locating mobile stations. The result aims towards a mobile communication system learning from the supplied information and adapting to modified conditions.

1 Objectives

The GSM Global System for Mobile Communications provides measurement reports for the radio resource management, delivering data about the radio propagation as seen by the mobile station. Currently these data are used for handover and power control decisions.

So already every mobile station in the GSM is a measurement device. It is desirable to process the provided data—the measurement reports—using intelligent techniques. The main problem of course is, *how* these measurement reports could be exploited.

The rationale behind the work presented here is that radio propagation is a very particular phenomenon, resulting both in complex behaviour of a mobile communication system and in radio propagation patterns characteristic for a specific location or event.

Two approaches and their feasibility will be shown in this paper: pattern recognition based on characteristic receive level patterns, and fuzzy control for radio resource management decisions.

Pattern recognition can be used to locate mobile stations in the network and to record and recognize/predict specific events and locations. Techniques used successfully for this purpose include a specifically designed neural network, fuzzy classification, and hidden Markov models.

Fuzzy Control is suitable as a replacement for the current GSM decision process, using rule-based processing of measurements, pattern classification, and location data.

Such an intelligent radio resource management could yield:

- Handover support: recognizing critical situations, avoiding unnecessary handover
- Locating mobiles using characteristic patterns
- Evaluation of radio propagation and traffic load

2 GSM

Radio Resource Management is a function of layer 3 (network) of the GSM. It comprises among others the functions of interest: *handover*, *power control*, and *measurement reports*.

GSM defines a framework for the implementation of the handover and power control process. The process itself is *not* a GSM standard. Most GSM network operators and equipment manufacturers use proprietary version of the basic algorithm specified in [1].

2.1 Measurement Reports

The common base for the algorithms are the measurement reports of the mobile station, which are sent periodically to the base station. The base station itself provides measurements at the uplink with the same time interval.

Mobile Station measurements on the downlink: receive level and quality of the serving cell, receive level and ID code of up to 6 neighbour cells.

Base Station measurements on the uplink per TCH: receive level and quality, distance mobile to base station (timing advance), transmit power of mobile and base station.

The measurement reports of the mobile station are delivered via the SACCH *Slow Associated Control Channel* to the base station. Due to the frame length of a traffic channel multi-frame and the 1:4 interleaving of control channels, the measurement reports are available every 480 ms at the base station controller.

3 Pattern Recognition

Evaluation of measurement data recorded in a GSM network has shown that the receive level traces are to a high degree characteristic for the vehicle's route. Above all, long-term fades due to shadowing obstacles cause typical receive level traces.

Furthermore, the distributions of the receive level of the serving cell and the surrounding neighbour cells as seen by the mobile station are characteristic for a specific sector of the cell.

3.1 Sectorization

This is a method to locate a mobile station in the network. The area of a cell is divided into several smaller sectors (e.g. 800×800 km in size), for each sector a database stores the receive level distributions of the serving cell and the neighbour cells. Fig. 1 shows an example of such distributions obtained by simulation (left diagram).

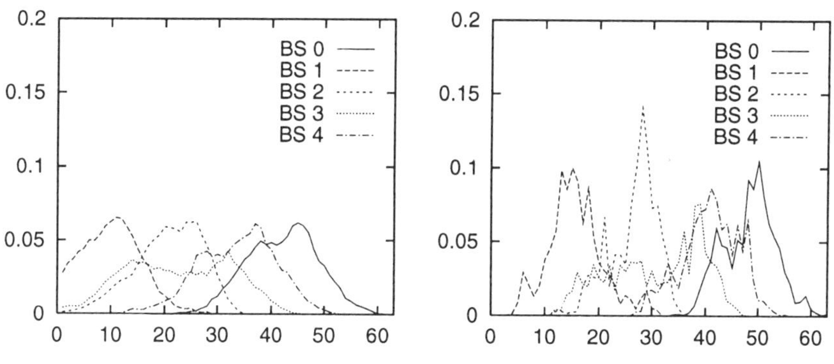

Figure 1: Sector receive level distributions

During an established call the current receive level distribution of the measurement data delivered by the mobile station is monitored and compared to the ones stored in the database.

For the actual classification process the maximum value of a sliding time window of n values is used for the distribution. This results in narrowed distributions with better classification results. Fig. 1 shows the maximum distributions (right diagram) for $n = 30$. Each class (= sector) is represented by a k-dimensional fuzzy set, with k the number of base stations and the level distributions as the membership functions. The classification fuzzifies the current maximum values of the measurement report as singletons.

The resulting recognition rate is 56 % for sector sizes of 800×800 m. Errors occur mostly at the border areas of sectors, where receive level distribution are very similar. Taking that into account and extending the sector location area to 1600×1600 m, the recognition rate is 94 %.

3.2 Receive Level Traces

The measurement reports delivered by a mobile station constitute a receive level trace, which is characteristic for the location of the mobile station. Pattern recognition techniques can used to classify receive level traces. Fig. 2 shows an example of such a trace.

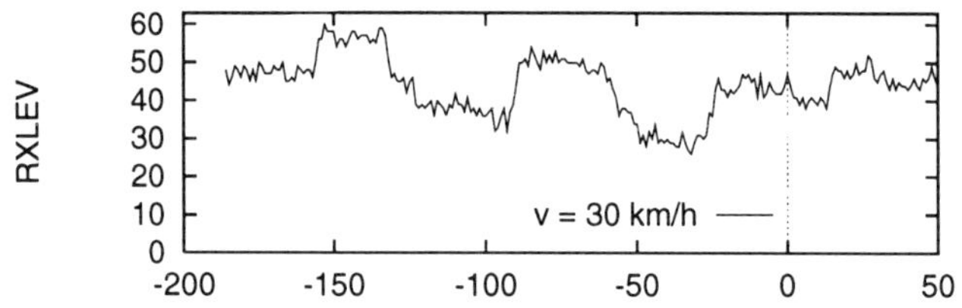

Figure 2: Receive level trace

The specific techniques presented here include a neural network and a hidden Markov model classifier.

3.3 Neural Network NOSDIC

NOSDIC — Node Splitting Distance Classifying Network — is a neural network specifically designed for the level trace pattern recognition problem [3, 4]. The network realizes a distance classifier by comparing input pattern vectors to exemplar vectors stored in the network. Input pattern vectors are the receive level traces, smoothed by an input filter.

The resulting classification has an error rate of 6 % for a sample with 68 classes and a varying number of training patterns, i.e. the recognition rate is 94 % for the sample case, above 90 % for all cases simulated so far.

The major drawback of this classifier is the velocity-dependent classification, which requires different classes for different mobility. An increasing number of classes will result in a classification with higher error rate.

Using a sufficient database of radio measurements, the classification of receive level traces can be used to locate a mobile station with a much finer resolution (approx. 100 m) than the sectorization technique. A combination of sectorization and trace pattern recognition is useful, because it reduces the number of classes for the trace pattern recognition.

The classification process can also be used in an event-oriented context, e.g. classifying traces before a handover event. In this case the database stores the history of all handover events, allowing future prediction of handover behaviour based on the currently measured receive level trace. The most useful application is to avoid ping-pong handover.

3.4 Hidden Markov Models

For the selection of a suitable method for pattern classification, one has to consider that the speed of the mobile is only known to be within a certain range and the speed cannot be assumed to be constant. According to that are the properties of the stream of measurement data. One method that takes those properties into account is based on hidden Markov models. They allow an appropriate mathematical description of the sequences of measurement data.

For simply locating mobiles at certain places like those proposed for handover a hidden Markov model like the one shown in Fig. 3 can be used.

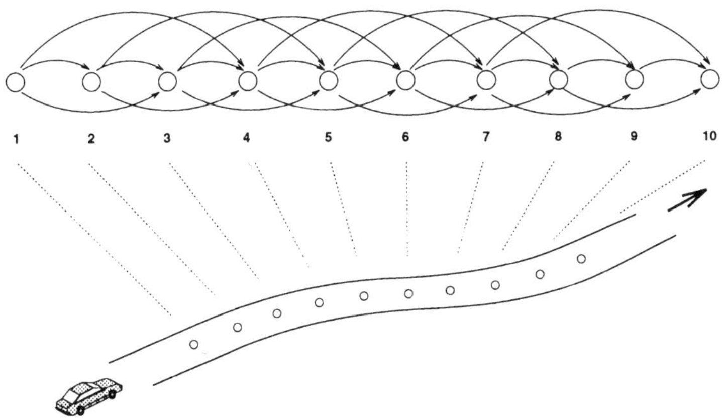

Figure 3: Description of a section of the road by a hidden Markov model

Each state of the model is assigned to a discrete location on the street or road. Transitions represent the motion of the mobile, i.e. each interval between two measurement values is represented by a transition into another state according to the velocity of the mobile. The transition probabilities can be calculated from the velocity distribution on the respective road. In addition an initial state probability has to be determined that describes the probability for each state to observe the first measurement value of the sequence under consideration in this state. This distribution can also be calculated from the velocity distribution.

What makes the model however be specific for a certain section of the road, is the probability distribution of the observation values for each state. Those values can only be determined from measurements in the real environment. From theoretical considerations it is known that those values can be assumed to be normally distributed and the variance can be determined from theoretical considerations. Thus reference measurements have to supply the mean or expectation value at each location that has been assigned a state of the model.

Evaluation is then done by calculating the probability that an incoming sequence of measurement data is represented by the model. If this probability exceeds a given threshold, then the mobile can be expected to be on the section of the road described by the model. A more precise determination of its position can be obtained by calculating the optimal state sequence, which means the sequence of states within the model that best represents the sequence of measurement values. The last state in the optimal state sequence can be expected to be the most probable state at the end of the sequence. Thus the location that has been assigned to this state is the most probable location of the mobile at the end of the sequence of measurement values.

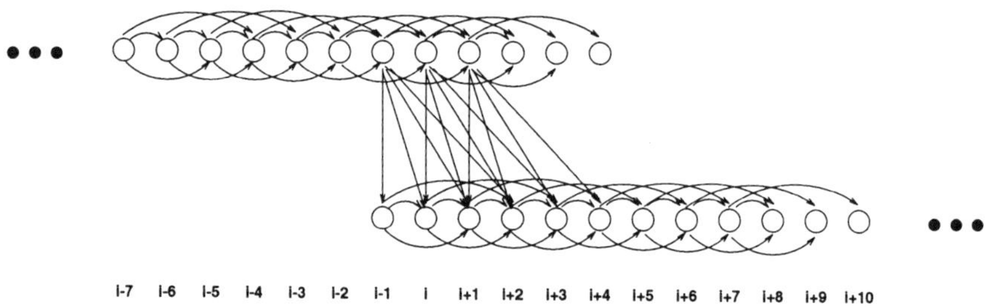

Figure 4: Continuous location by joining hidden Markov models

As described in [9], continuous location can be realized by continuously extending the hidden Markov model in the direction of the mobile's motion. This can be done by appropriately calculating the initial state probabilities of the following model allowing transitions for all states of the preceding model that have a non-disappearing probability to be the final position of the mobile and using the new model in the next step. The principle is shown in Fig. 4.

4 Integrated Intelligent System

Exploiting the available degree of freedom for the BSS and MSC decision process, an intelligent system for handover decisions can be constructed (Fig. 5).

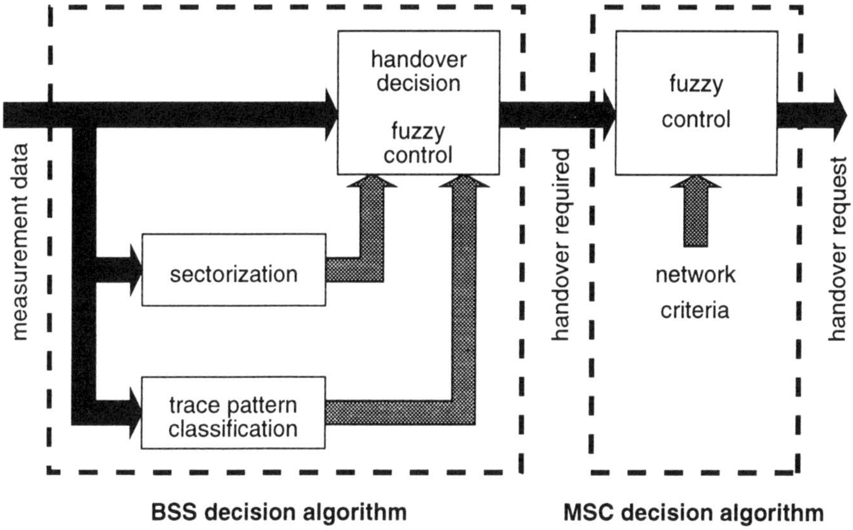

Figure 5: Integrated intelligent system

For both decision processes fuzzy control is an appropriate technique, as it has been shown in [5]. The rule-based design allows easy adaption to varying demands, such as implementation of network-directed handover.

5 Conclusion

Intelligent techniques such as fuzzy logic and neural network have been successfully applied to a variety of technical system. The work presented here has shown that the GSM radio resource management can be improved by implementing fuzzy logic and pattern recognition.

The handover decision process can be replaced with an intelligent system, without changing the GSM standard. The resulting process improves handover performance, reducing the total number of handover and avoiding ping-pong handover.

References

[1] ETSI/TC GSM, "Recommendation GSM 05.08, radio sub-system link control," January 1991.

[2] H. Hüning, "Cognition and neural network modelling," in *ICANN '92 Proceedings, Artifical Neural Networks 2* (I. Aleksander and J. Taylor, eds.), (North-Holland), 1992.

[3] H. Hüning, "A node splitting algorithm that reduces the number of connections in a hamming distance classifying network," in *International Workshop on Artifical Neural Networks IWANN '93*, (Sitges, Spain), June 1993.

[4] H. Hüning and M. Junius, "A node splitting neural network with integer inputs and application to a pattern recognition problem in mobile radio communications," in *First European Congress on Fuzzy and Intelligent Technologies EUFIT '93*, (Aachen, Germany), pp. 431–435, September 7–10 1993.

[5] M. Junius, "Intelligent GSM radio resource management using neuro-fuzzy technology," in *Proceedings The Sixth Nordic Seminar on Digital Mobile Radio Communications DMR VI*, (Stockholm, Sweden), pp. 184–191, June 13–15 1994.

[6] M. Junius, "Intelligentes Radio Resource Management — Mustererkennung mit GSM-Funkmeßdaten und Anwendung," in *ITG-Fachbericht 124 Mobile Kommunikation*, (Berlin, Offenbach), pp. 487–501, VDE-Verlag, September 1993.

[7] M. Junius and O. Kennemann, "New methods for processing GSM radio measurement data: Applications for locating, handover, and network management," in *Proceedings 44th Vehicular Technology Conference, Creating Tomorrow's Mobile Systems*, (Stockholm, Sweden), pp. 338–342, IEEE, June 7–10 1994.

[8] M. Junius and D. Marger, "Simulation of GSM handover and power control based on propagation measurements in the german d1 network," in *Fifth Nordic Seminar on Digital Mobile Radio Communications DMR V*, (Helsinki, Finland), pp. 367–372, December 1–3 1992.

[9] O. Kennemann, "Continuous location of moving gsm mobile stations by pattern recognition techniques," in *Proc. Fifth IEEE International Symposium on Personal, Indoor and Mobile Radio Communications (PIMRC'94)*, (The Hague), pp. 630 – 634, Sept. 1994.

Graph Coloring Algorithms for Assignment Problems in Radio Networks

Francesc Comellas and Javier Ozón

Departament de Matemàtica Aplicada i Telemàtica
E.T.S.E. Telecomunicació, Universitat Politècnica de Catalunya
Campus Nord C-3, Gran Capitán s/n, Barcelona, Catalonia, Spain
comellas@mat.upc.es   ozon@mat.upc.es

Abstract

Assignment problems in radio networks, like the channel assignment, may be solved by graph coloring algorithms. In this paper we compare the performance of several recent techniques on a simple coloring problem: the search for a bipartite subgraph with the maximum number of edges of a given graph. The algorithms considered are a neural network, a genetic algorithm, simulated annealing and two heuristics. The results show that one heuristic, a new proposed multiagent system, and the standard simulated annealing technique are faster outperforming the algorithms in the literature. We also show how the algorithms may be easily adapted to deal with the k-coloring of a graph and can therefore be used for solving assignment problems in telecommunications.

1 Introduction

Several issues related to the design of radio networks, as well as other problems in telecommunications, may be formulated as graph coloring problems. Consider, for example, the construction of a packet radio network which uses time division multiplexing. Since a user cannot transmit and/or receive more than one packet at a time, time slots must be associated to channels (one channel for each pair of users) in such a way that all channels that correspond to a particular user get different time slots. If this problem is modeled by graphs, users correspond to vertices, channels to edges and assigning time slots corresponds to coloring the edges so that incident edges have different colors. In a similar way we may deal with the task of assigning channels to the radio base stations in a spectrum efficient way. Particular instances of these problems are very often NP-complete optimization problems [6] and finding an optimal solution is a computationally hard task. In these cases, heuristic methods and general combinatorial optimization methods are used to obtain an acceptable answer in a reasonable time, see [5, 9, 11].

In this paper we have considered the instance of the coloring problem when only two colors are involved, and we have compared the methods described in the literature (a heuristic search [8] and a neural network [11]) with the techniques introduced in this work (simulated annealing, a genetic algorithm and a multiagent system). In Section 2 we give a short introduction to the terminology used and the exact description of the problem. In Section 3 we describe the algorithms considered. Finally, in Section 4, we present the details of the implementations, the results obtained and the extension of the algorithms to the general case.

2 Graph coloring

A *proper coloring* of a graph $G = (V, E)$ is a function from the vertices of the graph to a set C of *colors* such that any two adjacent vertices have different colors. If $|C| = k$, we say that G is *k-colored.* The minimum possible number of colors for which a proper coloring of G exists is called the *chromatic number* $\chi(G)$ of G. If $\chi(G) = 2$ then G is bipartite.

The problem of finding the chromatic number and a proper coloring of a graph is of great interest for its widespread applications in areas such as scheduling and timetabling and particularly in telecommunications. As many other problems in graph theory, it is NP-complete. In this work a variation of the graph coloring problem, the *bipartite subgraph problem*, has actually been considered. This particular problem, which is also NP-complete, consists of removing the minimum number of edges of a given graph in such a way that the remaining graph is a bipartite graph. Efficient algorithms for this problem are known only for particular graphs, e.g. triangle-free graphs with maximum degree three [2]. A heuristic algorithm, introduced by Hsu in [8], can be applied to any graph and more recently Lee, Funabiki and Takefuji [11] proposed a neural network model for this problem.

On the other hand, simulated annealing and genetic algorithms have outperformed neural network approaches when applied to several NP-complete graph problems, see [3, 4]. This fact encouraged the authors to look for the application of these techniques to the coloring problem, object of this paper.

3 Combinatorial optimization algorithms

The first technique that we have applied to the bipartite subgraph problem belongs to a general class of algorithms, known as genetic algorithms, which are inspired on the mechanics of natural selection and genetics.

In a genetic algorithm, see Goldberg [7], the starting point is a collection of possible solutions generated at random, known as *population*. A suitable encoding of each solution in the population is used to compute its *fitness* through a *cost function*. At each iteration a new population, or *generation*, is obtained by *mating* the best of the old solutions with one another. To create the next generation, new solutions are formed through *reproduction*, *crossover* and *mutation*. The solutions that will be considered for crossover are probabilistically selected according to the fitness values from the set that constitutes the current generation. This new population becomes the *parent pool*. In the Goldberg's approach a constant number of solutions are selected so that a fixed size population is maintained. Crossover creates two new *child* solutions from two solutions sampled from the parent pool. In this way, fitter parents have a better chance of producing children. This is done for the whole population. Children solutions are obtained by interchanging random parts of their parents. Some randomness is also introduced through the mechanism called *mutation* to ensure that the algorithms avoid getting stuck at local minima. Mutation changes selected parts of a solution. The crossover and mutation operations are done with fixed probabilities, thus ensuring that some solutions from the current generation will be kept in the new generation.

Once a new generation is created, the fitness of all solutions is evaluated and the best solution is recorded. The process is repeated until either the results stabilize or the optimal solution, when it can be identified, is reached.

The main aspects to decide are the representation of the solutions, the cost function and the crossover and mutation operators. Important parameters are the population size and the probabilities of crossover and mutation.

The second algorithm considered uses the simulated annealing (SA) technique which comes from the analogy made between the states of a physical system, e.g. a liquid, and the configurations of a system in a combinatorial optimization problem. If the temperature of the interacting molecules in a liquid is suddenly reduced below its freezing point, the result is a disordered glassy state with an energy higher than the true crystalline ground state. In fact the molecules are in a local energy minimum. On the other hand, if the temperature of the liquid is reduced slowly (annealing), waiting for equilibrium to be reached before a new reduction is made, the liquid freezes to the solid state through a cooling process that leads to the crystalline state, which is the global energy minimum. In the analogy with the combinatorial optimization problem the parameters being varied are equated with atomic positions in the liquid and its energy is identified with the cost function being optimized. The temperature is then defined as a control parameter related to the probability of accepting changes that worsen the state of the system, thus ensuring a more comprehensive search of the state space. Hence, in the simulated annealing, a change of state that decreases the energy is always accepted, whereas if energy increases, the change is accepted with a certain probability $e^{-\Delta E/T}$, where T is the temperature of the system. At a given temperature, a number of attempts large enough to obtain a good statistical set of trials is performed and thereafter the temperature decreased. This process is repeated and the system is gradually cooled until it is stopped according to some criteria such as a small number of changes accepted and/or a non significant reduction of the energy.

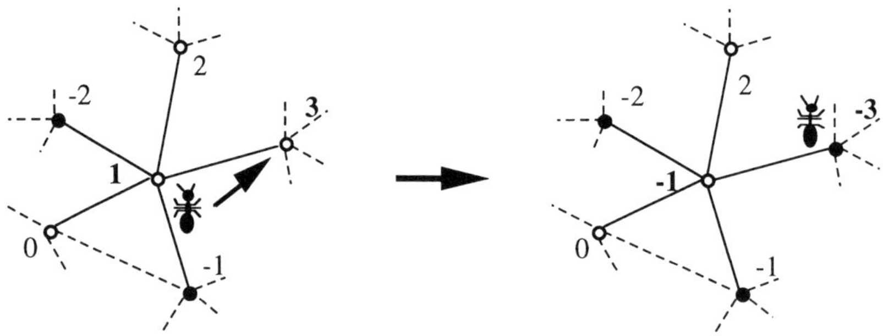

Figure 1: An ant acting on a graph.

The third algorithm we have considered is `ants`, a multiagent system where several autonomous agents work together to obtain good solutions of the bipartite subgraph problem. The `ants` algorithm starts by generating an arbitrary random coloring of the graph which is, in general, non proper, and by randomly allocating a set of agents, called *ants*, on different vertices of the graph. Once it has been initializated, the process begins to evolve by assigning to every vertex a certain value, calculated as the difference between the number of adjacent vertices with the same color and the number of adjacent vertices with different color. After that, every ant moves to the adjacent vertex that has maximum value, changes the color of this vertex and recalculates the values for it and its adjacent vertices (see Figure 1), all that with a certain probability to avoid local minima. Once each ant has moved, the process is repeated until the algorithm converges to a near-optimal solution.

4 Results and conclusions

As an illustration of the performance of the three proposed algorithms note that we may obtain from the graph with 30 vertices and 50 edges considered by Lee *et al.* [11] a bipartite subgraph with 43 edges, whereas Hsu's heuristic algorithm [8] found only a subgraph with 38 edges and Lee *et al.* with a neural network found a subgraph with 42 edges.

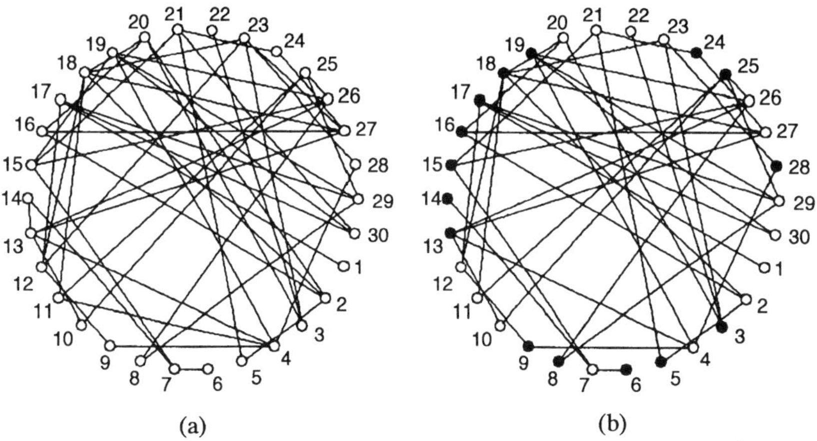

Figure 2: The graph proposed by Lee *et al.* and the solution with 43 edges found in this paper.

For our study, a large number of instances have been simulated to test the three algorithms with random graphs of orders ranging from 10 to 250 vertices and densities of 5%, 15% and 20% (the density of a graph is the ratio between the number of edges that actually has the graph and the maximum number that may contain). For each case 20 simulation runs were performed and compared with the results given by Lee *et al.* in their paper [11]. All simulations were programmed in C (less than 500 lines) and executed on a HP Apollo 715/75.

Possible solutions has been coded as lists where each position represents a vertex of the graph and has values 0 or 1 according to the color assigned to it. The cost function calculates the number of edges that must be removed from the associated graph to have a proper coloring.

In the genetic algorithm, we have used the same crossover and mutation operators than Goldberg [7]. Figure 3 shows an example of their application. Values for the relevant parameters were taken as follows: a population size of 400 individuals and probabilities of crossing and mutation of 0.95 and 0.01.

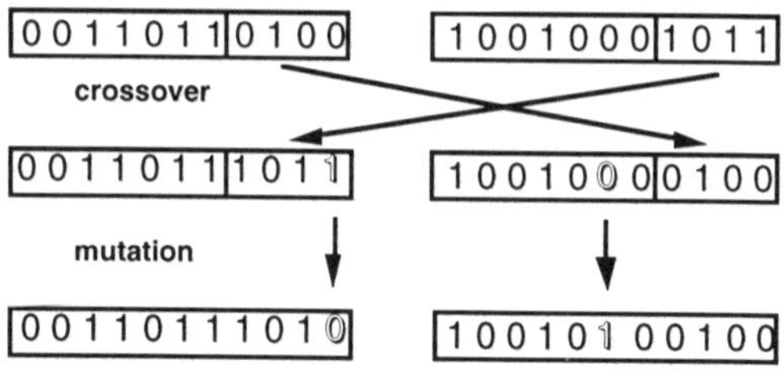

Figure 3: Crossover and mutation operators in the genetic algorithm.

For simulated annealing we have used a standard implementation as described in [1] and [10]. Simulated annealing requires careful tuning of its control parameters to achieve good results. Typical values considered were: initial temperature, T_0, ranging from 0.5 to 1.5, number of iterations for a given temperature $N_k = \alpha|E|$, where $|E|$ is the number of edges of the graph and α is a factor ranging from 1 to 10, and cooling rate exponential, $T_k = 0.9T_{k-1}$.

The third method, our multiagent system, used up to 10 agents (or *ants*) and a probability of 0.9. Simulations show that the total number of agents is not critical for the solution when the number of vertices of the graph is relatively small (less than 1000).

Table 1 shows the results corresponding to graphs with a density of 15%. Similar performances were obtained with different densities.

Finally, the new algorithms presented in this paper may be very easily adapted to solve the problem of finding a k-coloring of a given graph. For the genetic algorithm and the simulated annealing, all that is needed is to change the set of values associated to each vertex from $\{0, 1\}$ to $\{0, 1, \ldots, k-1\}$ in the list that codes a solution. The multiagent system may be also modified to deal with more colors, but in that case the cost function needs to be redefined. We have tested with success the modifications and generated easily a 3-coloring of the Lee graph that uses all its edges.

As a conclusion, the results show that the multiagent system, the simulated annealing and the genetic algorithm presented in this work perform better that the neural network approach. All three algorithms are simple and easy to implement and may be adapted to deal with the general k-coloring problem. Moreover, the multiagent system, besides finding better solutions, runs around ten times faster than simulated annealing and fifty

Node Size	**Hsu**	**NN**	**GA** max.	**GA** mean	**SA** max.	**SA** mean	**ants** max.	**ants** mean
10	6	6	6	6.0	6	5.8	6	5.8
20	24	25	24	22.5	26	24.2	26	24.7
30	49	50	54	52.1	58	52.9	58	53.4
40	90	90	94	89.9	94	91.0	96	91.1
50	128	135	139	135.4	140	137.9	142	137.4
60	191	195	199	191.1	198	193.5	200	193.7
70	246	254	257	251.9	261	257.6	265	258.4
80	311	330	330	326.2	338	331.4	339	331.6
90	390	405	406	402.9	419	412.4	421	413.1
100	478	494	492	485.0	508	502.8	514	503.4
150	1032	1060	1024	1015.0	1089	1080.7	1092	1084.8
180	1476	1502	1437	1420.5	1541	1527.9	1547	1532.2
200	1824	1860	1850	1812.1	1881	1867.4	1883	1870.1
250	2817	2854	2832	2807.6	2881	2866.3	2884	2865.1

Table 1: Comparison between Hsu's algorithm, neural networks, genetic algorithm (GA), simulated annealing (SA) and multiagent algorithm (`ants`) on graphs with density 15%.

times faster than the genetic algorithm. It may be also directly implemented on a parallel computer, thus improving its performance.

Acknowledgment

This work was supported by the *CICYT*, Spain, under grants TIC-92-1228-E, TIC94-0592 and the EU-HCM program ERBCHRX-CT920049.

References

[1] E. Aarts and J. Korst. *Simulated Annealing and Boltzmann Machines*, Chichester: John Wiley and Sons, 1989, ISBN 0-471-92146-7.

[2] J.A. Bondy and S.C. Locke. "Largest bipartite subgraph in triangle-free graphs with maximum degree three", *J. Graph Theory*, vol. 10, pp. 477–504, 1986.

[3] F. Comellas. "Using genetic algorithms for planarization problems", Computational and Applied Mathematics I, Eds. C. Brezinski and U. Kulish, Elsevier Science Publishers B.V. (North Holland), pp. 93-100, 1992, ISBN 0-444-89701-1.

[4] F. Comellas and E. Pallarès. "Optimització combinatòria i disseny de xarxes d'interconnexió", *Butll. Soc. Cat. Ciènc.*, vol. XIV (2), 1994, pp. 221–234.

[5] M. Duque-Antón, D. Kunz, and B. Rüber. "Channel assignment for cellular radio using simulated annealing", *IEEE Trans. Vehicular Technology*, vol. 42, pp. 14–21, 1993.

[6] M.R. Garey and D.S. Johnson, *Computers and Intractability: A Guide to the Theory of NP-Completeness*, New York: W.H. Freeman, 1979, ISBN 0-7167-1044-7.

[7] D. E. Goldberg. *Genetic Algorithms in Search, Optimization, and Machine Learning*, Addison-Wesley, 1989, ISBN 0-201-15767-5.

[8] C.P. Hsu, "Minimum-via topological routing", *IEEE Trans. Computer-Aided Design*, vol. 2, pp. 235–246, 1983.

[9] D. S. Johnson, C. R. Aragon, L. A. McGeoch and C. Schevon. "Optimization by Simulated Annealing: an Experimental Evaluation; Part II, Graph Colouring and Number Partitioning", *Operations Research*, vol. 39, pp. 378–406, 1991.

[10] S. Kirkpatrick, C.D. Gelatt and M.P. Vecchi, "Optimization by Simulated Annealing". *Science*, vol. 220, pp. 671–680, 1983.

[11] K. C. Lee, N. Funabiki and Y. Takefuji. "A Parallel Improvement Algorithm for the Bipartite Subgraph Problem", *IEEE Trans. Neural Networks*, vol. 3, pp. 139–145, 1992.

Fast Channel Assignment in Cellular Radio Systems

Matthias Oliver Berger

RWTH Aachen, Department of Computer Science IV
52056 Aachen, Germany
berger@i4.informatik.rwth-aachen.de

Abstract

This paper proposes a new algorithm using a nonpositive neural network model to solve channel assignment problems in arbitrary cellular mobile radio systems. Unlike traditional neural nets used in optimization problems, the proposed network is guaranteed to deliver valid solutions after every update and needs no parameter tuning. The power of the new method to solve this NP-complete problem will be shown in a number of simulations.

1 Introduction

Over the recent years, an increasing number of subscribers joined cellular mobile radio systems. This trend will continue in the future requiring efficient usage of the limited radio spectrum available to those services.

The channel assignment problem in cellular mobile radio systems is the task to assign a radio channel to each call made, while obeying interference constraints between channels adjacent in the spectrum [3, 7]. Conventional algorithms suffer from the fact that efficiently finding a valid channel assignment is known to be very difficult: channel assignment belongs to the class of NP-complete problems [2].

Other approaches to neural channel assignment have shown the general feasability of solving the problem, but suffer from slow convergence (or unknown convergence behaviour), thus employing numerous heuristics and parameters. In this paper, a neural net is proposed that runs autonomous to solve channel assignment problems. The performance is tested in a number of simulations, showing the improvement in speed and convergence.

2 Channel Assignment

In a cellular mobile radio, the world is assumed to be a plane, divided into n cells z_i. To communicate, each mobile user needs to be assigned a radio channel in the cell where it is located. The number of channels demanded in each cell is indicated by a demand vector $D = (d_i)_{i=1}^n$ where $d_i \in \mathbf{N}$ denotes the number of channels demanded in cell z_i. The overall number of channels demanded is $\hat{d} = \sum_{i=1}^n d_i$.

There are k radio channels available, represented by positive integers $1, \ldots, k$. To preserve radio spectrum, a channel assigned to a mobile phone in cell z_i may be reassigned to a second mobile phone in cell z_j, if interference constraints that ensure acceptable quality of communication are met.

A compatibility matrix is used to formulate these constraints: $C_{n \times n} = ((c_{ij}))$ with $c_{ij} = 0$ if a channel assigned in cell z_i may be reassigned in cell z_j. If $c_{ij} > 0$, then this denotes the reuse distance. For $c_{ij} = 1$ this is called the co-channel constraint: for each pair of cells, it is known whether they may use the same channel or not. To determine cell z_x where call j, $1 \leq j \leq \hat{d}$, was made, an array `cell` of size $\hat{d}$ is used. For call j, array element `cell[j]` denotes cell z_x, where it was demanded.

$K = (k_i)_{i=1}^{\hat{d}}$, $k_i \in [0 : k]$ is called a *feasible channel assignment* iff

$$|k_x - k_y| \geq c_{\texttt{cell[x]},\texttt{cell[y]}}, \forall x, y \in [1 : \hat{d}]$$

holds. Note that usually there exists more than one solution of finding a feasible channel assignment for a given problem $P = (C, D, k)$. To summarize:

Definition 1 (Channel Assignment Problem) *INPUT: A problem $P = (C, D, k)$, consisting of a compatibility matrix C for a cellular radio system with n cells, a demand vector D and the number of available channels k. PROBLEM: Find a feasible channel assignment. OUTPUT: A sequence K, indicating the channel assigned to each demand.* □

3 Neural Representation

To solve the channel assignment problem, an interference graph $G_P(V, E)$ is constructed in such a way, that its independent sets of vertices correspond with feasible solutions of the problem. The important terms from graph theory are defined in appendix A.

3.1 Channel Assignment as an Independent Set Problem

V consists of vertices v_i^x with $i \in [1:k]$ for every demand $x \in [1:\hat{d}]$. Two vertices v_i^x and v_j^y, $j \in [1:k], y \in [1:\hat{d}]$ are connected by an edge, iff

1. $x = y \wedge i \neq j$ or
2. $x \neq y \wedge |i-j| < c_{\mathtt{cell[x]},\mathtt{cell[y]}}$.

This construction results in a complete graph with k vertices for every demand $x \in [1:\hat{d}]$. Vertices v_i^x and v_j^y, $x \neq y$, are connected iff the distance between channels i and j is smaller than what is prescribed by the compatibility matrix.

Definition 2 (Interference Graph) Let $K_k^x(V,E)$ be complete graphs $\forall x \in [1:\hat{d}]$ with

$$\begin{aligned} V(K_k^x) &= \{v_1^x, \ldots, v_k^x\}, \\ E(K_k^x) &= \{(v_i^x, v_j^x) | i \neq j \wedge v_i^x, v_j^x \in V(K_k^x)\}. \end{aligned}$$

Then graph $G_P(V,E)$ is called interference graph of P with

$$\begin{aligned} V(G_P) &= \bigcup_{x=1}^{\hat{d}} V(K_k^x), \\ E(G_P) &= \bigcup_{x=1}^{\hat{d}} E(K_k^x) \cup \bigcup_{x=1}^{\hat{d}} \bigcup_{\substack{y=1 \\ x\neq y}}^{\hat{d}} \{(v_i^x, v_j^y) | \forall i,j \in [1:k], \\ & \quad v_i^x \in V(K_k^x), v_j^y \in V(K_k^y) : \\ & \quad |i-j| < c_{\mathtt{cell[x]},\mathtt{cell[y]}}\}. \end{aligned}$$

□

This graph is mapped on a two-dimensional neural net with $k \times \hat{d}$ neurons.

3.2 Nonpositive Neuron Model

This paper uses a modified version of the two-dimensional Hopfield-Tank neural network [4] consisting of k clusters each with $\hat{d}$ neurons, resulting in $k \times \hat{d}$ processing elements. The following notation is used: Let V_{ix} denote the output of the neuron with row index i and column index x. $\mathbf{V} = (V_{ix})$ is called the state of the network. The connection weights are given by a weight matrix $\mathbf{W} = (w_{ix,jy})$. Thresholds are coded using matrix $\mathbf{\Theta} = (\theta_{ix})$.

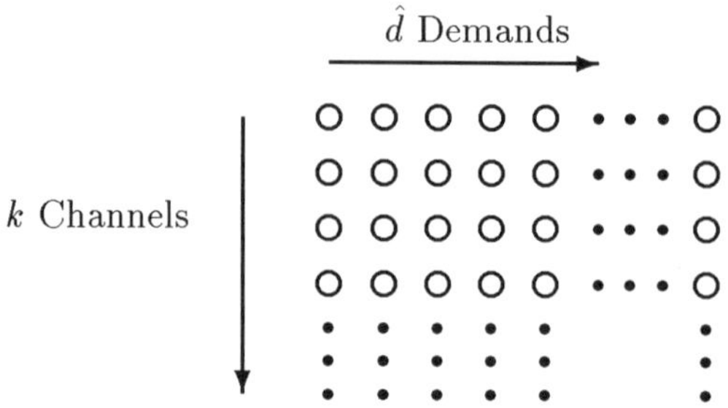

Figure 1: Mapping the channel assignment problem on a two-dimensional neural net.

Nonpositive neural networks are symmetric and use only weights $w_{ix,jy} \in \{0, -1\}$ with zero diagonal $w_{ix,ix} = 0$ and zero threshold $\theta_{ix} = 0$. States are binary only ($V_{ix} \in \{0, 1\}$). They are calculated using the common *sgn* function:

$$\operatorname{sgn}(x) = \begin{cases} 1 & \text{if } x \geq 0 \\ 0 & \text{if } x < 0 \end{cases} \tag{1}$$

This class of Hopfield networks was first analysed by Shrivastava et.al. [8]. They showed that all stationary points have a one-to-one correspondence with independent sets of certain undirected graphs. Moreover, they prooved that the nonpositive network always converges to a stationary point in at most $2k\hat{d}$ steps, provided each element is updated once in the first $k\hat{d}$ steps and again in the following $k\hat{d}$ steps.

3.3 Mapping

The state of a neuron represents the following:

$$V_{ix} = \begin{cases} 1 & \text{if channel } i \text{ is assigned to call } x \\ 0 & \text{otherwise} \end{cases} \tag{2}$$

for all $i \in [1, \ldots, k], x \in [1, \ldots, \hat{d}]$.

Neuron V_{ix} correspondends with vertice v_i^x of the interferencegraph with $x \in [1 : \hat{d}], i \in [1 : k]$. The weights are determined by the negated adjacency matrix of the graph G_P. They can be expressed using a weight matrix $\mathbf{W}_{(k\hat{d})\times(k\hat{d})}$ with $x, y \in [1 : \hat{d}], i, j \in [1 : k]$ and

$$w_{ixjy} = -\delta_{xy}(1 - \delta_{ij}) - (1 - \delta_{xy})(|i - j| < c_{\texttt{cell[x]},\texttt{cell[y]}})$$

where δ denotes Kroneckers delta function. No heuristics and no parameter tuning are needed.

The energy function has the form

$$E = \sum_{i=1}^{k}\sum_{j=1}^{k}\sum_{x=1}^{\hat{d}}\sum_{y=1}^{\hat{d}}(-\delta_{xy}(1-\delta_{ij}) - (1-\delta_{xy})(|i-j| < c_{\texttt{cell[x]},\texttt{cell[y]}}))V_{ix}V_{jy}$$

Now the new states are calculated sequentially starting from random binary states using the *sgn*-transition function. Convergence occurcs after at most $2k\hat{d}$ steps.

4 Simulation Results

The discrete network model described in the previous section was simulated on a sequential computer using sequential update. Several channel assignment problems were computed to measure the performance of the proposed system. All problems were taken from Funabiki and Takefuji [1]. Only P_2 was used by Kunz [5]. Simulation results can be found in tables 1, 2 and 3.

Table 1: Average number of satisfied demands (in percent).

Problem	New	Funabiki
P_1	100,0%	100,0%
P_2	95,3%	100,0%
P_3	95,7%	100,0%
P_4	92,4%	100,0%
P_5	93,8%	100,0%
P_6	94,3%	100,0%
P_7	98,9%	100,0%
P_8	95,4%	100,0%

The values given are derived from 25 simulation runs for each problem, as experience showed that at least 25 runs were needed to achive statistically reasonable results. They compare the average number of satisfied demands by the net, the convergence rate and the average number of iterations needed until the algorithm solved the problem against the nets proposed by Funabiki [1] and Kunz [5]. Lochtie's approach [6] is very similar to the one by Kunz and does not publish solutions on any problems of realistic size.

Obviously, Funabiki's and Kunz's algorithm are both slower than the new one. Comparing the number of demands satisfied and the convergence

Table 2: Convergence rates (in percent).

Problem	New	Funabiki
P_1	100,0%	100%
P_2	100,0%	9%
P_3	100,0%	93%
P_4	100,0%	100%
P_5	100,0%	100%
P_6	100,0%	79%
P_7	100,0%	100%
P_8	100,0%	24%

rate, the Funabiki's trade-off becomes clear. Although the net can satisfy 100% of the demands in Problem P_2, it converged only in 9% of the simulation runs. Kunz does not give any convergence rate or average number of demands satisfied.

Table 3: Average number of iterations.

Problem	New	Funabiki	Kunz
P_1	2,0	21,2	-
P_2	2,0	294,0	2450
P_3	2,0	147,8	-
P_4	2,0	117,5	-
P_5	2,0	100,3	-
P_6	2,0	234,8	-
P_7	2,0	85,6	-
P_8	2,0	205,6	-

5 Conclusion

This paper introduced a new algorithm to solve channel assignment problems in cellular mobile radio systems. Constraints were coded in a compatibility matrix, channel demands in a demand vector. Nonpositive networks have guaranteed convergence. As can be seen from the simulation results, the new network is much faster than the competition. Future research will include dynamic neural channel assignment.

A Graph theory

The following standard definitions were used.

Definition 3 *A simple graph $G(V, E)$ consists of a finite set V of vertices and a specified two element subset of edges $E \in V^{(2)}$, where each edge corresponds to a different pair of distinct vertices. To simplify notation, the two element set $\{u, v\} \in E$ is denoted by uv and u and v are called adjacent vertices in G. An adjacency matrix $A_{|V| \times |V|} = ((a_{ij}))$ of a graph G describes the set E of G: $a_{ij} = 1$ if vertex i is adjacent to vertex j, 0 otherwise. $W \subset V$ is called independent if $W^{(2)} \cap E = \emptyset$.* □

It is assumed that G is undirected, i.e. the order between the two vertices of an edge is not important. A class of graphs called the complete graphs will be introduced next.

Definition 4 *A simple graph $G = (V, E)$ with n vertices is called complete graph or K_n if uv belongs to E whenever $u \neq v$.* □

References

[1] Funabiki, N., Takefuji, Y., "A Neural Network Parallel Algorithm for Channel Assignment Problems in Cellular Radio Networks", *IEEE Trans. Veh. Tech.*, vol. 41, no. 4, pp. 430–437, 1992.

[2] Garey, M.R., and Johnson, D.S., *Computers and intractability*, New York: W.H.Freeman, 1979

[3] Hale, W.K. "Frequency Assignment: Theory and Applications", *Proc. IEEE*, vol. 68, pp. 1497–1514, 1980.

[4] Hopfield, J.J., and Tank, D.W., "Neural computation of decisions in optimization problems", *Biol. Cybern.*, vol. 52, pp. 141–152, 1985.

[5] Kunz, D., "Channel Assignment for Cellular Radio Using Neural Networks", *IEEE Trans. Veh. Tech.*, vol. 40, no. 1, Pt. 2, pp. 188-193, 1991.

[6] Lochtie, G.D., "Frequency channel assignment using artificial neural networks", in *Proc. 8th Int. Conf. Antennas and Propagation*, vol. 2, pp. 948-951, Edinburgh, 1993.

[7] MacDonald, V.H., "The Cellular Concept", *Bell. Syst. Tech. J.*, vol. 58, No. 1, pp. 15–41, 1979.

[8] Shrivastava, Y., Dasgoupta, S., Reddy, S.M., "Guaranteed Convergence in a Class of Hopfield Networks", *IEEE Trans. Neur. Net.*, vol. 3, no. 6, pp. 951–961, 1992.

Link Bandwidth Allocation in Multiservice Networks Using Neural Technology

M.B. Zaremba[(1)], K.-Q. Liao[(1)], G. Chan[(2)], M. Gaudreau[(2)]

*[(1)] Département d'informatique, Université du Québec à Hull
Hull, Québec J8X 3X7 Canada*

*[(2)] Industry and Science Canada, 300 Slater Street
Ottawa, Ontario K1S 3O6 Canada*

Abstract

Application of neural techniques to the optimization of link bandwidth allocation in Integrated Services Digital Networks is considered in this paper. The problem is presented as a decision problem defined in a multidimensional space. The decision to accept or reject a call for service is based on a performance measure. In this paper, the optimal decision is determined using a dynamic trunk reservation policy. A model of accurate approximation for an integrated services system is given. A variety of cases, particularly involving different statistical characteristics of the integrated services network, have been investigated and are discussed in the paper.

1 Introduction

Two transfer modes (switching and multiplexing), STM (Synchronous Transfer Mode) and ATM (Asynchronous Transfer Mode), have been proposed for the implementation of ISDN (Integrated Services Digital Networks). STM is based on circuit switching while ATM is basically a packet switching technique. Irrespective of the transfer mode used, call admission control (acceptance or rejection of a request to establish a communication) is an important aspect of maximizing the network resources utilization while providing a high grade of service. In this paper, we consider the application of neural techniques to the optimization of link bandwidth allocation in ISDN [4, 9, 10], a sub-problem of network resources allocation.

In a circuit-switched network (STM mode), a link is composed of a certain number of channels (for example, channels of 64 Kbps). Different services request different numbers of channels. For example, a telephone call requests one single 64 Kbps channel while a video-conference call demands 24 channels for a total capacity of 1.5 Mbps. In an ISDN, since the number of channels requested by a video-conference call is much larger, video-conference calls are blocked with a higher probability if there is no protection scheme for them.

We have the same problem in an ATM ISDN. At call admission control, when a call arrives, we have to estimate the network performance at the cell level (QOS, quality of service) making the assumption that the call is accepted [2, 6]. If the estimated performance is satisfactory, the call can be accepted if the factor considered is performance at the cell level (cell delay or loss). However, at the call level, it is also necessary to have an adequate grade of service (GOS). As the blocking probability of wide-band (WB) services is high without some protection mechanism, we have to protect the wide-band services against the overload of narrow-band (NB) services. With such a mechanism, we do not always accept a call when the estimated QOS is acceptable.

In general, we should develop control mechanisms to protect services requiring a large number of channels in the STM case and services with a large effective bandwidth [2, 6] in the ATM case. The objective is to maximize network resources utilization rate while guaranteeing a certain blocking probability for each wide-band service. A well designed control mechanism not only protects wide-band services but also eliminates or decreases the effect of garbage channels [10].

Different techniques have been proposed for link bandwidth allocation in ISDN. In trunk reservation [1], call acceptance or rejection is only a function of the number of free channels. This technique is not optimized if we consider the link utilization rate, although it can be very successful in protecting WB traffic against the overload of NB traffic. In [10], Miyake developed an optimization method based on Markov decision process theory. In this method, the acceptance or rejection of a call is a function of the detailed system state which is described by the numbers of calls in service of all traffic types. Although this method gives the optimal solution, the computation is so time-consuming that it cannot be used in practice.

In order to develop an efficient algorithm, a method called dynamic link bandwidth allocation was proposed in [9] where a queuing model was also presented. Although this method gives only a sub-optimal solution, the computation takes far less time than the method presented in [10]. Several numerical studies have shown that the solution obtained by this method is very close to the optimal solution in most cases. The proposed queuing model, which is very efficient as far as numerical calculations are concerned, produces highly accurate results and has been adopted in this paper.

The link bandwidth allocation problem can be presented as a decision problem defined in a multidimensional space. As such, it is easily amenable to a form that can be efficiently handled by neural networks. The main advantage of applying neural technology is the possibility of real-time quasi-optimal operation of the neural system under varying operating conditions of an integrated service network. The decision to accept or reject a call for service has to be based on a performance measure. In this paper, the optimal decision is determined using a dynamic trunk reservation policy. A model of accurate approximation for an integrated services system is presented.

2 Dynamic Trunk Reservation Policy Model

The system we consider has M channels. In order to simplify the presentation, we investigate the case with two call types, where the decision generated by the neural network is to accept or reject an incoming narrow-band (NB) call, which requires a smaller number of channels (without sacrificing generality, the number is assumed to be 1). A wide-band (WB) call requests N channels. The arrival processes are assumed to be Poisson with arrival intensities λ_1 and λ_2 respectively. The service times are exponentially distributed with mean holding times $1/\mu_1$ and $1/\mu_2$.

In the dynamic trunk reservation policy [9], a control parameter r_2 is introduced to protect the WB traffic against the overload of NB traffic. When the number of WB calls in service is less than or equal to r_2, N channels are reserved for WB traffic, i.e., when the number of free channels is less than or equal to N, the NB calls are blocked.

In [9], the system is analyzed by decomposition. The analysis is based on the realistic assumption that WB service time is significantly longer on average than NB service time. Consequently, the NB traffic will have sufficient time to approach the equilibrium conditions for each state of WB state. Based on this practical system behavior, the model assumes that the transient behavior of the system between the instants where the number of WB calls in service changes can be neglected.

The periods during which the trunk reservation operates, with less than N free channels, are called transient periods. These transient periods are distributed in a complex fashion. In order to make the WB process tractable, this complex distribution is approximated with an exponential distribution having the same mean. This assumption, with the previous assumptions of Poisson arrivals and exponential service times, enables us to model the WB process as a Markov process. After the calculation of WB state probabilities, we can easily compute the blocking probabilities (PB_1, PB_2) of NB and WB traffic (for details, see [9]). In effect, the approximate model is based on the following two assumptions:

- The narrow-band occupancy achieves a stationary distribution while there are a fixed number of wide-band calls in service except during the transient periods, since in that case, the NB process is a pure death process;
- The transient times are exponentially distributed.

We now proceed with the determination of the control parameter r_2. The objective is to:

maximize

$$U = \frac{\frac{\lambda_1}{\mu_1}\left(1 - PB_1\right) + \frac{\lambda_2}{\mu_2}N\left(1 - PB_2\right)}{M}$$

subject to the constraint

$$PB_2 \leq C$$

Here U denotes the utilization rate of the system which is a function of r_2. For each set of system parameters, we can determine the optimum value of r_2 (denoted by r^*_2). We then have the sub-optimum solution for each system state (x_1, x_2) where x_1 denotes the number of NB calls in service and x_2 denotes the number of WB calls in service. More specifically, the solution will reject a NB call, if $x_2 \leq r^*_2$ and if $M-(x_1+x_2N) \leq N$. Here, $M-(x_1+x_2N)$ gives the total number of free channels in state (x_1, x_2).

Remark: We here assume that the network dimensioning takes into account the possible traffic variations and that by a proper choice of the parameter r_2, the constraint $PB_2 \leq C$ can always be satisfied.

A number of studies have shown that the utilization rate is a quasi-convex function of r_2, and it is easy to see that PB_2 is a decreasing function of r_2. This allows us to devise a more efficient algorithm to find the optimal solution instead of searching all possible values of r_2, from -1 to r_0-1 (r_0 is the maximum number of WB calls that can be in service simultaneously).

The algorithm is:

1) set $r_2 = r_0$ and $U(r_2)=0$.

2) set $r_2=r_2-1$; calculate PB_1 and PB_2.

3) Calculate $U(r_2)$. If $U(r_2) \geq U(r_2+1)$ and if $PB_2 \leq C$, return to step 2; otherwise $r^*_2 = r_2 + 1$.

3 Neural Implementation

3.1 Network Architecture

Feedforward, multi-layer architectures have been primarily investigated in this paper. The input vector presented to the network was constituted by the following data:

M	- total number of channels
N	- trunk size of the wide-band call
λ_1/μ_1	- ratio of the arrival rate to the service rate of narrow-band calls
λ_2	- arrival rate of wide-band calls
μ_2	- service rate of wide-band calls
C_2	- maximum loss probability for wide-band calls
x_1	- number of narrow-band calls
x_2	- number of wide-band calls.

The desired binary output of the network, which was required for the supervised learning of the network, was calculated in the following way. For a given pair of the number of calls requested (x_1, x_2) for the two types of calls, the decision is calculated such that it minimizes the total loss or maximizes the link utilization rate subject to the condition $PB_2 \leq C_2$, where PB_2 is the blocking probability of WB traffic. The (x_1, x_2) space is covered according to a quasi-uniform distribution. The program that computes the required values gives greater space, at a pre-declared ratio, to the subspace close to the decision surface.

3.2 Scope of the Tests

The input elements that are inherently subject to large variations are λ_1 and λ_2. They represent conditions that may change over a longer period of time, for instance during a day, as well as quite instantaneously over a short time interval. Our analysis concentrated, therefore, mainly on the impact of these input variables on the performance of the neural controller. In order to gain better insight into the structure of the problem, the decision surfaces were found by calculating the desired service acceptance signal using algorithms in Section 2. The resulting surfaces are illustrated in Figures 1 and 2.

The graphs were obtained as a result of calculation of 5,000 pairs (x_1, x_2). The points that still belong to the space of accepted calls were retained. It can be noticed that the surface obtained for variable λ_2 exhibits additional nonlinearities in the area of smaller values of the arrival rate of wide-band calls. This non-monotonic variation was also observed by Gimpelson [5] in a loss system, and by De Serres, Mason [3] and Liao, Mason [8] in a combined queueing and blocking system. This oscillatory variation is due to the effective bandwidth reservation of an NB call. An NB call can keep a WB call out of service by leaving an insufficient number of channels.

3.3 Test results

The architecture that most consistently provides a high ratio of success over a broad range of operating conditions is a three-layer architecture 8-10-3-1.

For a steady-state situation the neural network is able to operate at a success rate of 99.0-99.5%. The success rate diminishes when we try to cover a broader spectrum of operating conditions. For variable arrival rate of narrow-band calls, λ_1, it is situated between 96.8% (for $\lambda_2 = 0.3$) and 97.8% (for $\lambda_2 = 0.1$). The value of λ_1 changed in our tests between 20 and 100. Variations in λ_2, from 0.05 to 0.4, have a more pronounced impact on the network performance. The lowest rate, equal to 94.1%, was observed for $\lambda_1 = 40$, and the highest for $\lambda_1 = 100$, and equal to 97.2%. Generally, the neural network performs better for higher values of λ_1. A single network covering both variables (λ_1 and λ_2) generated 95.7% of correct responses.

4 Conclusions

Neural networks were shown to perform favorably in the process of the acceptance or rejection of a call in multiservice communication networks. Networks that have learned the patterns for all different call arrival rates perform at 95% of the desired result obtained using the dynamic link bandwidth allocation method. In practice, the network performance can only be increased, since in many situations, where the pattern of traffic changes comparatively slowly, at a rate that is slower than the neural network learning rate, the performance will reach the level of 99%.

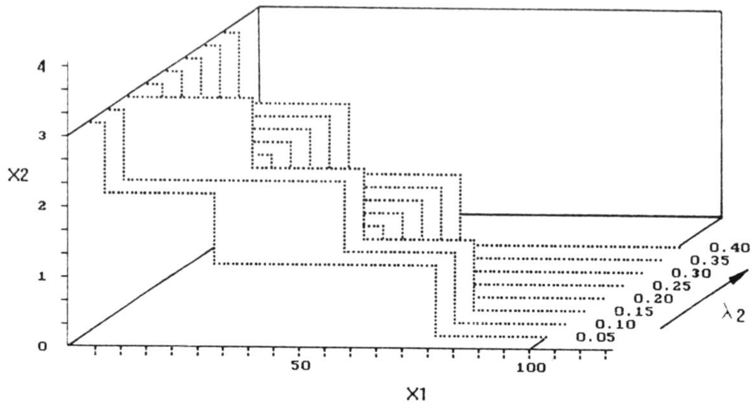

Fig. 1 Decision surface for λ_1 = constant

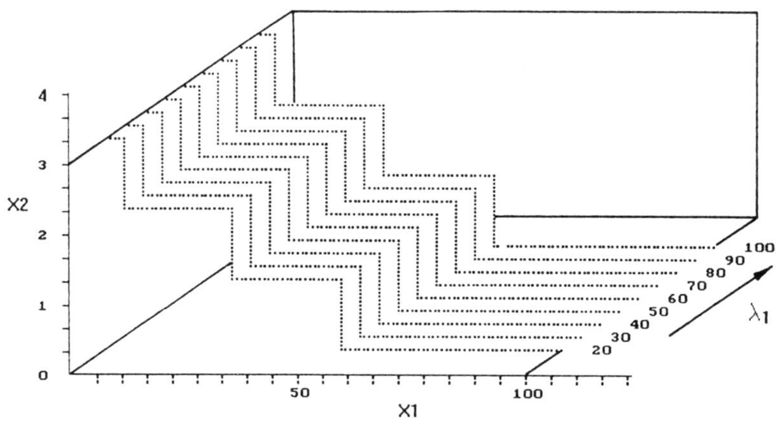

Fig. 2 Decision surface for λ_2= constant

Taking into account the shape of the decision hypersurface, which can be rapidly evaluated on-line for some cases, we are currently considering the applicability of constructive methods for fast generation of feed-forward neural networks optimized for changing operating conditions.

5 Acknowledgment

The authors acknowledge financial support of consecutively Department of Communication and Industry and Science Canada through their Program of Centres of Excellence (PCE).

References

[1] J.M. Akinpelu, "The Overload Performance of Engineered Networks with Nonhierarchical and Hierarchical Routing", *International Teletraffic Congress ITC-10*, Paper 3.2.4, June 1983.

[2] J.J. Bae and T. Suda, "Survey of Traffic Control Protocols in ATM Networks", *Proceedings of IEEE*, vol. 79, Feb. 1991.

[3] Y. De Serres and L. Mason, "A Multiserver Queue with Narrow and Wide Band Customers and Wide Band Restricted Access", *IEEE Trans. on Communications*, vol. COM-36, pp. 675-684, June 1988.

[4] Fochini, B. Gopinath and J.F. Hayes, "Optimum Allocation of Servers to Two Types of Competing Customers", *IEEE Trans. on Communications*, vol. COM-29, July 1981.

[5] L.A. Gimpelson, "Analysis of mixtures of wide- and narrow-band traffic", *IEEE Trans. on Communications*, vol. COM-13, pp. 258-266, Sept. 1965.

[6] R. Guerin, H. Ahmadi and M. Naghshineh, "Equivalent Capacity and its Application to Bandwidth Allocation in High Speed Networks", *IEEE JSAC*, vol. 9, Sept. 1991.

[7] A. Hiramatsu, "ATM Communications Network Control by Neural Networks", *IEEE Trans. on Neural Networks*, vol. 1, March 1990.

[8] K.-Q. Liao and L. Mason, "An approximate Performance Model for a Multislot Integrated System", *IEEE Trans. on Communications*, vol. COM-37, pp. 211-221, March 1989.

[9] K.-Q. Liao, Z. Dziong and L. Mason, "Dynamic Link Bandwidth Allocation in an Integrated Services Network", *IEEE ICC'89*, Paper 31.5, Boston, June 1989.

[10] Miyake, "Optimal Trunk Reservation Control for Multi-Slot Connection Circuits", *Transactions of IEICE* (Japan), Nov. 1988.

Dimensioning of Telephone Networks using a Neural Network as Traffic Distribution Approximator

AP Engelbrecht and I Cloete

Department of Computer Science
University of Stellenbosch
Stellenbosch, 7600
South Africa
ian@cs.sun.ac.za

Abstract

A feedforward neural network is used to approximate the distribution of both primary (first-offered) traffic and overflow traffic in a telephone network. We show that a neural network accurately approximates both primary and overflow traffic distributions, and if utilized in an automated dimensioning model, reduces the complexity of that model.

1 Introduction

Dimensioning is the process of determining link sizes in communication networks such that dimensioning costs are minimized and subscriber demands are satisfied. Increases in the subscriber population of a telephone network and increases in the telphone usage per subscriber cause traffic demand on links between switches to increase. In order to account for these growths in traffic demand, the number of circuits in affected links must be increased. However, there is a trade-off between network expansion costs and performance of a telephone network: a large number of circuits implies a higher financial cost, while a smaller number of circuits implies an increase in overflow traffic and congestion. Methods for automated dimensioning are therefore essential for cost-effective management and planning.

Traditionally dimensioning was done manually using traffic tables and graphs [2,7]. In the past two decades, automated dimensioning models were developed which simulate these traffic tables and graphs [3]. In this paper we investigate the utilization of a feedforward neural network (FFNN) in an automated dimensioning model. We show that a neural network (NN) is an effective and reliable traffic distribution approximator. We also show that the utilization of a NN in an automated dimensioning model reduces the complexity of the dimensioning model.

2 Traffic models

A traffic model is a mathematical model which characterizes the traffic in a telephone network. The traffic model is used to derive a relation among the traffic offered to a link, the traffic carried on the link and the traffic which overflows from that link.

Two offered traffic types are encountered in a telephone network, i.e. primary (first-offered) traffic and overflow traffic. The statistical characteristics of these two traffic types differ, in the sense that the former is modelled as a Poisson process, the latter as an Interrupted Poisson Process [5]. Under the assumptions of full-availability and that lost calls are removed, the Erlang B distribution is used to characterize primary traffic, whereas the equivalent random theory (ERT) is used to model overflow traffic [3,6,8]. Given the traffic load on a link, the required grade-of-service and the type of traffic offered to the link, the respective distribution formulas are used to calculate the number of circuits needed to satisfy the specified demand.

3 Dimensioning models

Historically, dimensionining is performed manually. With manual dimensioning, telephone engineers use tables and graphs to plan network expansion [2]. In addition to being time consuming, manual dimensioning has the drawback that only small telephone networks, or a small part of a larger telephone network can be considered, due to the enourmous complexity of the dimensioning problem. The introduction of computers into the management of telephone networks presents the advantages of automated dimensioning models. With such automated models, larger telephone networks can be dimensioned as a whole, instead of dividing the network into small subnetworks and manually dimensioning each subnetwork. These dimensioning models implement formulae to simulate the tables and graphs used in manual dimensioning.

The automated dimensioning process consists of two phases. During the first phase the number of circuits in each link is calculated using an appropriate traffic model. The second step is an optimization process, where the objective is to minimize the dimensioning cost. In this paper we concentrate on the first dimensioning phase.

Two types of links are identified in a telephone network, i.e. primary links which carry first-offered traffic only, and overflow links which carry traffic that overflows from congested routes. Since the two traffic types are characterized by different traffic distributions, the dimensioning model must employ different traffic models. The Erlang B and ERT formulas are respectively used by the model to dimension primary and overflow links.

Let M and N respectively denote the traffic offered to a link and the number of circuits in that link. The Erlang B function (1) expresses the probability that an arriving call is rejected because no circuit is available:

$$B = E(M,N) = \frac{M^N/N!}{\sum_{i=0}^{N} M^i/i!} \tag{1}$$

Given the required blocking probability B, also referred to as the grade-of-service, and the traffic load M, equation (1) is solved for N – the number of circuits in the primary link.

On the other hand, the ERT model determines an approximate probability distribution for the overflow traffic which is offered to an overflow link. The basis of the ERT model is the assumption that the overflow traffic offered to an overflow link with N circuits can be represented by a single overflow from a fictitious equivalent link with N_E circuits, offered a Poisson traffic stream A_E. The equivalent parameters A_E and N_E are chosen such that the overflow from the equivalent link has the same mean and variance as the actual overflow traffic offered to the overflow link:

$$A_E \approx V + 3Z(Z-1) \tag{2}$$

$$N_E = A_E\left(\frac{M+Z}{M+Z-1}\right) - M - 1 \tag{3}$$

where $Z = \frac{V}{M}$, and V and Z are respectively the variance and peakedness of the overflow traffic. Equation (4) is solved for N to obtain the number of circuits in the overflow link:

$$B = \frac{E(M, N+N_E)}{E(M,N_E)} \tag{4}$$

Several algorithms exist to find solutions for N from equations (1) and (4) [3]. For the purpose of this paper we have used the iterative algorithms developed by Jagerman which uses Newton's interpolation formula [6].

4 A neural network dimensioning tool

Neural networks have been shown to be excellent computational tools for non-linear modelling and function approximation. Research literature shows that a significant niche exists for the combination of telecommunications and neural networks, and that the capability of neural networks to learn from data allows them to be used on function approximation problems [1,9].

We investigate the use of a multi-layer FFNN to approximate the distribution curve of both primary and overflow traffic. Only one NN is used to approximate both traffic distributions. The utilization of a NN to aid dimensioning consists of two phases, i.e. the training of the NN and the application of the trained network to find the correct link sizes. Training is performed once only using Jagerman's algorithms to simulate the traffic tables proposed in the CCITT (Consultative Committee of the International Telegraph and Telephone) recommendations.

For training purposes, the NN recieves three input variables which describes the state of the link and the service requirements, i.e. the traffic load offered to a link, the peakedness of the offered traffic and the grade-of-service. The peakedness input variable yields an indication of the traffic type. Since primary traffic is Poisson, it will have a peakedness factor of one, while overflow traffic exhibits a peakedness greater than one. The output of the NN is a single value which is the number of circuits necessary to satisfy the specified demand. After training of the NN, only one forward pass is needed to dimension a single link.

4.1 Complexity analysis

The utilization of a NN in an automated dimensioning model reduces the complexity of that model. We show this by comparing the NN approach with a dimensioning model which utilizes the algorithms developed by Jagerman to solve equations (1) and (4) [6]. We use the number of arithmetic calculations as measure of complexity.

To calculate the complexity of the Jagerman dimensioning algorithms, we have to determine the number of calculations needed to solve both equations (1) and (4). The dimensioning of a primary link which is offered M erlang traffic with a grade-of-service requirement B, requires in total

$$T_x^{\text{Erlang}} = 293 + 30x \tag{5}$$

arithmetic calculations, with $x = (int)[(1 - B)(1 + M)]$. To dimension an overflow link, using the ERT traffic model, requires

$$T_x^{\text{ERT}} = 1332 + 90x \tag{6}$$

calculations, where $x = A_E\left(\frac{M+Z}{M+Z-1}\right) - M - 1$. It is clear form equations (5) and (6) that the number of calculations to dimension a link is dependent on the traffic load M: larger traffic loads result in more calculations.

Consider a three-layer FFNN with I inputs, J hidden units and K outputs. Training is accomplished using the delta rule with sigmoid activations for the hidden and output units. Since only one forward pass is needed to dimension a single link, we need to compare the complexity of only the forward pass with the complexity of the Jagerman algorithms. The total number of calculations in one forward pass for a I-J-K neural network,

$$T^{\mathrm{FFNN}} = 3(JI + JK + J + K) \tag{7}$$

shows that the number of calculations is independent of the traffic load and traffic type.

In the next section we use a 4-21-1 neural network[1] to approximate the two traffic distributions. From equation (7) a total of 381 calculations are performed to dimension any link using the NN approach. Table 1 shows the complexity of Jagerman's solutions to the Erlang B formula (1) as a function of M, respectively for a grade-of-service $B = 0.001$ and $B = 0.2$. The complexity of Jagerman's solution to the ERT model is not illustrated, since $T_x^{ERT} > T_x^{Erlang}$. We see from Table 1 that $T_x^{Erlang} > T^{FFNN}$ even for traffic loads as low as $M = 2$ erlang and $M = 3$ erlang for a grade-of-service of respectively $B = 0.001$ and $B = 0.2$.

Table 1: Complexity of Jagerman's Erlang B solution

	B = 0.001				*B = 0.2*			
M (erlang)	1	2	3	4	1	2	3	4
T_x^{Erlang}	353	383	413	443	341	365	389	413

Further analysis of the design complexity of the respective dimensioning models shows that the NN dimensioning approach is less complex than the Jagerman approach: because the statistical characteristics of primary and overflow traffic differ, existing automated dimensioning models employ a different traffic model for each traffic type [3]. This results in a different dimensioning model for each one of the traffic types. In contrast to traditional dimensioning models, only one NN model is needed for the dimensioning of both primary and overflow links.

[1]Both the input and hidden layers have an additional bias unit.

4.2 Experimental results

We have used a 4-21-1 neural network architecture to approximate both traffic distributions. The traditional error back propagating training algorithm, also referred to as delta learning, is used for training [10]. Sigmoid activations are used in the hidden and output units. All weights are initialized as random variables in the range $[\frac{-1}{\sqrt{fanin}}, \frac{1}{\sqrt{fanin}}]$. A data set of 85 000 patterns was generated using Jagerman's algorithms for primary and overflow traffic. For experimental convenience we have restricted the input parameters B, M and Z and the output parameter N to the following ranges

$$B \in [0.001, 0.2], \quad M \in [0.001, 200], \quad Z \in [1, 4], \quad N \in [0, 236]$$

All inputs are linearly scaled to the range $[-\sqrt{3}, \sqrt{3}]$ which is the active area of the sigmoid activation function, while the output variable N is scaled to $[0, 1]$. A training set of 40 000 patterns was randomly selected from the generated data set. The remaining 45 000 pairs were used as test set. The NN was trained for an average error per training pair of 0.000005, corresponding to a rescaled error of 0.27848 [4]. Training stopped after 660 epochs.

To establish the performance of the NN, the floating point output of the NN was rounded to the nearest integer value and compared to the rounded output of Jagerman's algorithms. Our results indicate that the NN provides a sufficiently accurate approximation of both traffic distributions. Table 2 summarizes the results compared to the results of Jagerman's algorithms. Table 2 gives the average dimensioning error $\overline{N}$ in number of circuits, the standard deviation s, and the coefficient of determination R^2 for the training and test sets. The results on the test set show a good generalization performance. The average error $\overline{N}$,

$$\overline{N} = \frac{\sum_{p=1}^{P} |t_p - N_p|}{P}$$

with P the total number of patterns, and t_p and N_p respectively the rounded values for the target and actual NN output for pattern p, shows a very low average dimensioning error for both the training and test sets.

We have found that the performance of the NN degrades for extreme cases, i.e. when the number of circuits is near the maximum range for which the NN was trained. For example, the NN underdimensions a primary link which is offered 200 erlang traffic for a grade-of-service of 0.001 by 15 circuits. In this case the output of the NN is 220 circuits instead of 235 circuits. This explains the relatively high standard deviation. These large errors are due to the fitting of a flat surface over the data points which does not accommodate the curved surface at the extreme values. A solution to this problem may be to increase the number of hidden neurons or the number of hidden layers to learn the curved structure of the traffic distribution.

The coefficient of determination R^2 is calculated as

$$R^2 = 1 - \frac{(n-1)}{(n-m)} \left[\frac{\sum_{p=1}^{P}(t_p - N_p)^2}{\sum_{p=1}^{P}(t_p - \bar{t})^2} \right]$$

where $\bar{t}$ is the average target value; n represents the degrees of freedom (i.e. total number of weights) and m the number of parameters which need to be estimated (i.e. number of input parameters). The coefficient of determination is included as an indication of how well the NN has approximated the traffic distributions. In this case we have an approximation accuracy of 99.96% on both the training and test sets.

Table 2: Results of neural network dimensioning approach

	Training set	**Test set**
Average error ($\overline{N}$)	0.483175	0.47745
Standard deviation (s)	0.741539	0.732874
Coefficient of determination (R^2)	0.999576	0.999586

5 Conclusion

We have shown in this paper that a neural network is an excellent tool to approximate both primary and overflow traffic distributions. As a dimensioning tool, the NN presents sufficiently accurate estimations of link sizes. We have also shown that the NN dimensioning approach reduces the complexity of the dimensioning model compared to Jagerman's algorithms. The NN approach requires one forward pass to dimension a link, while the number of iterations of Jagerman's algorithms depends on the traffic load.

Equation (7) shows that the complexity of the NN dimensioning approach depends on the number of hidden units in the hidden layer. Since the primary and overflow traffic distributions are smooth functions, we may find that the approximation can be accomplished using less training patterns and consequently less hidden units. Further investigation is needed to find the optimal NN architecture for the approximation of traffic distributions.

Because of the necessity to scale traffic data to the active range of the sigmoid activation function, the NN dimensioning approach has the disadvantage of limited extrapolation capabilities. We are currently investigating automatic scaling learning algorithms which will eliminate the need to scale traffic data before learning [4]. Such automatic scaling algorithms will enable the NN approximation tool to incrementally learn examples which are not in the range covered by the original data set.

We conclude that a NN can be used as an effective and reliable traffic distribution approximator in an automated dimensioning tool.

References

[1] J Alspector, R Goodman, TX Brown (eds.), "Proceedings of the International Workshop on Applications of Neural Networks to Telecommunications," Lawrence Erlbaum Associates Publishers, 1993.

[2] G Bretschneider, "The Calculation of Trunkgroups for Overflow Traffic in Dial Offices," *Nachrichtentechnische Zeitschrift,* vol. 9, no. 11, pp. 533–540, 1956.

[3] AP Engelbrecht, "Modelling Algorithms for the South African Telephone Network," M.Sc. Thesis, Department of Computer Science, University of Stellenbosch, 1993.

[4] AP Engelbrecht, I Cloete, J Geldenhuys, JM Zurada, "Automatic Scaling using Gamma Learning for Feedforward Neural Networks," submitted to IWANN'95.

[5] A Girard, "Routing and Dimensioning in Circuit-Switched Networks," Addison-Wesley, 1990.

[6] DL Jagerman, "Methods in Traffic Calculations," *AT&T Technical Journal,* vol. 63, no. 7, pp. 1283–1310, September 1984.

[7] Siemens, "Tables for Planning and Operation of Telephone Systems", Siemens & Halske, München, November 1961.

[8] RI Wilkinson, "Theories for Toll Traffic Engineering in the USA," *Bell System Technical Journal,* vol. 35, no. 2, pp. 421–514, March 1956.

[9] B Yuhas, N Ansari (eds.), "Neural Networks in Telecommunications," Kluwer Academic Publishers, 1994.

[10] JM Zurada, "Introduction to Artificial Neural Systems," West Publishing Company, 1992.

Theory and Implementation of an Analog Network that Solves the Shortest Path Problem

Tzi-Dar Chiueh and Ling-Kai Bu

Room 511, Department of Electrical Engineering
National Taiwan University, Taipei, Taiwan 10617
chiueh@cc.ee.ntu.edu.tw

Abstract

In this paper we propose an analog network that solves the shortest path problem between any two nodes in a given graph. This network is constructed by elements called variable turn-on-voltage diodes (VTD) and this network has the same topological structure as the graph at hand. Each VTD has a threshold voltage proportional to the length of the corresponding edge in the graph. To find the shortest path between two nodes, one needs only to ground one node and gradually raise the voltage at the other node. As soon as the network starts draining large current, the shortest path — made up of those VTDs that conduct currents — emerges. SPICE simulations confirm that the proposed network can indeed solves the shortest path problem efficiently.

1 Introduction

The shortest path problem is one of the important problems in combinatorial optimization. Given a graph composed of nodes and undirected edges, one is to find the shortest path between two nodes. Traditionally, this problem is solved by the Dijkstra's algorithm [1], which requires $O(N^2)$ time complexity. Recently, following Tank and Hopfield's pioneer work on traveling-salesman-problem neural network, there have been results on using the Hopfield neural network to solve the shortest path problem [2,3]. However, this approach suffers from the same difficulty as in most applications of the Hopfield network to combinatorial problems: extensive parameter tuning is needed to achieve acceptable results. In this paper we first introduce a physical system that solves the shortest path problem in $O(1)$ time complexity, using as many elements as there are edges in the graph. We then proceed to introduce an analog VTD network implementation of this physical system.

2 A String System

Given a graph, one can construct a string system that is made up of the strings whose lengths are proportional to those of the edges in the graph. Hence each edge in the graph is represented by a string in the string system. For each node in the graph, use a small ring to represent that node. If two nodes are joined by an edge in the graph, the corresponding rings are connected by a piece of string whose length is proportional to that edge (see Figure 1(a)).

To find the shortest path between two nodes in the graph, one needs only to pick up the two corresponding rings in the string system and pull them until the strings are tight. Instantly the shortest path stands out as a sequence of tight strings (see Figure 1(b)) [4]. All the rings in this path represent the nodes that are along the shortest path between the two nodes in the graph.

Since this string system is a physical one, the above proposition can be proved by repetitive experiments or though application of known physical laws and properties of the string system. Here we will only attempt to explain this phenomenon heuristically. Once the two rings are pulled tight, their physical distance can not be shorter or longer than the length of the shortest path between them (assuming that rings have zero radius). If the distance is shorter, then the strings are not tight. If the distance is longer, then some strings along the shortest path must have been broken. Therefore, only those strings that are in the shortest path will remain tight and stand out, all other strings will be slack due to the gravity force exerted on them.

3 Analog Network Realization

After carefully examine the proposed physical string system, we find that the major components of the system are strings that cannot be stretched beyond their maximum lengths. To realize this element using electronic circuits, we represent length as voltage across the proposed analog two-terminal device. So we need a two-terminal element whose voltage drop can not rise higher than a predefined value. To detect "tight strings", we choose to let the current along the element signify "string tightness." Hence, the element is tight only if it carries large current, at which juncture voltage drop across it is equal to the maximum allowable voltage.

We can also construct an analog network using the proposed element to represent each edge in the graph and connect all these elements as the edges are joined in the graph. To locate the shortest path between two nodes in the graph, we first ground one of the corresponding node in the network and gradually raise the voltage at the other corresponding node until large current starts to flow. The shortest path can be located by following those elements with large currents.

The I-V curve of this element is illustrated in Figure 2(a). Since it is basically a diode with variable turn-on voltage, we call it Variable Turn-on-voltage Diode (VTD). We will introduce two VTD designs. The first uses a floating-gate diode-connected MOS transistor. By introducing high-voltage pulses into the controlling gate, one is able to pump charges into the floating gate and effectively alter the threshold voltage of that transistor [5]. Even though the saturation current in the strong inversion region grows quadratically instead of exponentially as in diodes, this floating-gate diode-connected MOS transistor can still function as the VTD in our network.

The second implementation of VTD is more elaborate. It is based on a pass transistor and a biasing circuit. The goal of the biasing circuit is to bias the gate voltage of the pass transistor to V_q. The voltage V_q is near V_{dd} when V_{ds} is greater than or equal to the maximum allowable voltage across the VTD, and V_q is near zero otherwise. Circuit diagram of the proposed VTD is depicted in Figure 2(b). V_{dto} represents the diode turn-on voltage of the VTD. V_{dto} and V_2 are added by a voltage adder and the result is fed to the negative input of a wide-range operational transconductance amplifier (OTA). We use a wide range amplifier to ensure that the gate voltage can assume full swing of the power supply.

4 Simulations and Discussions

SPICE simulations of the VTD whose turn-on voltage is set to 0.7 V is presented in Figure 3. As the V_{ds} of the pass transistor increases from 0 to 1 V, the current through VTD initially stays at around order of $10^{-10}A$, it then rapidly rises to $10^{-5}A$ as soon as V_{ds} reaches 0.7 V. This amounts to an increase of five orders in VTD current for 50 mV difference in the voltage across the VTD.

We then use the VTDs to construct a graph as Figure 1(a) with turn-on-voltages specified. Currents flowing through edges AB, BG, and GI turn significant when the voltage across nodes A and I is around $1.9V$ (see Figure 4), which is just the voltage equivalent of the length of the shortest path between nodes A and I (0.7+0.8+0.4). Currents flowing through other edges stay relatively small, four to five orders less than the currents in the edges on the shortest path. There are two points that should be mentioned about our simulations. Firstly, since edges in the graph are undirected while the VTDs are directed, hence two VTDs in parallel are required to implement an edge in an undirected graph. By using two VTDs of different turn-on voltages (V_{dto}), one can construct a directed graph and solve its shortest path problem, while the Dijkstra's algorithm solves shortest path problems for only undirected graphs. Secondly, in dealing with larger graphs, one may encounter cases when the voltage equivalent of a particular shortest path is larger than the supply voltage, V_{dd}. In this case, the V_{dto} of the VTDs need to be rescaled so that the pass transistors work as they are designed.

Another simulation is conducted to locate the shortest path between nodes A and E. As we increase the voltage, V_{AE}, the shortest path that consists of edges AC and CE surfaces when V_{AE} is around 1.1 V. Further increase in V_{AE} results in large currents along all edges in the paths from node A to node E (see Figure 5). This would cause some difficulty if the second shortest path is almost of the same length as the shortest path. To circumvent this problem, a dummy node E' can be inserted and connect nodes E and E' by a VTD with zero turn-on voltage. This technique has the effect of constraining the current sunk by node E to around tens of μA and thus preventing the currents along edges on other paths to rise to the same level as that of the currents along edges AC and CE (see Figure 6). With this technique, finding the shortest path becomes easier since, instead of gradually increasing voltage V_{AE}, all one needs is to supply a high enough voltage across the two nodes and find those VTDs that conduct significant currents or whose OTA outputs are at V_{dd}.

We have also notice that the voltage drop across a nonideal diode when it conducts is higher than its threshold voltage. Therefore there are times when a second shortest path with less hops comes out the winner in our network. This feature can be exploited in a communication network where the link delay as well as the node delay need to be taken into account. The ratio of the link delay and the node delay can be set by the current level since, for example, in a MOS-based diode

$$V_{gs} - V_{th} = \sqrt{2I_d/\beta}.$$

So when one set the current threshold for detecting the shortest path, $V_{gs} - V_{th}$ in each turned-on VTD is set and thus the ratio of the node delay and the link delay is determined.

5 Conclusion

In this paper, we introduced an analog circuit network that solves the shortest path problem in real time. The hardware complexity of the network is the same as the complexity of the graph in the problem. We have also shown that with proper adjustment of the current level the network can also solve fastest path routing problems in a communication network with node delays as well as link delays.

Acknowledgments

The authors would like to acknowledge the support of the National Science Council, Taiwan, ROC under Grant No. NSC84-2215-E002-023.

References

[1] T. C. Hu, *Combinatorial Algorithms,* Reading, MA: Addison-Wesley, 1982.

[2] L. Zhang and S. C. A. Thomopoulos, "Neural Networks Implementation of the Shortest Path Algorithm for Traffic Routing in Communication Networks," in *Proc. Int. Joint Conf. Neural Net.*, June 1989, p. II-591.

[3] , M. K. M. Ali and F. Kamoun, "Neural Networks for Shortest Path Computation and Routing in Computer Networks," *IEEE Trans. on Neural Net.,* Vol. 4, Np. 6, Nov. 1993, pp. 941–954.

[4] A. K. Dewdney *The Armchair Universe,*New York, NY: W. H. Freeman and Company, 1988.

[5] T. H. Borgstrom, M. Ismail, S. B. Bibyk, "Programmable Current-Mode Neural Network for Implementation in Analog MOS VLSI," *IEE Proceedings-G,* Vol. 137, No. 2, April 1990, pp. 175-184.

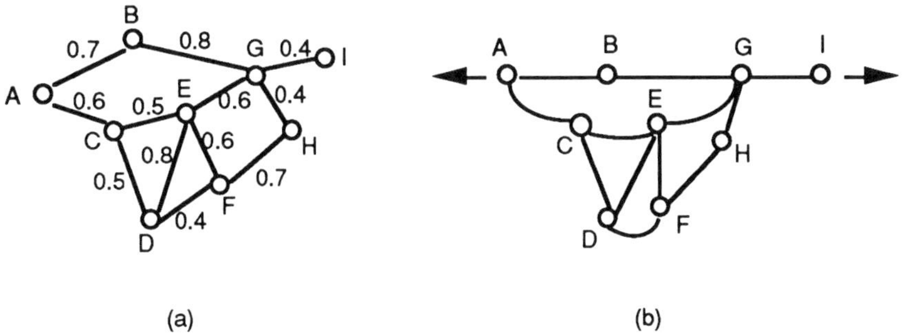

Figure 1: (a) A string system that represents a graph. (b) Finding the shortest path between nodes A and I by pulling these two nodes.

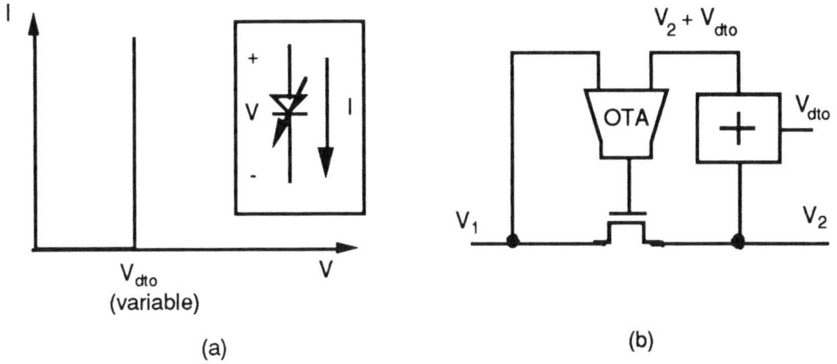

Figure 2: (a) The symbol and I-V curve of an ideal Variable Turn-on-voltage Diode (VTD). (b) Circuit diagram of the proposed VTD implementation.

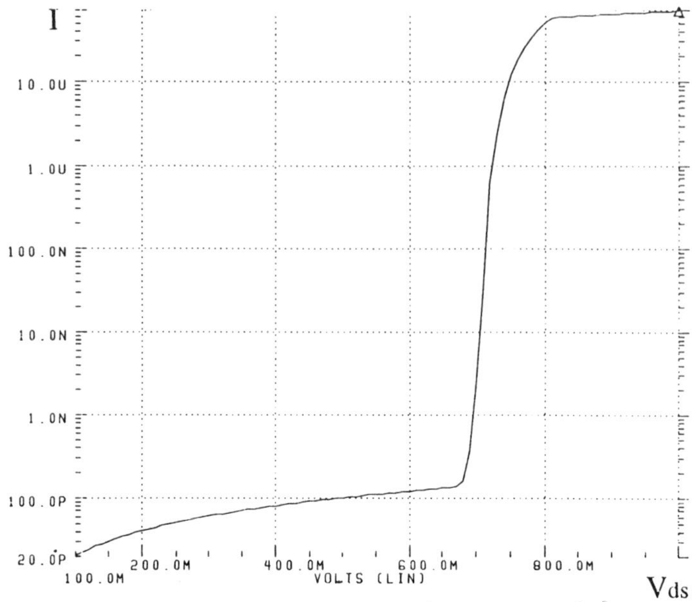

Figure 3: I-V curve of a VTD with $V_{dto} = 0.7V$.

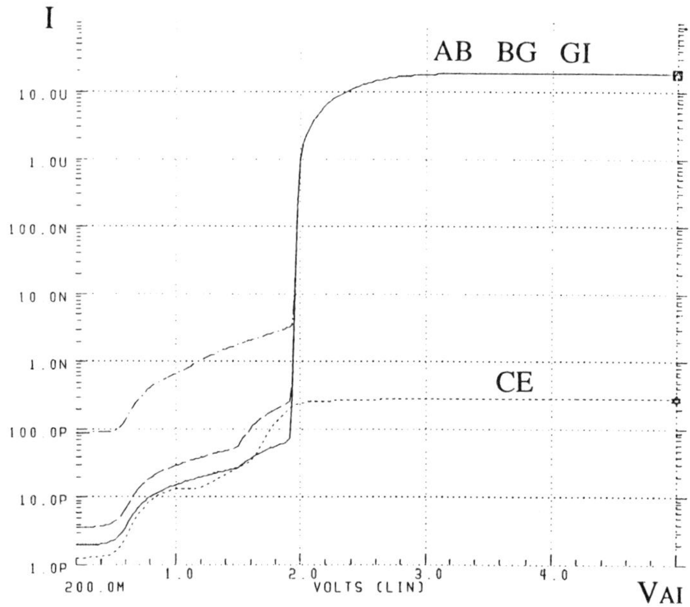

Figure 4: Currents along some VTDs when finding the shortest path between nodes A and I by increasing the voltage across nodes A and I, V_{AI}.

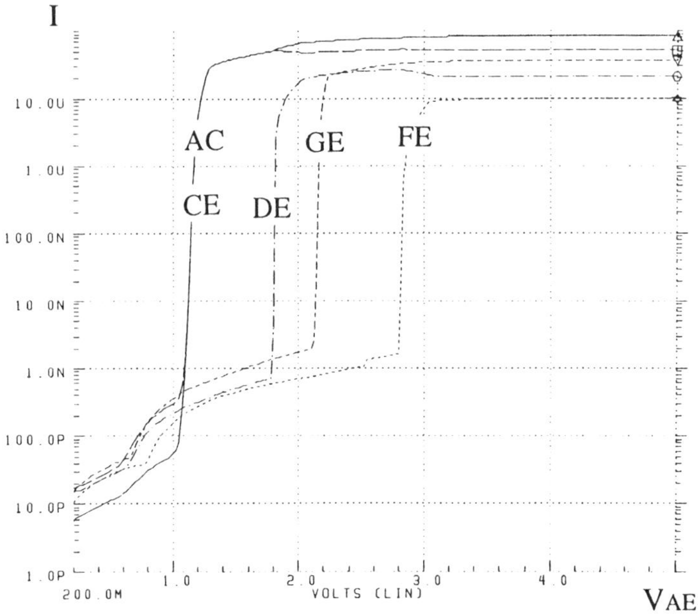

Figure 5: Currents along some VTDs when finding the shortest path between nodes A and E by increasing V_{AE}.

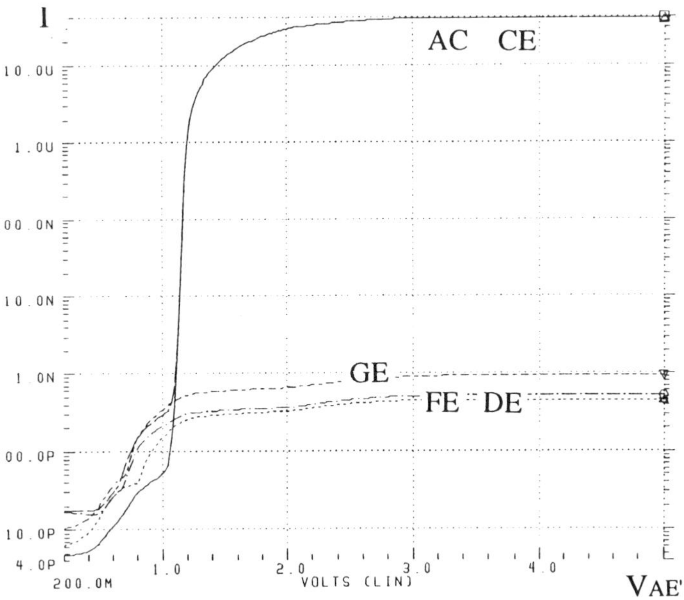

Figure 6: Currents along some VTDs when a dummy node E' is added and a VTD with $V_{dto} = 0V$ is inserted between nodes E and E'.

A Reinforcement Learning Scheme for Adaptive Link Allocation in ATM Networks

Ernst Nordström, Jakob Carlström

Department of Computer Systems, Uppsala University,
Box 325, S–751 05 Uppsala, Sweden
Fax: +46 18 55 02 25,
`<http://www.docs.uu.se/docs/ann/>`
`jakobc@DoCS.UU.SE, ernstn@DoCS.UU.SE`

Abstract

This paper presents an adaptive scheme for a sub-function in Asynchronous Transfer Mode (ATM) network routing, called link allocation. The scheme adapts the link allocation policy to the offered Poisson call traffic such that the long-term revenue in maximized. It decomposes the link allocation task into a set of link admission control (LAC) tasks, formulated as semi-Markov Decision Problems (SMDPs). The LAC policies are directly adapted by reinforcement learning. Simulations show that the direct adaptive SMDP scheme outperforms static methods, which maximize the short-term revenue. It also yields a long-term revenue comparable to an indirect adaptive SMDP method.

1 Introduction

Routing in public Asynchronous Transfer Mode (ATM) networks has two objectives: maximizing the operator revenue and maintaining the network availability for different call types. Adaptive routing techniques are efficient when the traffic demand varies over time. The approach presented in [1], views the routing task as an adaptive semi-Markov Decision Problem (SMDP). The method selects a route from a set of candidate routes, the objective being to maximize the long-term revenue. It uses an indirect algorithm, which adapts a model of the underlying controlled Markov Process, and computes control policies based on the latest model. In order to simplify the revenue analysis, the call traffic load and revenue generation on successive transmission links are assumed to be independent.

In this paper, we assume that two adjacent switches are interconnected by a set of parallel transmission links. The adaptive routing problem is decomposed into a set of adaptive link allocation problems,where the task is to select the link within a link group that maximizes the long term revenue. An adaptive link allocation scheme, based on a direct (model-free) SMDP approach is presented. A near-optimal link allocation policy is found by solving a series of simple link admission control (LAC) tasks, formulated as direct SMDPs. The link admission controllers use reinforcement learning [2] [3], in form of the actor-critic method [4], to find optimal state-dependent LAC policies. In particular, the controllers should detect link states where blocking of narrow-band calls leads to higher long-term revenue. A set of functions that measure the relative merit of accepting a call in a particular link state, controls the link allocation after adaptation.

The experimental results show that the proposed scheme has comparable performance with the indirect adaptive SMDP method, both in terms of long-term revenue and in terms of adaptation rate.

2 The Link Allocation Problem

In the link allocation problem, a group of M links with capacities C_i [units/s], $i \in I = \{1, ..., M\}$, is offered calls from K different classes. Calls belonging to a class $j \in J = \{1, ..., K\}$ have the same bandwidth requirements b_j [units/s], and similar arrival and holding time dynamics. As in [1], we assume that type-j calls arrive according to a Poisson process with intensity λ_j [s^{-1}], and that the call holding time is exponentially distributed with mean $1/\mu_j$ [s]. In this work, the parameter b_j is given by the peak ATM cell transmission rate, since deterministic cell multiplexing is assumed.

The task is to find a link allocation policy π that maps *request states* $(j,n) \in J \times N$ to *allocation actions* $a \in A$, $\pi: J \times N \rightarrow A$, such that the long-term revenue is maximized. The set N contains all feasible link group states, and the set A contains the possible allocation actions, $I \cup \{REJECT\}$. The set of feasible link group states is given by the Cartesian product of the sets of feasible link states N_i,

$$N_i = \left\{ n_i : n_{ij} \geq 0,\ j \in J; \sum_{j \in J} n_{ij} b_j \leq C_i \right\},\quad i \in I\ ,$$

where n_{ij} is the number of type-j calls accepted on link i.

The network availability constraint (limited call blocking probabilities) is not considered in the present work. Moreover, we assume an uniform call charging policy, which means that the long-term revenue is proportional to the cell throughput at the call level.

3 An Adaptive Link Allocation Scheme

In order to speed up the adaptation process, the link allocation task is decomposed into a set of link admission control (LAC) tasks with actions $a_i \in A_i = \{ACCEPT,$

REJECT}, see Figure 1. The link admission controllers adapt to a constant-rate call flow, during a number of periods. The call flows are kept unchanged during each period, which ends when an optimal LAC policy has been found for each link. Then, new call flows are determined for the following period, based on the performance of the LAC policies determined during the previous period.

A load sharing link allocation policy with constant load sharing coefficients h_{ij} maintains the LAC task during the policy adaptation period. That is, a type-j call is offered to link i with probability h_{ij} (Figure 1). The selected link admission controller can then accept or reject the call. The load sharing coefficients used during period p are determined by:

$$h_{ij,p} = \frac{\bar{\lambda}_{ij,p-1}}{\sum_{k \in I} \bar{\lambda}_{kj,p-1}}, \quad i \in I, \ j \in J, \tag{1}$$

where $\bar{\lambda}_{ij,p-1}$ denotes the measured rate of accepted type-j calls on link i during period $p-1$. Hence, a link which has a relatively high admission rate will be offered more calls during the next adaptation period. The adaptation stops when the new coefficients $\{h_{ij,p}\}$ are sufficiently close to the old coefficients $\{h_{ij,p-1}\}$.

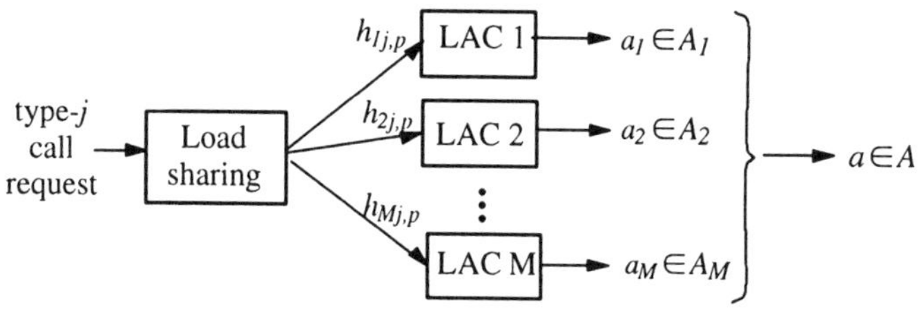

Figure 1: Link allocation during adaptation.

In the course of LAC adaptation, each link admission controller i estimates merit functions $m_{ACCEPT,i}(j,n_i)$, which measure the relative merit of accepting a type-j call in link state n_i. The accept merit functions control the link selection after the adaptation phase.

When a type-j call request arrives, each link is checked to see if it has sufficient free capacity to accept the call. Provided this is the case, the controller selects an action $a_i \in A_i$, with higher probability for the action which yields higher long-term revenue (see section 4). The controller outputs the resulting action a_i along with the accept merit value $m_{ACCEPT,i}$. The link allocator then selects the link with the highest accept merit value (among the links that accept the call), see Figure 2. If all a_i = *REJECT*, the link allocator rejects the call.

In certain link states, called "intelligent blocking" link states, rejecting calls of some types yields a higher long term revenue than accepting them. They typically occur when the link has a free link capacity that is equal to the size of a wide-band call. By rejecting a narrow-band call request, the controller reserves bandwidth to the wide-band class, and so increasing the long-term revenue. However, if many narrow-band calls are accepted on the link, at least one of them is likely to depart before the next wide-band call arrives. Hence, narrow-band calls can be accepted, although the free capacity equals the size of a wide-band call.

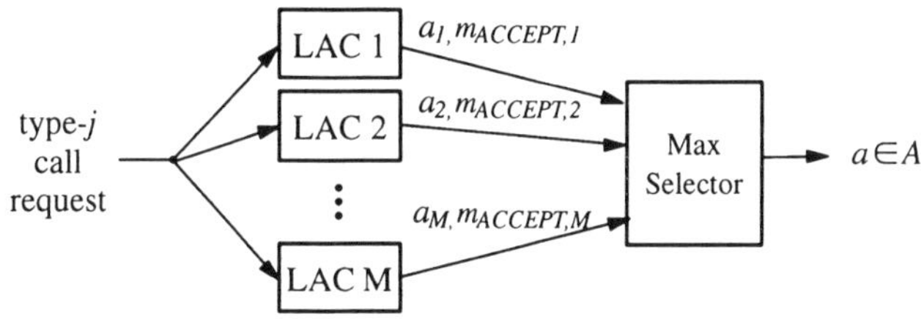

Figure 2: Link allocation after adaptation.

4 Reinforcement learning of the LAC policy

Within each link i, a link admission controller constructs a policy π_i: $X_i \rightarrow A_i$, $A_i = \{ACCEPT, REJECT\}$. $\pi_i(x_i)$ indicates what action to $a_i \in A_i$ to select at each SMDP state $x_i \in X_i$. X_i is defined by $X_i = N_i \times E \times J$, where the two possible types of events, an arrival or a departure of a call, are the elements in $E = \{ARRIVAL, DEPARTURE\}$.

The objective of the link admission controller of link i is to find a policy π_i which maximizes the long-term revenue, expressed as the expected (infinite horizon) discounted reward. This "utility" is denoted $V_{i\pi}(\xi_i)$, for a SMDP state $\xi_i \in X_i$:

$$V_{i\pi}(\xi_i) = E\left\{ \int_{t=0}^{\infty} e^{-\beta t} r_i(x_i(t), a_i(t)) dt \right\} \quad (2)$$

where the reward $r_i(x_i(t),a_i(t))$ is the continuous-time total cell transmission rate on the link, $x_i(t)$ and $a_i(t)$ denote the SMDP state and action at time t, respectively, and $x_i(0) = \xi_i$. This maximization is performed by a delayed reinforcement learning method, which is a modification of the actor-critic method [4], with its redefinition for SMDPs [3].

The actor-critic method solves the task using two separate function approximators (Figure 3): an evaluation function $V_i(x)$ which models $V_{i\pi}(x)$ and a policy function $\pi_i(x)$. In our modification, π_i is divided into two sub-policies: an arrival policy π_{ia}, which is adaptive, and a departure policy π_{id}, which is deterministic. A sub-policy selector chooses what sub-policy to employ, according to

$$\pi_i(n_i, e, j) = \begin{cases} \pi_{ia}(n_i, j), & e = ARRIVAL \\ \pi_{id}(n_i, j), & e = DEPARTURE \end{cases}, \text{ where} \tag{3}$$

$$\pi_{ia}(n_i, j) \in \{ACCEPT, REJECT\}, \tag{4}$$

$$\pi_{id}(n_i, j) \equiv ACCEPT. \tag{5}$$

The motivation for Equation 5 is that the link admission controller must accept all call departure requests.

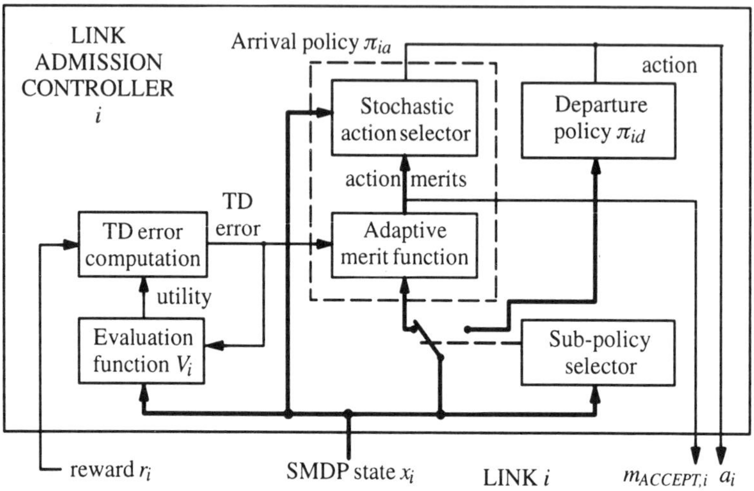

Figure 3: The architecture of the modified actor-critic method

π_{ia} uses an adaptive merit function (Figure 3), which indicates the relative merits $m_{ACCEPT,i}$ and $m_{REJECT,i}$, for accepting or rejecting a requested call, respectively. The accept merits $m_{ACCEPT,i}$ are also output to the link allocation algorithm. A stochastic action selector chooses among the actions, with higher probability for actions with higher merits. The probability of selecting an action a_i in state x_i is determined by the action merits and the SMDP state, by choosing action $a_i(x)$ as in [4]:

$$a_i(x) = \arg\max_{u \in A_i} \left(m_{u,i}(x_i) + e_u \right) \tag{6}$$

where $m_{u,i}(x_i)$ is the merit of action u, and e_u are independent random numbers, drawn from an exponential distribution with mean $1/T(x_i,u_i)$. The "temperature" $T(x_i,u_i)$ adjusts the randomness of action selection. After adaptation, $T(x_i,u_i)$ is set to zero for all (x_i,u_i).

The discounted cumulative reward $q_{i,xy}$ received between two state transitions, from a SMDP state x entered at time t_x, to another SMDP state y entered at time t_y, is defined by

$$q_{i,xy} = \int_{t_x}^{t_y} e^{\beta(t_x - t)} r_i(t - t_x) dt \tag{7}$$

The link admission controller learns from interacting with the link in repeated trials. By definition of the evaluation function and (Equation 2), the desired evaluation function $V_{i\pi}(x)$ must satisfy

$$V_{i\pi}(x) = q_{i,xy} + e^{\beta(t_x - t_y)} V_{i\pi}(y) \tag{8}$$

During learning, this may not be true. The difference between the two sides of the equation is called the temporal difference (TD) error. This is used to update both $V_i(x)$, according to the TD(0) rule [2], and $\pi_i(x)$:

$$\Delta V_i(x) = \eta_V [q_{i,xy} + e^{\beta(t_x - t_y)} V_i(y) - V_i(x)] \tag{9}$$

$$\Delta m_{u,i}(x) = \eta_\pi [q_{i,xy} + e^{\beta(t_x - t_y)} V_i(y) - V_i(x)] \tag{10}$$

where η_V and η_π are "learning rate" parameters, and $u \in A_i$ is the action chosen in state x.

It should be noted that although an effect of using a deterministic departure policy is that the policy is not updated after call departures, the evaluation function is updated, which leads to better estimates of V_i, resulting in faster and safer convergence of the arrival policy. The non-zero probability of choosing and evaluating actions with low merits (Equation 6), allows the link admission controller to improve its policy.

In reinforcement learning, neural networks, for example multi-layer perceptrons, are often used to approximate the evaluation and policy functions. This is beneficial when the state space is too large to explore completely, since the neural network allows generalization between states. Neural networks also allow incorporation of other environment parameters, providing the link admission controller with information which may improve its performance, for example in cases where the Poisson call model does not hold. However, in this work, lookup tables were used for function approximation.

5 Results

The proposed adaptive link allocation scheme was tested on simulated Poisson call traffic. Results for three other methods are presented for comparison: the indirect adaptive SMDP method [1] and the static First Fit and Best Fit methods. The static methods maximize the short-term revenue, using the following algorithms:

- First Fit: Search the links in a predefined order, and allocate the call to the first link found with sufficient capacity.
- Best Fit: Choose the link with least, but sufficient, capacity.

The simulations were done for a link group of 3 links with capacities $C_i = C = 24$ [units/s] for all i. The link group was offered calls from two classes, characterized by bandwidth requirements $b_1 = 1$, $b_2 = 6$ [units/s] and call holding times $1/\mu_1 = 1/\mu_2 = 1$ [s]. The arrival intensities λ_1 *and* λ_2 [s^{-1}] were varied so that:

$$\frac{b_1\lambda_1}{C\mu_1} + \frac{b_2\lambda_2}{C\mu_2} = 1.5 \tag{11}$$

The temperature $T(x_i,u)$ of the actor/critic-method was set using prior knowledge of the intelligent blocking states, introduced in section 3. In particular, intelligent blocking should be possible for the narrow-band class, at link states where the free capacity equals the size of one wide-band call, that is, for the link states $n_i \in \{(0,3), (6,2), (12,1)\}$. In the corresponding SMDP states x_i, different temperatures were used for accept and reject actions: $T(x_i, ACCEPT) = 0.4$, and $T(x_i, REJECT) = 0.3$. For all other $(x_i,u) \in X_i \times A_i$, the temperature $T(x_i,u)$ was set to zero.

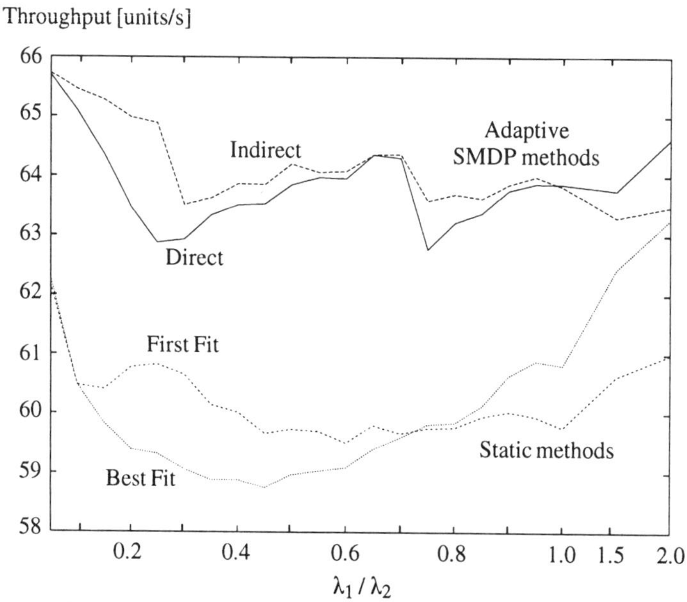

Figure 4: Call level throughput versus arrival rate ratio for different methods.

Some prior knowledge was also needed to complement to the load-sharing policy during adaptation. Experiments with the indirect SMDP scheme showed that one link will always reject narrow-band calls. When $0.25 < \lambda_1/\lambda_2 < 0.75$, this occurred for two links, and when $\lambda_1/\lambda_2 < 0.25$, all narrow-band calls were rejected. The direct scheme did not succeed in finding these complete blocking links, so it had to be predefined in the simulations. A uniform load-sharing policy, set according to the prior knowledge, was used during the initial adaptation period.

The actor/critic parameters were set to $\beta = 0.74$, $\eta_V = 0.1$ and $\eta_\pi = 0.2$. Also, the merit values were initialized to favor *ACCEPT* actions for all $x_i \in X_i$.

The results for the indirect and direct SMDP schemes presented in the diagram in Figure 4 were obtained after 4 adaptation periods, where each adaptation period contained 1 000 and 15 000 simulated call events, for the indirect and direct SMDP method, respectively. The throughput values in the diagram are based on measurements on 300 000 calls events after policy convergence.

The diagram shows that the adaptive SMDP methods yields up to 7% higher long-term revenue than the static methods. The diagram also shows that direct SMDP scheme yields a performance similar to the indirect scheme's.

6 Conclusion

This paper has presented an adaptive scheme, based on reinforcement learning, for a sub-function in ATM network routing called link allocation. The scheme adapts the link allocation policy to the offered Poisson call traffic such that the long-term revenue in maximized.

The experimental results show that the proposed scheme outperforms the static methods and yields a long-term revenue similar to the indirect adaptive SMDP method [1]. The results also show that the adaptation rate of the reinforcement scheme is comparable to the indirect method's.

In our future work, we will consider link allocation of non-Poisson traffic, exploiting the advantages of neural networks as function approximators.

Acknowledgements

The authors would like to thank Mats Gustafsson, Olle Gällmo and Lars Asplund for stimulating discussions. This work was financially supported by ELLEMTEL Telecommunication Systems Laboratories and by NUTEK, the Swedish National Board for Industrial and Technical Development.

References

[1] Z. Dziong and L. Mason,"An Analysis of Near Optimal Call Admission Control and Routing Model for Multi-service Loss Networks", *INFOCOM'92*, Session 2A.1.1, Florence, Italy, May 1992.

[2] R.S. Sutton, "Learning to Predict by the Methods of Temporal Difference", *Machine Learning*, vol. 3, Kluwer Academic Publishers, 1988, pp 9–44.

[3] S.J. Bradtke and M. O. Duff, "Reinforcement Learning Methods for Continuos-Time Markov Decision Problems, in *Advances in Neural Information Processing Systems 8*, D.S. Touretzky, ed., MIT Press, 1995.

[4] A. Barto, R. Sutton and C. Watkins, "Learning and Sequential Decision Making", Report COINS 89–95, Dept. of Computer and Information Science, University of Massachusetts, Amherst, USA, September 1989.

Modular Neural Control of Buffered Banyan Networks

A. Garcia-Lopera, A. Ariza Quintana y F. Sandoval Hernandez

Departamento de Tecnologia Electronica
Escuela de Ingenieros de Telecomunicacion
Universidad de Málaga
Plaza del Ejido, 29013 Malaga, ESPAÑA
E-Mail AGL@TECMA1.CTIMA.UMA.ES

Abstract

A distributed Neural Control (NC) for Banyan type Multistage Interconnection Networks (MIN) with input buffering at each switch node and his application to Asynchronous Time Mode (ATM) communications is considered. The neural control is independently done at each stage level and for each switching node. Simpler and smaller NC are used at the switching node level instead of only one NC with higher number of neurons and interconnections. Potential VLSI applications are considered. Simulation results on throughput and buffer sizes are presented considering Constant Bit Rate (CBR) and Variable Bit Rate(VBR) traffic types.

1 Introduction

The applications of neural networks to solve the switching fabrics control problem in ATM communications has been widely studied in the few last years. In [1] an extended bibliographical reference is given. The buffered Banyan MIN are accepted like one of the best suited technologies to meet the requirements of the ATM for the B-ISBN. These switches suffer of input (Head Of Line) blocking, output blocking and, additionally to the non blocking switches, internal blocking too. To avoid all these blocking, buffers can be placed at the input ports, output ports and at the internal stages switches ports (internal buffers). For these blocking networks, performance analysis and simulation have shown, under different arrival traffic types and different non uniform destination distributions, that buffering at the internal level stages produces considerable improvement in their performance [2]. They offer better results in parameters like throughput, latency and cell loss ratio than the only input-buffered switches, but they require a more complex control. In [3] an NC is proposed for an only input buffered Banyan Network with N inputs and N outputs that uses N^2 neurons arranged in a matrix of NxN. They are interconnected in a *Multiple Overlapped Winner Take All* way (MOWTA). The number of connections is $O(N^2 log_2 N)$. Our paper extends this approach simplifying the NC. We introduce here a modular NC architecture for a multibuffered Banyan switch with input and internal buffers. The neural control is carried out at the Switching Element (SE) level and independently for each stage that is considered like a non blocking network with output buffering.

2 Banyan Networks and the Proposed Modular NC

The Banyan network can be defined by the number of communication paths of each SE, the number of the stages and the interconnection paths between SE described by a set of *topology describing rules* [4]. We consider networks of NxN inputs/output with $log_2 N = m$ stages of $2x2$ SE, with $N/2$ SE in each stage. In the Figure 1 is represented one of these networks. In 2.1 we will consider dxd SE MIN. The Banyan networks are deterministic in the sense that there is only one path for a given input/output pair. Also, they are selfrouting and have the unique path property. All the information needed to route a cell through the network is the output line address. The sequence of nodes connecting an input to an output is unique for all the input-output pairs. Let $A = (a_m a_{m-1} .. a_1)$ the input line address, $B = (b_m b_{m-1} .. b_1)$ the output line address, and $P = (p_m p_{m-1} ... p_1)$ the link address that the cell defined for the pair (A/B) visits in each stage, all in binary representation with m bits. In the selfrouting scheme, we can determinate the logical position of a cell through the network applying succesives bit-wise operations to the input address [3]. The switching node at the stage k performs the replacement operation:

$$R_k (p_m \, p_{m-1} \cdots p_2 \, p_1) \setminus (b_m \, b_{m-1} \ldots b_2 \, b_1) = (p_m \, p_{m-1} \cdots p_2 \, b_{m-k+1}) \tag{1}$$

The *Link Operator* is a mathematical mapping of the interconnection network between the stages k and $k+1$ into a bit-wise operation upon the address of the links corresponding to the outlets of the switching nodes of the stage k, to get the address of the corresponding links at the inlets of the switching nodes in the stage $k+1$. For an Omega Network the *Link Operator* is the Shuffle operator $L=S$.

$$S (p_m \, p_{m-1} \cdots p_2 \, p_1) = (p_{m-1} \cdots p_2 \, p_1 \, p_m) \tag{2}$$

For the SW-Banyan network the *Link Operator* is the Banyan Exchange $L=B$.

$$B_k(a_m a_{m-1} \ldots a_2 a_1) = (a_m a_{m-1} \ldots a_{m-k+2} a_1 a_{m-k} a_{m-k-1} \ldots a_2 a_{m-k+1}) \quad (3)$$

The proposed architecture is oriented to control internal buffered Banyan network. We assign an NC to every switching node. This node is considered like a *non blocking* network (*cross-bar network*). Therefore the neural control is carried out "locally" for each stage of the MIN. The NC have all the same size (2x2 neurons if the SE are 2x2) *independently* of the size of the MIN. This small number of neurons will simplify the VLSI design, and provide small times of convergence which is assured for all the cases.

Next we describe the neural architecture. To clarify the description we will refer to the example of the Figure 2. For this example, in the Figure 3 we represent the matrixes defined as follow.

The cells waiting in the N input buffers of length W at each stage k can be represented by N/2 Traffic Matrixs $T_{KV}[b_{ij}]$ of dimension $Wx2$ associated to each SE numbered V in this stage k. In this matrix, each element is the bit b_{n-k+1} of the output address of the cells waiting in the input buffers of S_V. Each S_V has two input buffers with the addresses $(p_m p_{m-1} \ldots p_2 0)$ and $(p_m p_{m-1} \ldots p_2 1)$ respectively.

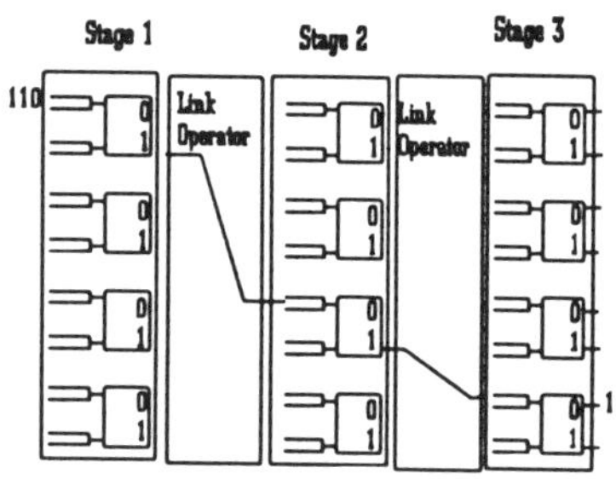

Fig 1: Buffered MIN Scheme

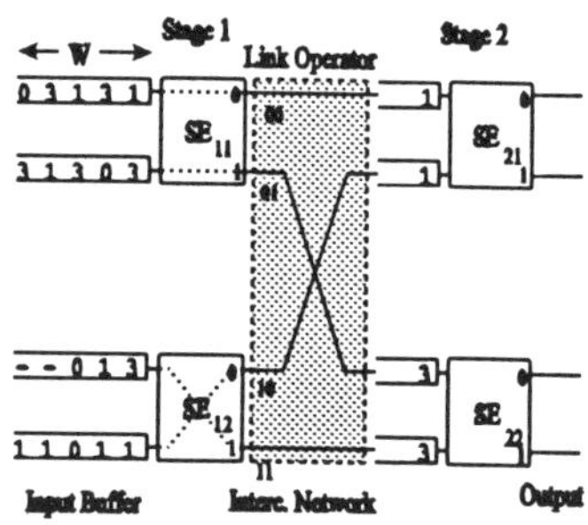

Fig 2: 4x4 Banyan Buffered

From equation (1), two cells at the inlets of the S_V having the same b_{m-k+1} bit in their output addresses will compete for the same outlet of S_V with the address $(p_m p_{m-1} \ldots p_2 b_{m-k+1})$.

From the Traffic Matrix $T_{KV}[b_{ij}]$ is derived the Request Matrix $R_{KV}[r_{ij}]$. This is a 2x2 binary matrix. Each element represents a request of cell $C(ij)$ from input i to output j. If the cell $C(ij)$ exists in the buffers, r_{ij} will be 1. If not, r_{ij} will be *0*. . Because they are two inlets and two outlets in the 2x2 SE, this Matrix will have dimension *2x2*.

The Request Matrix is the input to an NC. This is a square neural array of *2x2* neurons mapped with R_{KV}.They can be totally interconnected like a Hopfield

Network, or, more efficiently, in the MOWTA way, bearing in mind that the SE is a non blocking network. In the Figure 4, the NC for each SE is represented. Each neuron is activated or not, depending if the corresponding element r_{ij} is *1* or *0*. The function *tangh* is considered for the transfer function of the neurons. The neural network will converge to the Output Matrix $O_{KV}[r_{ij}]$. At most, only two neurons will be in *on* state, one per row and one per column. The corresponding cell *C(ij)* is scheduled to be transmitted to the outlets of S_V in the current time slot, not existing conflict at internal link level. The cells selected at the stage *k* will be stored in the

$$T_{11}=\begin{bmatrix}0&1&0&1&0\\1&0&1&0&1\end{bmatrix}\; R_{11}=\begin{bmatrix}1&1\\1&1\end{bmatrix}\; O_{11}=\begin{bmatrix}1&0\\0&1\end{bmatrix}\; T_{12}=\begin{bmatrix}-&-&0&0&1\\0&0&0&0&0\end{bmatrix}\; R_{12}=\begin{bmatrix}1&1\\1&0\end{bmatrix}\; O_{12}=\begin{bmatrix}0&1\\1&0\end{bmatrix}$$

Fig 3: Generated Matrixes for the 4x4 Network of Figure 2

input buffers of the stage *k+1* if they are not full. A back pressure mechanism is implemented to avoid this situation. The buffer with not rooms enough will generate an external inhibitory input to the neurons corresponding to their addresses. They are two neurons for each buffer, corresponding to each column in the R_{KV} matrix. These neurons are inhibited, so, not cells will be sent to the corresponding buffer in the *k+1* stage. This situation is represented in the Figure 5.

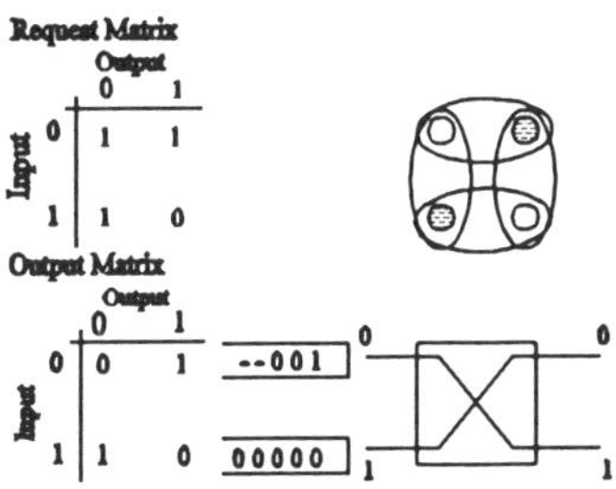

Fig4: Neural Controller for 2x2 SE

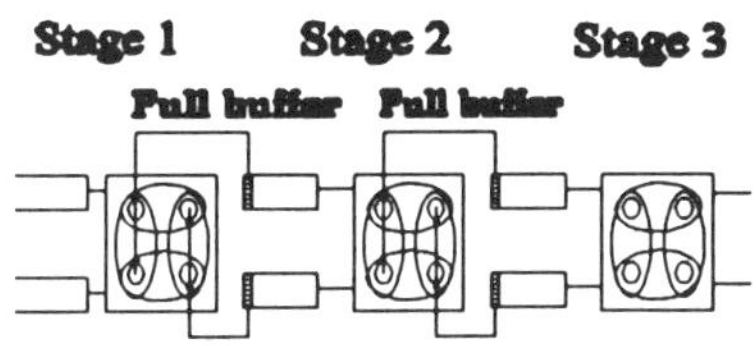

Fig 5: Back Pressure Mechanism

The new Traffic Matrix $T_{k+1}[b_{ij}]$ of the stage *k+1* is obtained applying to the addresses of the stage *k* the MIN operators *B*, *L*, (Equations 1,2,3).The competition is repeated at the *k+1* stage with the corresponding Traffic Matrix. Therefore, we have avoided the internal blocking (at link level). However cells with output blocking can be selected in the input and internal stages but not in the output stage (supposing a non output buffered MIN) where they are contemplated like link level blocking cells. We point out that the internal buffers must be able to accept only one cell within a slot.

2.1 Modularity and Size of the Switching Elements

The proposed architecture is easily applicable to buffered Banyan networks built out of *dxd* SE where $d > 2$. In this case, the number of the stages needed for a NxN switching fabric reduces to $log_d N < log_2 N$. They maintain the selfrouting and unique path property, but all the addresses must be represented in a system base *d*. The reduction in the number of stages, will reduce the internal blocking possibility and the latency time for the cells. The SE is considered cross-bar non blocking network of *dxd* lines. We will need in this case NC having *dxd* neurons. The number of NC needed is $(N/d)\log_d N$. For instance, in the case of a non buffered 64x64 switching matrix with 2x2 SE, it will be necessary one NC with 4096 neurons having Winner Take All (WTA) with 64 neurons, if we apply the approach [3]. For one same size internal buffer Banyan network with 2x2 SE, will be required 192 NC of 4 neurons (768 total neurons), having two neurons WTA . If 4x4 SE is used, will be required 48 NC of 16 neurons (768 total neurons) having 4 neurons WTA. The convergence time of the NC necessary to arrive to stable state is very low for WTA with small number of neurons. To validate our proposal we has designed the full custom NC for *2x2* and *4x4* SE. The H-SPICE simulations show convergence time ranging from *8* to *12 ns*. We are using 2 μm. MIETEC CMOS technology.

2.2 Control Flow

In this section we discuss two different ways the cells can flow through the MIN. In ATM the time slot is divided in two intervals: The path setup interval or setup time (t_p), and the transmission interval or transmission time (t_x). In the setup time t_s, control signals are generated and passed across the Switch from the output stage toward the input stage, so that every waiting cell knows whether it can go to a next stage's input buffer or should stay in the current buffer. In the transmission time t_x, the selected cells may move forward through the switch.

In the first situation considered, during the setup time, the NC compete simultaneously in parallel "seeing" only the buffer one stage ahead. The last stage has not buffers at the output. The control information is exchanged only between two successive stages, and the selected cells can go forward only one stage. For the output stage, the selected cells will be that cells without output blocking condition. So, the minimum time required to get the output for one cell waiting at the inputs of a n-stages MIN is $m.T$ slots time.

In the second situation considered, multiple setups per slot time, we multiply the setup time by *m*. We assume that the information is exchanged between all the stages in *m* times t_s and the selected cells can go through the Switch from the input buffer to the output in one slot time. With this scheme, output buffering is required.

3 Performance Evaluation and Empirical Results

For simulation we used a discrete model of Hopfield Neural Network connected in MOWTA. Simulation was carried out in a HNC Balboa Neurocomputer installed into a Sun Sparcserver 330.

For Constant Bit Rate simulation, a Bernouilli process was assumed at chosen arrival rate λ for different buffers length, W, and different MIN sizes N. One run comprised three series of 10^4cell arrivals per input line with three different seeds used to randomly initialize the Neural Network after each convergence process. In any case we detected bad routing or convergence failure, needing always less than 50 iterations per neuron to converge. The simulation process was as follow. We obtained the input buffer queue distribution supposing W(input) = ∞ for different internal buffer size, W(internal). The distribution obtained was of geometric type. Applying on it the loglinear regression method, we estimated the cell loss probabilities as represented in the Figure 6. On this figure we can see that, for the input queues, an internal buffer of length 20 has the same effect that an internal buffer of length ∞. Shorter lengths need of bigger input buffer for the same cell loss probability. For all the following simulations, we used W(internal)=20. In the Figure 7 we compare the curves for MIN of several sizes. We verify that the input buffers lengths are bounded at values W(input)<30 in all the cases. In the Figures 8 and 9 we show the maximum throughput and the throughput of our proposal compared with the theoretical for a crossbar network given in [5]. In the Figure 9, are compared both the throughput values for a non buffered Banyan with strict FIFO policy (W=1) [5], and our proposal in the same condition. It is interesting to point out that the throughput obtained by our simulation for buffer input W(input)=1, FIFO policy, is 0.746. This value is agreed with the saturation throughput, 0.75, reported in [5] for a 2x2 crossbar with FIFO policy. It is a logic result considering that in our architecture we are controlling always 2x2 crossbar switching independently of the size of the MIN. The number of time slots that a cell need from her arrival time to the input buffer until the arrival time to the scheduled output buffer, is the mean waiting time shown in the Figure 9. It depends of the input buffer related with the cell loss probability shown in the Figure 6.

For the VBR simulations we have applied the Markov Modulated Poisson Process (MMPP) model. This is a doubly stochastic Poisson Process, with two states T_0 and T_1 having two Poisson parameters λ_0 and λ_1 respectively. It was simulated a bursty type of traffic with simultaneous arrival of several cells in the same time slot. We simulated the arrival of 10^6 cells repeated 10 times using different seed to generate the cell arrival. We chose the MMPP parameters for a hardly bursty traffic. The queue length distribution is shown in the Figure 11. As well it is represented the loglinear regression obtained to estimate the tail of the distribution. As we can see, the controller is able to absorb the bursty traffic but increasing the size of the input buffers.

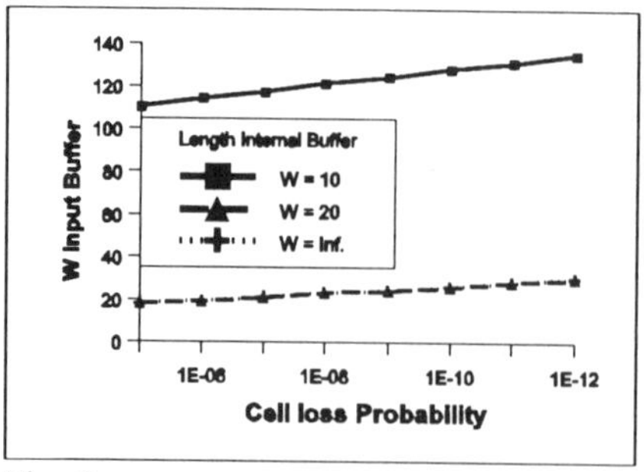

Fig. 6: 32x32 Banyan Matrix. Cell arrival rate λ= 0.95

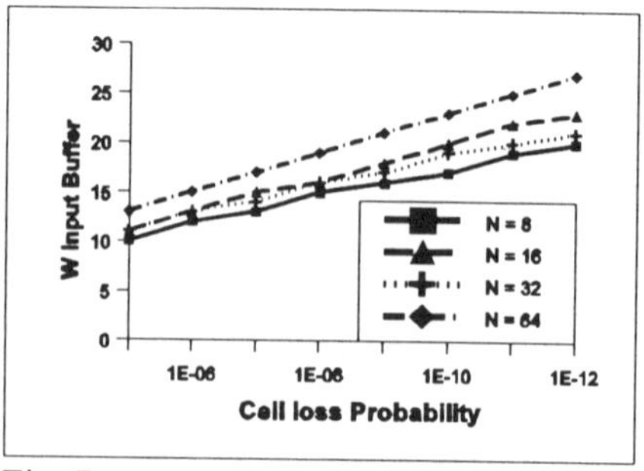

Fig. 7: Internal buffer length 20. Cell arrival rate λ = 0,9

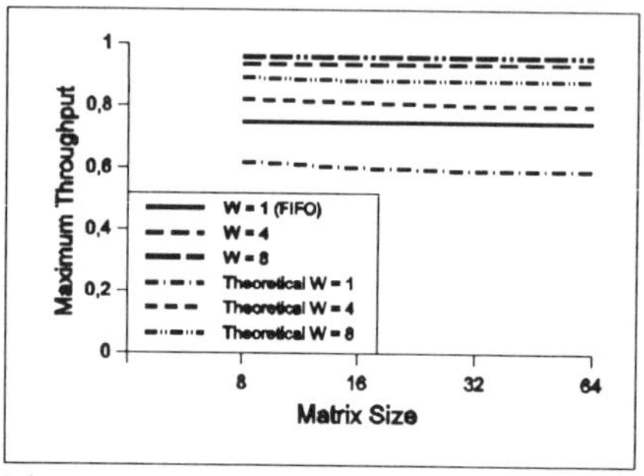

Fig. 8: Theoretical Cross bar and proposed NN model max throughput. Internal buffer length 20

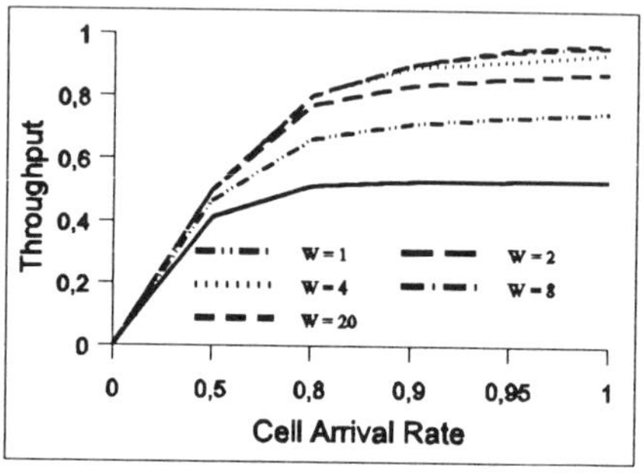

Fig. 9: 16x16 Buffered Banyan Matrix. Throughput for several input buffer compared with a FIFO (W = 1) non buffered Banyan ———

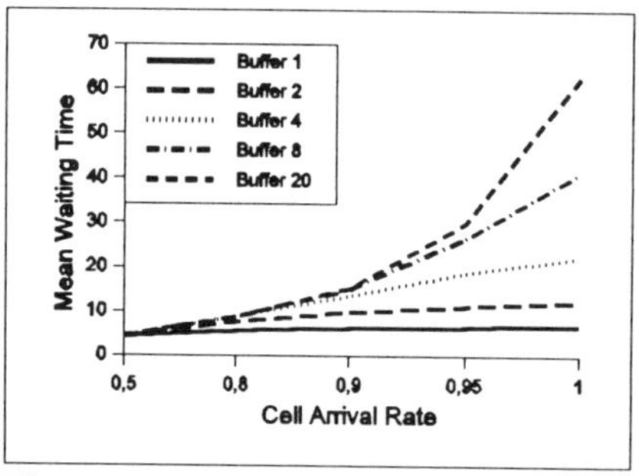

Fig. 10: 16x16 Banyan matrix. Mean Waiting Time for several input buffer. Internal buffer length 20.

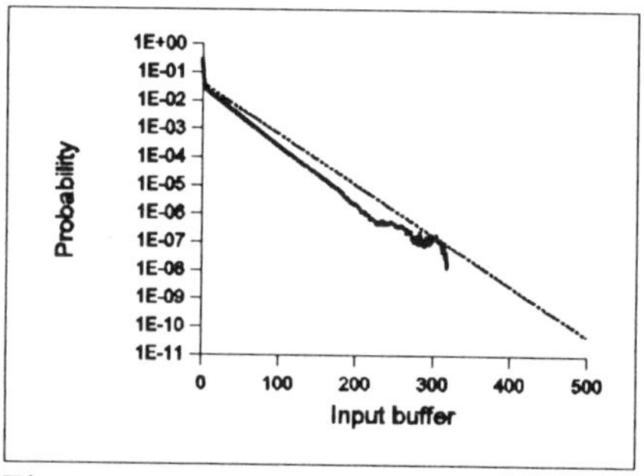

Fig. 11: 8x8 Banyan Matrix. Input queue length distribution probability and estimated linear regression for MMPP trafic type $\lambda_0 = 0{,}5$, $\lambda_1 = 8{,}5$, $T_0 = 1$, $T_1 = 42$. Internal buffer length 20.

4 Conclusions

The NC proposed is very simple in his architecture and fitted to the VLSI design. It is robust and need small time to converge. The total number of neurons is $2N\log_2 N$ gathered into arrays of 2x2 neurons, independently of the input buffer length. Each NC "see" one 2x2 crossbar network and the saturation throughput equals to the analytic results in [5]. Smaller buffers size and lower cell loss probabilities are obtained compared to other approaches [3,5]. The architecture is scalable in a straigth way. For a very bursty VBR traffic type the NC is able to absorb the arriving cells with low cell loss probability, but increasing significatively the input buffers size.

Acknowledgments

This work has been partially supported by the Spanish "Plan Nacional de Banda Ancha", Project TEMA, and the CICYT Project No. TIC92-1325-PB.

References

[1] Brown, T.X., "Neural Networks for Switching", in *Neural Networks in Telecommunications* (Yuhas, Ben and Ansari, Nirwan, Eds.), Kluwer Academic Publisher 1994, Chap. 2, pp. 11-35.

[2] Morris, D. Todd and Perros, Harris G., " Performance Modelling of a Multi-Buffered Banyan Switch Under Bursty Traffic", *IEEE Transactions on Communications,* Vol 42, No 2/3/4 , pp 891-895, February, March, April 1994.

[3] Brown, T.X. and Liu, K.H., "Neural Network Design of a Banyan Network Controller", *IEEE Journal of Selected Areas in Communications*, vol. 8, no 8, pp 1428-38, Oct. 1990.

[4] Wu, C.L., and Feng, T.Y., " On a class of Multistage Interconnection Networks", *IEEE Transactions on Computer,* vol. 29, no 8, pp. 694-702, Aug. 1980.

[5] Hluchyj, M.G., and Karol, M. J., "Queueing in High-Performance Packet Switching", *IEEE Journal on Selected Areas in Communications,* vol. 6, no. 9, pp. 1587-97, Dec. 1988.

ATM Traffic Neural Control: Multiservice Call Admission and Policing Function

A. D. Estrella, E. Casilari, A. Jurado, and F. Sandoval

E.T.S.I. Telecomunicación, Dpto. de Tecnología Electrónica
Univ. de Málaga, Plza. El Ejido, s/n, 29013 Málaga, Spain
estrella@tecmal.ctima.uma.es

Abstract

An ATM multiservice call admission control using backpropagation learning is presented. Its architecture allows different quality of service for each service and includes an appropriate policing function. The proposed neural control model is analyzed by computer simulations in a heterogeneous traffic environment and the results show its effectiveness compared with other methods.

1 Introduction

The asynchronous transfer mode (ATM) is a key technology in integrating multimedia communications services in the broadband integrated services digital network (B-ISDN). Because ATM involves a statistical multiplexing scheme, degradation of quality of service (QOS) may easily occur in cases of network congestion and, thus, some traffic controls are necessary. The Call Admission Control (CAL) is an algorithm which, following a call request, decides whether a call may be accepted while maintaining the QOS of existing calls. If the call request is accepted, the policing function monitors the traffic generated by that call and discards or marks those cells that violate the negotiated parameters of traffic when the call was established. There are a lot of types of policing function (leaky bucket, jumping or sliding window, state load, etc...) but, in most cases they only polices the mean and peak rate of the traffic source [1]. Figure 1 shows the situation of these controllers in an ATM network.

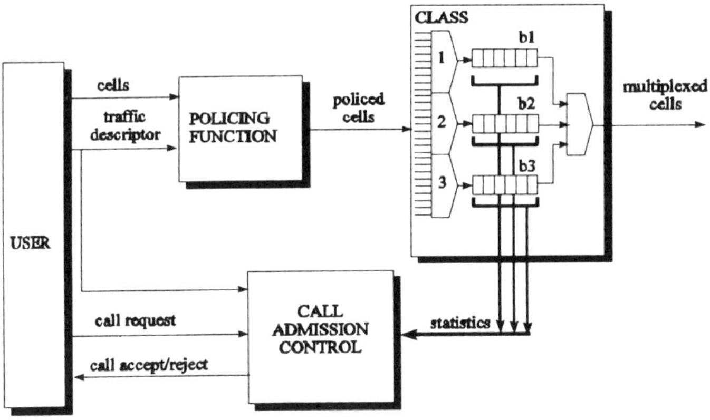

Figure 1: Traffic controls in an ATM network

The CAL in ATM using neural networks has been studied by several authors [2,3,4]. These approximations are based on the continuous monitoring of the network state and the use of a feedforward multilayer perceptron with backpropagation learning.

In [4], an admission controller with a new method of training data selection that supplies local learning and a good generalization is proposed. In this paper, the performance of this ATM admission neural controller with one QOS for each class of service is presented. Each class of service is controlled by a different neural network. Besides, a policing function, which is usually ignored, that maintains the high statistical gain obtained by the neural control is studied.

2 Admission control algorithm

The proposed algorithm assigns a buffer b_i and an elemental CAL to each class i of service (see figures 1 and 2). Each elemental CAL includes a neural network that is defined by the function $q_i = g_i(\boldsymbol{n}, \boldsymbol{W}_i)$. The input $\boldsymbol{n}$ is a call vector whose ith element n_i is the number of calls in progress for class i $(i=\{1,2,\dots,N\}$ $)$, the output q_i is the estimated QOS parameter by the neural network and $\boldsymbol{W}_i$ is the matrix of weights. The performance of the controller consists of three phases: observation, control and training.

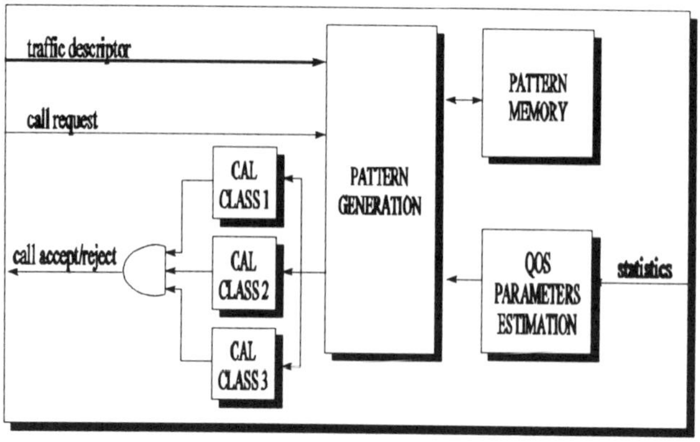

Figure 2: Block diagram of the neural CAL

The observation phase monitors the QOS parameters for each buffer. When there is a change in the number of calls in progress, the set of old connected calls number combined with its measured QOS parameters, is stored in a pattern memory as a training pattern. The pattern memory consists of a set of sliding windows, each of which represents a significant surface of the state space. The selected window is determined by a function that depends on $\boldsymbol{n}$, the traffic descriptor and the link capacity B [3].

The control phase is activated every time a user requests to setup a new call of class j. The function learned by each neural network q_i predicts a QOS parameter of that new call together with all the calls in progress. Then, the call request is accepted if $q_i = g_i(n_1, n_2, \ldots, n_j + 1, \ldots, n_N, \boldsymbol{W}_i) < t_i \quad \forall i$, where t_i is the objective QOS parameter for the class i.

The training phase is carried out, independently of previous phases, for each period T_{train}. The controller will randomly select a training pattern of some window. The selected training pattern will then be presented as an input to all the neural networks. We use an adaptive control algorithm based on feedforward multilayer networks with on-line back-propagation learning.

The ATM call admission neural control has been studied by several authors, however few of them [5] have investigated its associated policing function. As has been seen, the typical neural controller learns the relationship between the number of calls in progress of each class and a measure of the performance of the network, so it does not need to know in full detail the traffic characteristics. Its basic advantage is that we can obtain a very high statistical gain, unlike analytical algorithms. However, malicious users could easily modify their instantaneous bit rate and the conventional policing function (peak and mean) would not be able to control all statistical parameters of the source so, as a consequence, degradation of QOS may occur and the neural admission control would start to learn bad training patterns. In the following sections, this effect is studied and some solutions as leaky bucket array and police neural function are discussed.

3 Policing function

In a leaky bucket array, each leaky bucket controls one cell rate R_j and consists of a counter which a) is incremented every time a cell is generated by the source and b) is decremented at a fixed rate R_j. The incoming cells that find the counter at its threshold U_j are discarded or marked.

The neural control is based on a prediction of the generated traffic by the source. The neural network used can be defined by the function $y(k) = f(\lambda(k-1), V)$ where $\lambda(k)$ and $y(k)$ are the number of measured and predicted cells, respectively, in the interval k and V is the matrix of weights. The number of cells to be discarded or marked in the interval k is $\rho(k)$:

$$\rho(k) = \begin{Bmatrix} \lambda(k) - y(k) & if & \lambda(k) > y(k) + U \\ 0 & & otherwise \end{Bmatrix} \tag{1}$$

where U is a threshold related to the error tolerance of neural network. If cells are discarded, the next input to the neural network will be $y(k)$, and $\lambda(k)$ otherwise.

4 Simulation results

The simulation model is made up of several classes of traffic source generators with various policing controllers, a statistical multiplexer and the call admission neural controller. The selected models of traffic sources are ON-OFF (typical in voice) and first order autoregressive (typical in video). The model ON-OFF is defined by the peak bit rate V_{peak} and the active and silence mean periods (T_{ON} and T_{OFF}, respectively). The autoregressive model is defined by the equation $\lambda(k+1) = a\lambda(k) + bA(k)$ where $\lambda(k+1)$ is the bit rate for the frame $k+1$, a and b are constants and $A(k)$ is a gaussian noise.

To study the multiservice call admission control, the model generates calls of three classes of ON_OFF traffic (see table 1), with an exponentially distributed average holding time T_{hold} = *5s*, an average offered traffic load C = *60 erlangs* (variable between 0 and 100) and link capacity B = 150 Mbps. The pattern memory size is 640 patterns and is composed of 64 windows, T_{train} = 1 ms. Small neural networks with three input neurons, two hidden neurons and one output neuron are used.

Table 1: Simulations Conditions

Traffic characteristics	**Class 1**	**Class 2**	**Class 3**
V_{peak} (Mbps)	10	2	5
T_{ON}(s)	0.0001	0.001	always
T_{OFF}(s)	0.001	0.001	0
b(cells)	1000	64	80
q	10^{-5}	10^{-3}	10^{-5}

The learning time depends on the variations of C. In our case, after 1000 s, the neural controller has learned a stable admission boundary which can be seen in the figure 3. The axes indicate the number of calls in progress for each class. To show the advantages of this control, the simulation is repeated but with homogeneous QOS requirements, that is to say, b_i = 64, t_i=10^{-5} $\forall i$ Figure 4 shows a two-dimension map of the admission boundaries of both simulations. The admission region with heterogeneous QOS requirements is bigger because the control considers the characteristics of each class of traffic Besides, in this figure, the admission boundaries obtained by some analytical methods (peak bandwidth and virtual loss [6]) are also presented.

The neural control with a leaky bucket array is compared. For this, a first order autoregressive model as well-behaved source and a small ON-OFF model as interfering traffic are used. The reason for this, is that all attempts to predict ON_OFF traffic were unfruitful, owing to the model not being autocorrelated. The parameters of the autoregressive model are a = 0.8781, b = 0.1108 and a gaussian noise with mean 4.3Mbps and variance (normalized by the squared mean) 0.054. For these values, the mean bit rate and peak bit rate are 3.9 Mbps and 6 Mbps (with probability of 0.99), respectively. The neural network was trained with 400 patterns of video traffic and a sample interval of T = 1/30s and 15000 iterations were needed to obtain a mean squared error of 0.000016. The ON_OFF source has a peak bit rate of 1Mbps with T_{ON} = 0.042s and T_{OFF} = 0.425s. The leaky bucket array has two counters for controlling peak bit rate (R_1= 6 Mbps and bucket size of U_1=1) and mean bit rate (R_2= 3.9 Mbps and bucket size of U_2 = 2000).

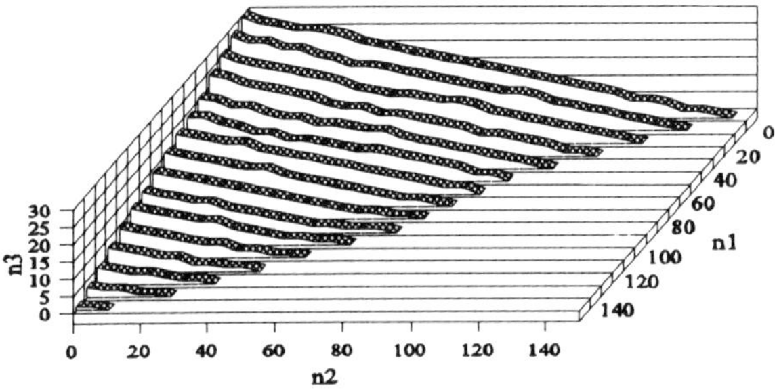

Figure 3: 3-classes admission boundary

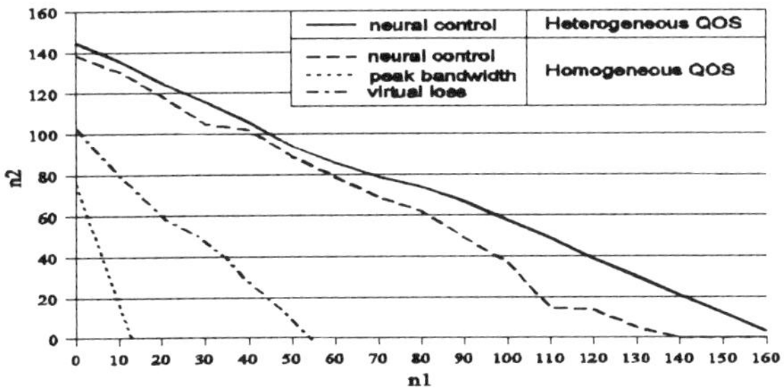

Figure 4: 2-classes boundaries for different QOS requirements

Figure 5 shows the generated traffic for 200 frames and the number of cells discarded or marked $\rho(k)$ by the neural control and the leaky bucket array. As can be seen, the performance of the neural control is superior. The reaction time of leaky bucket array is higher than that of neural control because the former needs to fill up the bucket to start to control. Besides, once full, the bucket is activated too easily and it can discard well-behaved cells.

Figure 6 shows discard errors and filtered traffic of both methods ((negative) positive discards corresponds to (well) bad-behaved sources). Again, the behavior of the neural control is better because it has less errors and the filtered traffic is more similar to video traffic. On the other hand, for ON-OFF traffic, it is observed that with leaky bucket array and similar controllers, the more traffic parameters are negotiated, the more exact the representation of the behavior of the service is and the policing function can react more accurately.

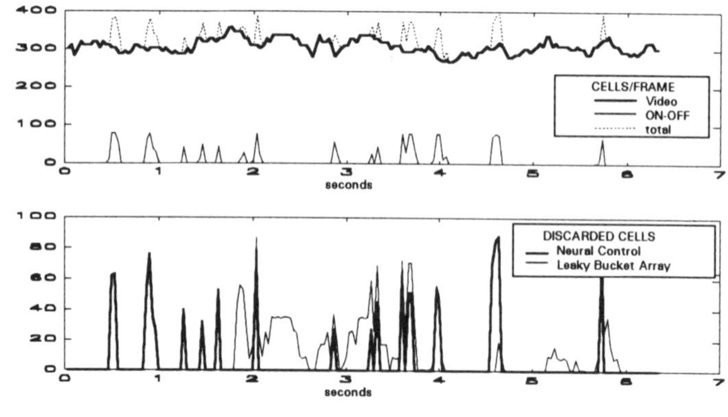

Figure 5: Generated traffic and discarded cells

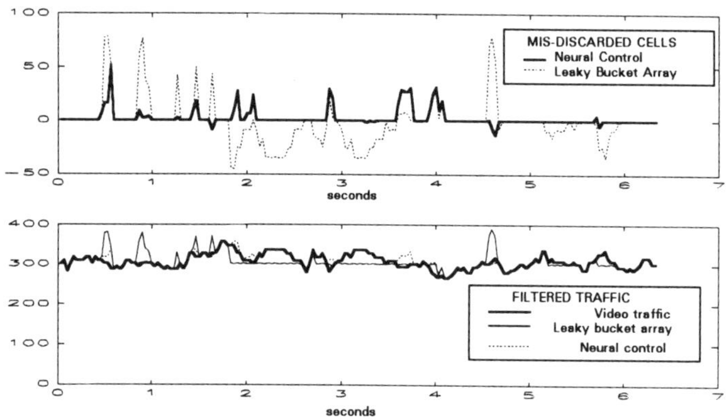

Figure 6: Discard errors and filtered traffic

5 Conclusions

An ATM multiservice call admission control has been obtained and the simulation results show that complicated admission boundaries can be learned. This control is suitable for any QOS objective and it can easily control different QOS objectives for different classes of calls. The design is modular and can be extended to further classes of service.

On the other hand, the high gain obtained by the neural controller can be lost if we do not use an optimum policing function. Unfortunately, with conventional controllers an optimum control is very difficult to obtain and can result too expensive. The studied neural policing function polices the global characteristics of the traffic against the leaky bucket array which only polices some parameters. However, this neural control is only useful for autocorrelated traffic. So, it is necessary to investigate new neural methods that obtain more general results.

Acknowledgements

This work has been partially supported by the Spanish Comisión Interministerial de Ciencia y Tecnología (CICYT), Project No. TIC92-1325-PB.

References

[1] H. Hemmer and P.T. Huth, "Evaluation of Policing Functions in ATM Networks," in *Queuing, Perfomance and Control in ATM (J.W. Cohen and C.D. Pack, Eds).* Elsevier Science Publishers, 1992, pp. 111-116.

[2] A. Hiramatsu, "ATM Communications Network Control by Neural Networks," *IEEE Transactions on Neural Networks*, vol.1, no. 1, pp. 122-130, 1990.

[3] R.J.T. Morris, "Prospects for neural networks in broadband networks resource management," in *Teletraffic and Data traffic (A. Jensen and V.B. Iversen, Eds)*, Elsevier Science Publishers, 1991, pp. 335-340.

[4] A. D. Estrella, A. Jurado and F. Sandoval, "New training pattern selection method for ATM call admission neural control," *Electronics Letters*, vol. 30, no. 7, pp. 577-579, 1994.

[5] A.A.Tarraf, I.W. Habib and T.N. Saadawi, "A Novel Neural Network Traffic Enforcement Mechanism for ATM Networks", *IEEE JSAC*, vol.12, no. 6, pp. 1088-1096, 1994.

[6] T.Murase, H. Suzuki, S. Sato and T. Takenchi: "A Call Admission Control Scheme for ATM Networks Using a Simple Qualite Estimate," *IEEE JSAC*, vol. 9, no. 9, pp. 1461-1470, 1991.

Adaptive Flow Control in Multistage Communication Networks Based on a Sliding Window Learning Algorithm

Alexandru Murgu

University of Jyvaskyla, Department of Mathematics
P.O. Box 35, FIN-40351 Jyvaskyla, FINLAND
murgu@julia.math.jyu.fi

Abstract

In this paper we consider the problem of adaptive flow control in a single-hop structured network having multiple layers and packet fragmentation between layers using a sliding window approach. To make tractable the control effort in a multistage network, we consider a reduced equivalent of the network to a single stage of switched queues. The control mechanism is implemented as learning algorithm based a gradient descent rule and mapped onto neural networks for nonlinear optimization. Due to the generalization feature of neural networks processing, only a local pattern of information is needed, in handling general flow patterns. The proposed control approach is analysed in terms of tracking performances of bursty Markovian flow patterns for a two port communication network.

1 Flow Control for Multistage Communication Networks

Adaptive flow control is a powerful approach for transmission of data or switching the resources in a communication network as compared with static routing schemes. The adaptive schemes have been adopted for packet routing, transmission capacity sharing of a link among several competing information sources or for flow control in a packet switched network [12]. In large networks, the control functions are based on a local information pattern and they have a distributed implementation. For a multistage communication network as shown in Fig. 1, a virtual circuit connection for bursty data sources can be translated into a reduced model consisting of only one stage network, as in the case of end-to-end connection models [1]. With one stage network model, the adaptive flow control scheme we want to design as a neural controller has the structure shown in Fig. 2. A packet fragmentation between successive stages or layers is a better approach toward a finer control scheme [7] when the packet flow is subject to some interfering traffic which will utilize part of the capacity on each link (or trunk for ATM networks) and occupy some buffer space at each node. In order to account for the interfering traffic, the usual approach has been to decrease the average service rate at each node [1], [6].

1.1 Single-Hop Flow Control in Multistage Networks

The flow patterns in multistage networks can be adaptively controlled according to the downstream link one hop away criterion [7]. The interconnection controller works cyclically and at the beginning of each time slot, the input ports (see Fig. 2) are scanned at the given discrete time periods (slots), while the routing/flow control parameters are on-line computed within the present slot to produce the flow assignments for the next one. The basic approach to realise this estimation task within a single hop is to used the deviations of the flow patterns from the corresponding multihop values to built a measure of similarity/dissimilarity among these patterns. More specifically, the estimation is performed by using a sliding window approach over the set of incoming flows while the goal is to release such amount of traffic into network to achieve the desired level of control performances measured in terms of delay–throughput balance. By considering such a control device, the average flow patterns at the input ports are to be detected from the stochastic incoming flows so that, the processing speed (or equivalently the processing priority) for different flows is to be computed. The above control problem is jointly realized by reconfiguring periodically the routing matrix and the speed of flow processing units (see Fig. 2). One more point to be mentioned is that the flow control strategy has to preserve the network stability against the congestion phenomenon.

1.2 The Reduced Equivalent of a Multistage Network

By using the Norton theorem for a multistage network which is modelled as a closed multiple queue system [6], the reduced equivalent model is synthetically given by the following dynamical system equation:

$$\dot{\mathbf{x}} = \mathbf{A}(t,\xi)\mathbf{x} + \mathbf{B}(t,\mathbf{x})\mathbf{u} \tag{1}$$

where $\mathbf{A}(t,\xi)$ is the routing matrix (a general stochastic matrix), $\mathbf{u}(t) = (\alpha,\phi)^T$ is the control vector (see Fig. 2) and $\xi(t)$ is an i.i.d. random process modelling the stochastic flow pattern. The packet fragmentation between the multiple layers is taken into account as a modified processing rate at the node processors given by

$$\mathrm{d}\mathbf{x}(t) = (\mathbf{A}(t,\xi)\mathbf{x}(t) + \mathbf{B}(t,\mathbf{x})\mathbf{u})\mathrm{d}t + \mathrm{d}\mathbf{w} \tag{2}$$

where $\mathbf{w}$ is a zero mean and unit matrix variance Brownian motion.

2 Sliding Window Approach of Flow Control

2.1 Sliding Window Function for Statistical Flows

The use of sliding control techniques in variable structure systems [10] like those represented by neural networks architectures improves considerably the richness of the dynamic behaviour and the control performance. The general idea in coping with the fluctuations of the input flow patterns is to use time dependent switching surfaces generated by using a sliding window. For a network with n input/output links, a sliding window function realises the mapping of the input variables $\mathbf{x}_i, i = 1,...,n$ into the corresponding level of the throughput performance represented by $y_i, i = 1,...,n$. This idea is naturally related to the function approximation by neural networks. A sliding window function satisfying the above requirements is given by

$$\Phi(\mathbf{x}) = \frac{N(\mathbf{x})}{D(\mathbf{x})} \tag{3}$$

where the quantities $N(\mathbf{x})$ and $D(\mathbf{x})$ are defined as follows

$$N(\mathbf{x}) = \sum_{i=1}^{n} y_i \exp\left(-\frac{(\mathbf{x}-\mathbf{x}_i(t))^T\mathbf{W}(\mathbf{x}-\mathbf{x}_i(t)}{2\sigma_i^2}\right) \tag{4}$$

$$D(\mathbf{x}) = \sum_{i=1}^{n} \exp\left(-\frac{(\mathbf{x}-\mathbf{x}_i(t))^T\mathbf{W}(\mathbf{x}-\mathbf{x}_i(t)}{2\sigma_i^2}\right) \tag{5}$$

In the expresion of $\Phi(\mathbf{x})$ the time dependent sliding surfaces are defined by

$$S(\mathbf{x}) = (\mathbf{x}-\mathbf{x}_i(t))^T\mathbf{W}(\mathbf{x}-\mathbf{x}_i(t)) \tag{6}$$

where $\mathbf{W}$ is the interaction matrix among different flow patterns. The idea behind this control strategy is to track the desired throughputs at the destination nodes of the multilayer network (represented by the vector $\mathbf{y} = (y_1, y_2, ..., y_n)$) by an inverse mechanism, i.e., controlling the incoming mean flow patterns $\mathbf{x}_i(t), i = 1, ..., n$.

2.2 Learning Sliding Window Control

The learning objective can de defined as minimization of a cost functional which is constructed based on the mean flow patterns $\mathbf{x}_i(t), i = 1, ..., n$, and the interconnection weights matrix $\mathbf{W}$. Such a cost functional is defined as follows

$$J(\mathbf{W}) = \sum_{i=1}^{n}\sum_{j=1}^{n}(\lambda - f_{ij}(\mathbf{x}))H(S(\mathbf{x}))(f_{ij}(\mathbf{x}) - \lambda) \tag{7}$$

and penalizes the large deviations from the reference level λ whose meaning will be more clear during the experimental part. This level is a reference level to be tracked and is depending upon the physical capacity of the links. If we neglect the link failures or multiple links between pairs of nodes which may be switched on/off at random time instances, λ can be taken to have a constant value. In more general situations, λ is a piecewise constant function determined by the link capacity changes. The functions

$$f_{ij}(\mathbf{x}(t)) = \sum_{k=1}^{n} w_{ij}x_{ki}(t)x_{kj}(t) \tag{8}$$

represents the scalar measure of the interactions between the interfering traffic patterns i and j and $H(z)$ is the step function penalizing the switching over the sliding surface $S(\mathbf{x})$

$$H(z) = \begin{cases} 0 & \text{if } z > 0 \\ 1 & \text{if } z < 0 \end{cases} \tag{9}$$

The learning mechanism through sliding window consists in approximating the probability distribution $f_{XY}(\mathbf{x}, \mathbf{y})$ of the stochastic variable (X, Y) : $\Omega \mapsto R^n \times R^n$ such that the distribution of $\mathbf{w}_i, i = 1, ..., n$ simulates the distribution of $\mathbf{x}_i, i = 1, ..., n$, where $\mathbf{w}_i$ denotes the i-th column of the interconnection matrix $\mathbf{W}$.

Theorem 1. *If* $|\mathbf{W}| \leq \beta$ *for* β *a suitable chosen constant, then there exists a solution* $\mathbf{W}$ *that stabilizes all the incoming flow patterns* $\mathbf{x}_i, i = 1, ..., n$.

The proof of Theorem one is based on the concept of linear separability and is omitted from this presentation. Instead, we point out that the above result depends only on the desired flow patterns and not on the rule of achieving these patterns.

3 Neural Networks Based Adaptive Control

3.1 Discontinuous Control Rule

The neural controller is implemented as feedback feedforward neural network for nonlinear optimization such that the state of the dynamic neural system is used to compute the control vector according to the following rule

$$\mathbf{u} = \begin{cases} \mathbf{u}^{+} & \text{if } S(\mathbf{x}) > 0 \\ \mathbf{u}^{-} & \text{if } S(\mathbf{x}) < 0 \end{cases} \tag{10}$$

Such a control rule realizes the tracking of any desired trajectory joining points of the sliding surfaces while minimizing the cost function $J(\mathbf{W})$. This cost function which is to be minimized constraints the control strategy to be implemented as a multilayer feedforward neural network (MFNN) [8].

3.2 Learning Neural Networks for Nonlinear Optimization

The recursive equations for computing the gradient components of the cost function with respect to the interconnection weights, generalizing the classical adjoint system equations of N-stage optimal control problems (windowing control) [10] are given by

$$w_{ij}(n+1) = w_{ij}(n) - \frac{\partial J(\mathbf{W})}{\partial w_{ij}} \tag{11}$$

If we are interested for a real time flow control algorithm as it is very often the case in large networks, the asynchronous/distributed fashion of neural networks operation appears as being particularly very attractive. We take advantage on this feature to use only a local pattern of information when we adapt the flow control variables for a given node. Beside of delay reaction of a sliding window control scheme, the asynchronous/distributed operation poses very delicate questions of when the overall convergence of the neural control algorithm has been achieved [8].

4 Numerical Results for Markovian Flows

The numerical experiment is designed to emphasize the adaptive properties in term of tracking ability for a two input ports ($n = 2$) network when subject to bursty Markovian flow patterns as represented in Figs. 3–4. The long-run flows are considered to be constant within a suitable time scale, although they can be any regular time dependent function. The long-run patterns explicitly depend on the link capacities, but for the specific goals of this research, all the links were taken to have the same capacity. The long-run flows are acting as references for the adaptive control mechanism, more specifically as references levels λ. The mean flows are to be retrieved by the adaptive flow control procedure through sliding window estimation. The traffic patterns attacking the input ports are the actual flows having a bursty structure (Figs. 3–4). These patterns are obtained by using a discrete time random disturbance of bounded magnitude which is imposed to the mean flows (preassigned during the experiment, but they are unknown to the traffic controller) and they are intended to mimic the bursty structure of Markovian flows in the multistage networks (or ATM group trunk). The corresponding simulation results obtained by using the window adaptive flow control procedure are presented in Figs. 5–6. In Fig. 7, the learning performance in terms of cost functional reduction is displayed, where the inner time scale of the learning algorithm is different from the outer time scale of the real controlled process (a unit time of the outer time scale comprises 10 time units of the inner time scale). The result reported in Fig. 7 is valid for the time period 8–10 for the joint control costs for port 1 and 2, where the most difficult control effort occurs, because of the periodic structure of disturbances in Fig. 3 and the large jump in Fig. 4.

5 Conclusion

A framework for mapping sliding window flow control in multistage communication networks onto a neural network algorithm has been developed. The central point of the adaptive control strategy is to built a sliding window learning algorithm which is amenable for approximating the distribution probability of the input traffic patterns. A numerical experiment has been designed to produce evidence on the quality of flow control which is obtained with such a scheme. A feedforward neural network has been used to implement the learning control algorithm as a nonlinear optimization problem.

References

[1] D.P. Bertsekas and R.G. Gallager, *Data Networks,* Englewood Cliffs, NJ: Prentice-Hall, 1987.

[2] D.P. Bertsekas, "An Auction Algorithm for Shortest Paths," in *SIAM J. Optimization,* vol. 1, no. 4, pp. 425–447, 1991.

[3] P.E. Gill et al., "Inertia-Controlling Methods for General Quadratic Programming", in *SIAM Review,* vol. 33, no. 1, pp. 1–36, 1991.

[4] A. Murgu, P. Neittaanmaki and V. Hara, "A Neural Networks Approach of Routing/Flow Control for Communications Networks," in *Proc. of the 1994 IEEE International Conference on Neural Networks ICNN'94,* vol. 4, pp. 2667–2672, 1994.

[5] T. Parisini and R. Zoppoli, "Neural Networks for Feedback Feedforward Nonlinear Control Systems," in *IEEE Trans. on Neural Networks,* vol. 5, no. 3, pp. 436–449, 1994.

[6] M. Schwartz, *Telecommunication Networks: Protocols, Modeling and Analysis,* Reading, MA: Addison-Wesley, 1987.

[7] G.W. Shapiro and H.G. Perros, "Nested Sliding Window Protocols with Packet Fragmentation," *IEEE Trans. on Communications,* vol. 41, no. 1, pp. 99–109, 1993.

[8] Special Issue on Neural Networks in Control Systems, in *IEEE Control Systems Magazine,* vol. 10, no. 3, pp. 3–125, 1990.

[9] P. Tseng, D.P. Bertsekas and J.N. Tsitsiklis, "Partially Asynchronous Parallel Algorithms for Network Flow and Other Problems," in *SIAM J. Control and Optimization,* vol. 28, no. 3, pp. 678–710, 1990.

[10] V.I. Utkin, *Sliding Modes in Control and Optimization,* Berlin: Springer-Verlag, 1992.

[11] T. Wang et al., "A Neuron-Weighted Learning Algorithm and Its Hardware Implementation in Associative Memories," in *IEEE Trans. on Computers,* vol. 42, no. 5, pp. 636–640, 1993.

[12] B. Yuhas and N. Ansari, eds., *Neural Networks in Telecommunications,* Boston: Kluwer Academic Publishers, 1994.

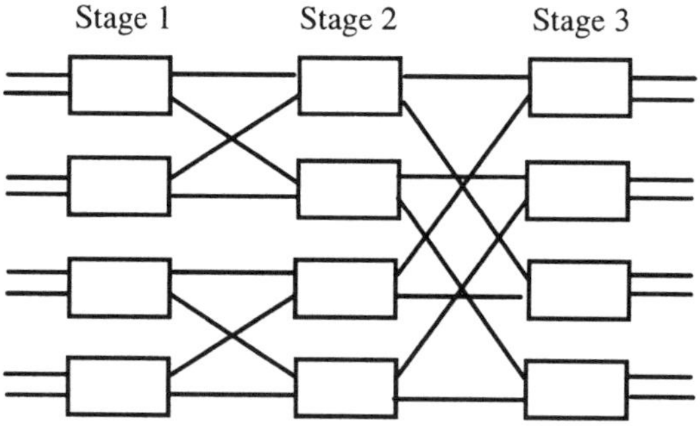

Fig. 1. A multistage network

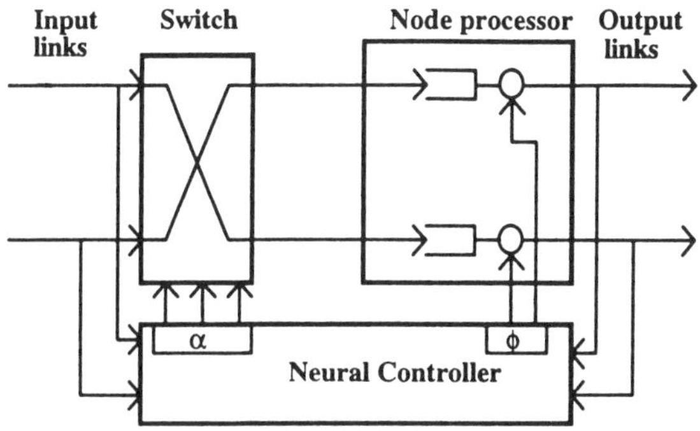

Fig. 2. The Nodal Controller Architecture

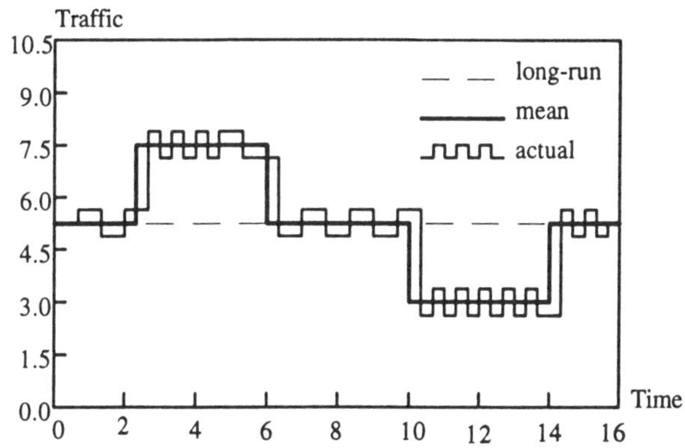

Fig.3. The squared traffic pattern at input port 1.

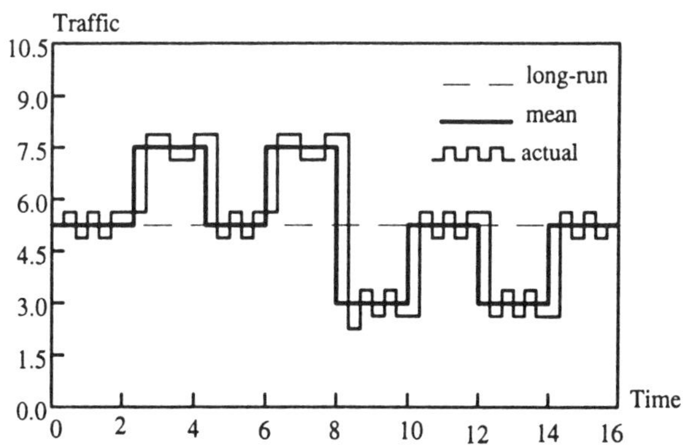

Fig. 4. The squared traffic pattern at input port 2.

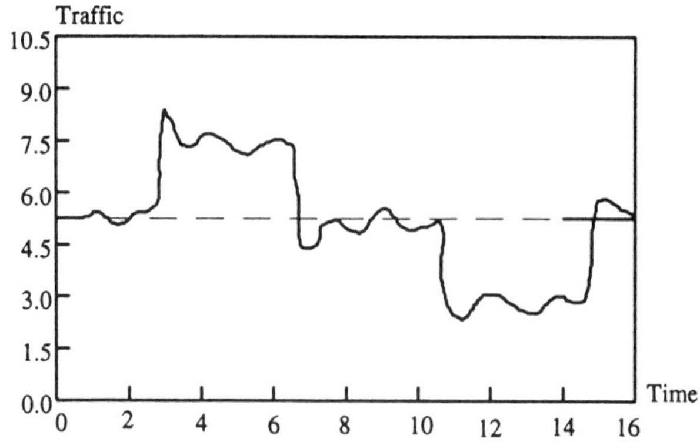

Fig.5. The tracking traffic performance for input port 1.

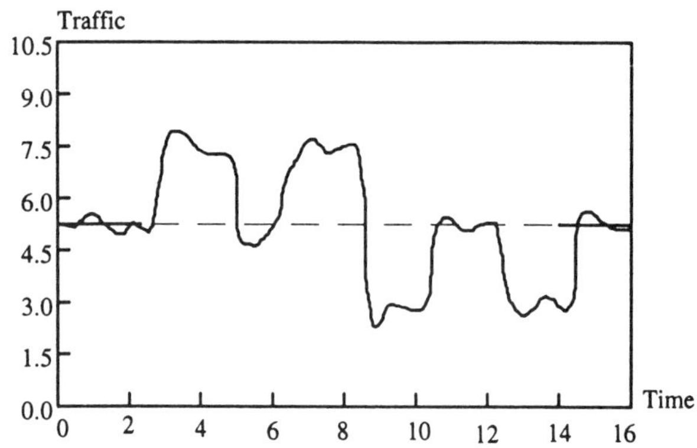

Fig. 6. The tracking traffic performance for input port 2.

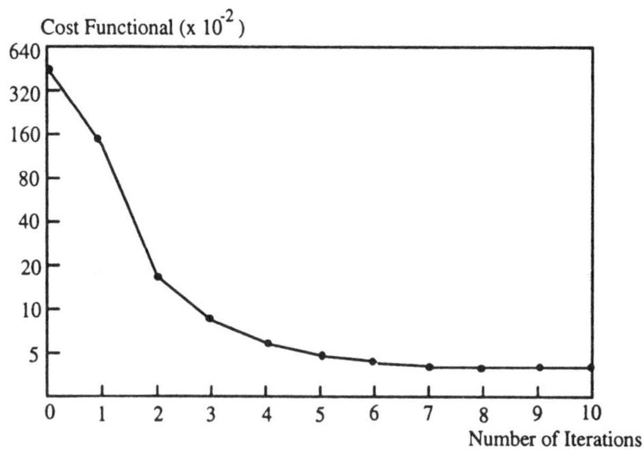

Fig. 7. The learning performance of NN for optimization.

A Neural-Network Controller for Scheduling Packet Transmissions in a Crossbar Switch

Anujan Varma and Robert Antonucci

Computer Engineering Department
University of California
Santa Cruz, CA 95064
varma@cse.ucsc.edu

Abstract

A Hopfield neural network is presented for scheduling packets in an input-buffered crossbar switch. The controller has a two-dimensional winner-take-all design and is driven by request signals representing the packets stored in the input buffers. Schemes are presented to support multiple priorities of traffic by changing the binary request signals to multi-valued signals that vary over a limited range. Simulation results show that the controller provides close-to-ideal throughput for switches as large as 32×32. In addition, by varying the relative strength of request signals, the scheme is shown to be able to control jitter effectively in a communication network consisting of multiple switches in series.

1 Introduction

Packet switches based on nonblocking crossbars are used extensively in communications switching. The problem of switching the maximum number of packets in a crossbar with windowed or randomly-accessible queues can be reduced to the matching problem in a bipartite graph [1]. Optimal algorithms exist for finding the maximum matchings in a bipartite graph, but are too computationally demanding to be implemented in the controller of a high-speed packet switch. In this paper, we explore an alternate approach, a controller based on a neural network for scheduling packets in an input-buffered crossbar switch.

Research supported by NSF Young Investigator Award MIP-9257103 and NSF Grant No. MIP-9111241.

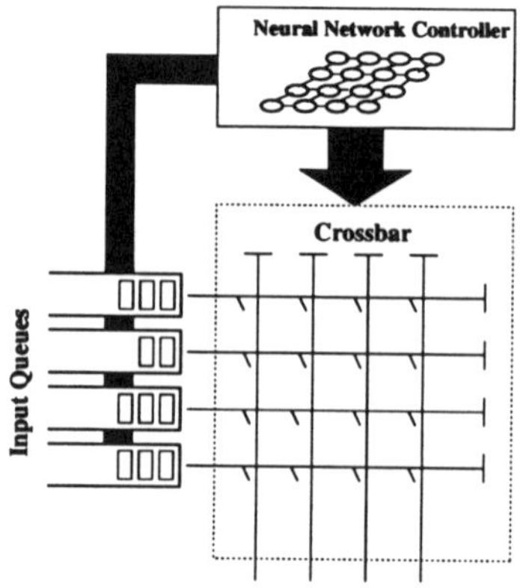

Figure 1: Architecture of the neural-network-controlled crossbar switch.

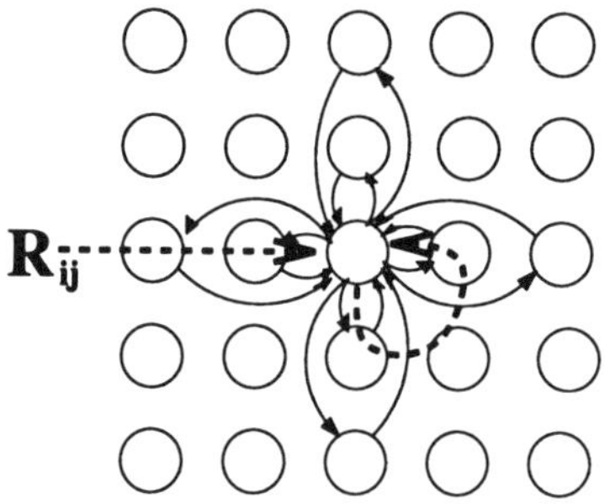

Figure 2: Architecture of the neural network, with the signals to and from one neuron shown. Dashed lines indicate excitatory signals, while normal lines indicate inhibitory signals.

The model of the input-buffered crossbar used in this paper is illustrated in Figure 1. Packets arriving at each input join the queue corresponding to the input port. The packets are assumed to be of fixed length, for example the 53-byte cells in networks based on ATM. Packet scheduling is performed in cycles. During each cycle the neural network controller chooses at most one packet from each input queue for transmission such that the destination output ports of the packets are all distinct. We assume that the state of the queues is presented to the controller at the beginning of each cycle in the form of a two-dimensional matrix R. The entry R_{ij} in the matrix is 1 if the queue at input port i currently holds a packet destined for output port j, and 0 otherwise.

2 Neural Network Architecture

Figure 2 shows the architecture of our switch controller. Each neuron receives an excitatory signal from the request matrix and produces a self-excitatory signal. It receives inhibitory signals from neurons in its row and column. This design was determined through gradient descent and mimics a two-dimensional winner-take-all network. The network is stable at states that represent maximal matchings and is unstable at ambiguous states or states that represent invalid matchings. By the use of a summing unit in each row and column, the network can be implemented with $\mathcal{O}(N^2)$ connections [3].

The energy function we used to express the value of a matching in our neural network controller needed to reflect the following three goals and constraints: (*i*) the maximum number of packets should be matched in each cycle, (*ii*) only one neuron in each row and column should be firing, and (*iii*) each neuron should be clearly firing or clearly idle. We designed the following energy function to account for these goals and constraints.

$$E = -A\sum_{ij} V_{ij}R_{ij} + \frac{B}{2}\sum_{i}(\sum_{k} V_{ik} + IR_i - 1)^2 + \frac{B}{2}\sum_{j}(\sum_{k} V_{kj} + IC_j - 1)^2 + \frac{C}{2}\sum_{ij} V_{ij}(1 - V_{ij}), \quad (1)$$

where Vij is the output of neuron (i, j) and A, B, C are constants. The first constraint is modeled by an energy drop for every firing, requested neuron. The second constraint is modeled by an energy gain for every row whose sum is not 1. To prevent neurons from firing in an unrequested row and possibly blocking viable connections, the terms IC_j and IR_i (*idle column* and *idle row*) are added to the column sums and row sums, respectively. If the buffer at input port i contains no packets IR_i is 1, otherwise IR_i is 0. With this term added, the energy will rise if the neurons in an unrequested row sum to a value above zero. Similarly, the term IC_j is used to prevent neurons in columns with no requests from interfering with the operations of the network. The second and third terms model this constraint. The final term raises the energy for every neuron with output between 0 and 1, thus discouraging ambiguous solutions. A, B, and C are parameters to be determined by experimentation.

The neuron we used was the sigmoidal neuron whose output V_{ij} is a function of the internal charge U_{ij} given by

$$V_{ij} = \frac{1}{1 - e^{-6U_{ij}}} \, . \quad (2)$$

The motion equation of the network is given by

$$\frac{dU_{ij}}{dt} = -\frac{\partial E}{\partial V_{ij}} = AR_{ij} - B\left(\sum_k V_{ik} + IR_i\right) - B\left(\sum_k V_{kj} + IC_j\right) + CV_{ij} + Bias, \tag{3}$$

where a bias is introduced, and all constants are absorbed by it. We used a genetic algorithm to tune the values of A, B, C, and $Bias$ in the motion equation of the network. Details of the algorithm are given in [3]. The values found by the algorithm for an 8×8 switch were $A = 4.0$, $B = 4.9$, $C = 6.4$, and $Bias = 1.5$.

An important feature of our design is the ability to assign weights selectively to the packets in the scheduling process. In our design, each request generates a signal to drive its neuron. These signals are not restricted to binary, as in previous designs, but can be varied over a continuous range, thus making certain packets more likely to be chosen over certain others in a matching. This capability could be used in various ways in the switching system, such as supporting multiple packet-priorities in the communication network, or improving the variance of the delay distribution by modifying the weights of individual packets as a function of their queueing delays. The biasing scheme can also be used to control delay jitter across multiple hops in a networks of switches, as described in Section 3.

3 Performance Evaluation

We performed extensive simulations to evaluate the neural network controller. Because of space constraints, only some representative results are discussed here. More detailed results can be found in [3]. Except for some experiments in which we tested the scalability of the switch controller, all experiments were performed on an 8×8 switch. Each switch was simulated with infinite, randomly accessible input queues, and packets were assumed to be of fixed size. The controller processed the state equation with a time step of 0.1. After 300 iterations the result was passed to the crossbar matrix. Should a neuron output an intermediate value between 0.1 and 0.9, the connection was left open. Likewise, if an invalid matching was provided by the controller, all rows and columns with multiple active neurons were disabled.

We first examine the performance of our switch in its basic operation. Figure 3(a) plots the efficiency of our neural network switch controller, as compared to that of an optimal algorithm that always finds a maximal matching, and that of the probabilistic iterative matching algorithm proposed in [1]. The *request density* on the x-axis is the probability of entry (i, j) of the request matrix being a 1. In each of these experiments,

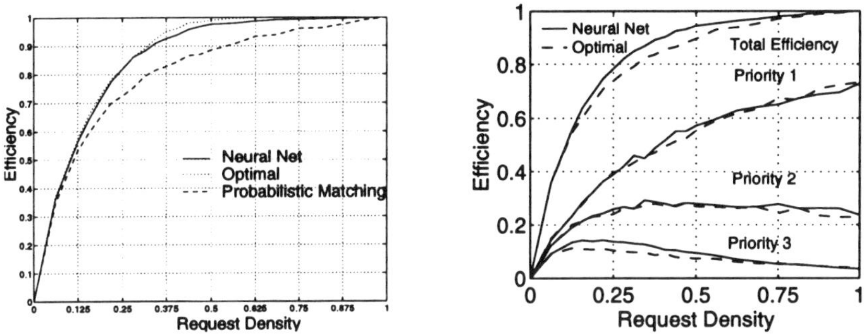

(a) (b)

Figure 3: (a) Efficiency of an 8×8 switch scheduled by the neural network, an optimal algorithm, and the probabilistic matching algorithm of [1]. (b) The total and per-priority efficiency of the neural network controller compared to that of an optimal scheduling algorithm. The separation between request signals of adjacent priority levels was 10%.

we simulated 22 distinct request densities that resulted in average queue lengths in the range of 0–4 packets per input queue. Each point in the plot is the average of the results for 500 randomly-generated request matrices. The neural network switch performed as well as the optimal algorithm up to 85% efficiency, and only then showed sub-optimal performance.

When multiple priorities are added to the network the designer must choose the relative request signal strengths between priorities. A larger separation approximates the optimal scheduling more closely, but can support only few distinct priority levels. A smaller separation may support more priorities, but the boundaries between priorities is not as strict, resulting in the controller often scheduling a lower priority packet over a higher priority one. Figure 3(b) shows the efficiency of both an optimal scheduling algorithm and the neural network handling three distinct priorities. The switch simulated in Figure 3(b) has priority signals separated by only 10% in their relative strengths. Both the overall efficiency and the efficiencies for packets belonging to individual priority levels are shown. It is interesting to see that the neural network was able to achieve higher overall throughput than the optimal scheduler for moderate loads. This is due to two reasons. First, because of the small separation between adjacent priority levels, the neural network occasionally held a higher priority packet so that several lower priority packets may be scheduled. Second, the optimal algorithm we implemented schedules the packets at each priority level independently, proceeding from the highest to the lowest. If several maximum matchings exist at any level, the algorithm chooses one randomly. The neural network, on the other hand, is more likely to choose a matching that also maximizes the total number of packets taking into account all priority levels.

The average efficiency of the matching produced by a scheduling algorithm on random request matrices may not always be indicative of the average queueing delay of packets in the switch. For example, if certain packet configurations lead to poor matchings by the algorithm, the average efficiency may still not be degraded significantly if such configurations occur rarely in the generated matrices. However, in a real system, such a configuration is likely persist for several cycles once it occurs, noticeably affecting the queueing delay of packets in the system. Indeed, our neural network controller suffered from this limitation when the request signals were restricted to binary levels. Although the efficiency of the matching produced by the neural network controller on random request matrices was superior, on the average, to that produced by probabilistic iterative matching, the average queueing delays did not show noticeable improvement. This problem was overcome by adding a small time-dependent bias to requests in the neural network controller, so that the effect of configurations leading to poor matchings do not persist over consecutive cycles.

To favor packets buffered for a long time, we used a bias that increased with the queueing delay of the packet. In addition, if multiple priority levels are supported by the network, the delay distribution of low-priority traffic can be improved by successively increasing the bias of packets to higher priority levels when they remain in the queue at a certain priority level for a set number of cycles. We evaluated the merit of such variable biasing and priority promotion on the queueing delay of switch supporting three distinct priority levels. Each priority level has associated with it a base signal, $Base_p$, a signal range within the priority level, $Range_p$, and a threshold in terms of the number of cycles after which the packet is promoted, $Threshold_p$. The algorithm used to compute the bias of each packet is summarized below. T_{arr} is the arrival time of the packet, T the current time, and p the priority.

if $(T - T_{arr}) > Threshold_p$ and $p > 0$
(promote packet to next higher priority level)
$T_{arr} = T_{arr} + Threshold_p$; $p = p - 1$;

$$R_{ij} = Base_p + Range_p \cdot \mathcal{F}\left(1.65 \times \frac{T - T_{arr}}{Threshold_p}\right). \tag{4}$$

We used base values of 0.54, 0.74, and 1.0 for priorities 3, 2, and 1 respectively. A fourth priority level, priority 0, was added so that priority 1 packets can be promoted to this level if they stay in the system for too long; the base value of the bias for this level was set at 1.34. The range was 0.05%

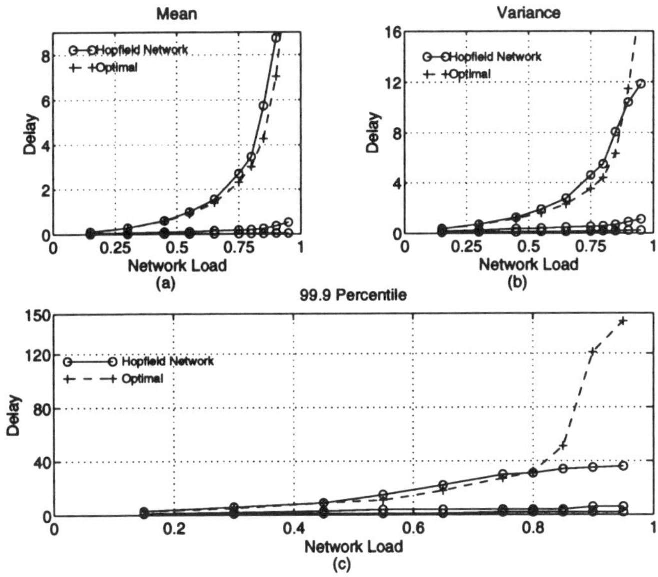

Figure 4: Queueing delay in an 8 × 8 switch for three priority levels with a neural network controller employing variable biasing of requests and priority promotion.

of the base value, and the thresholds were 30, 5, and 3 for priorities 3, 2, and 1 respectively. Although no packet is promoted beyond priority zero,it still needs a threshold to scale the continuous bias; thus, a value of 6 was used in the denominator of the second term in Eq. (4). $\mathcal{F}$ is the squashing function used in the neurons, scaled to keep signals between 1 and −1.

$$\mathcal{F}(x) = \frac{2}{1 - e^{-x}} - 1. \tag{5}$$

The constant 1.65 in Eq. (4) scales the time such that as the packet approaches its threshold for promotion, the neural network may take a small efficiency hit to send the packet, in order to avoid promotion which may cause a larger efficiency hit.

Figure 4 shows the results of implementing variable biasing and priority promotion in a communication network supporting three priority levels. Both the average delay and the variance are comparable to those produced by optimal scheduling, and the 99.9 percentile delay is better than that of the optimal algorithm at heavy load.

The packet-biasing scheme can also be used to control jitter across multiple hops in a network of switches. To evaluate the effectiveness of this approach, we designed a jitter control scheme where a packet arriving at a switch is assigned a priority level based on its accumulated jitter. This way a late packet may be transmitted immediately, even though several packets may be waiting ahead of it in the queue. We simulated a network configuration consisting of four 8×8 switches in series, each controlled by a neural network. At total of 112 communication sessions were simulated, out of which 16 were 4-hop sessions that traveled through all the switches, while the remaining 96 were divided equally among one-hop, two-hop and 3-hop sessions. To minimize jitter, each packet in a switch was biased by the difference between the current time and its *expected* arrival time. We compared the end-to-end delays of packets from different sessions for this scheme with three other schemes — a neural network that biases packets based on only the local queueing delay, probabilistic iterative matching, and an optimal matching algorithm. Detailed results from these simulations are given in [3]. The jitter control scheme increaseds the average delay of packets with shorter paths in favor of more control over packets with longer paths. For long paths, the 99.9 percentile delay of the jitter control scheme was significantly better than even that of the optimal algorithm.

References

[1] T. E. Anderson, S. S. Owicki, J. B. Saxe, and C. P. Thacker, "High-speed switch scheduling for local area networks," Technical Report 99, Digital Systems Research Center, Palo Alto California, April 1993.

[2] T. X. Brown and K. Liu, "Neural network design of a banyan network controller," *IEEE Journal on Selected Areas in Communications*, Vol. 8, No. 8, October 1990, pp. 1428–1438.

[3] Y. Park, V. Cherkassky, and G. Lee. "ATM cell scheduling for broadband switching systems by neural network," in *Proceedings of the International Workshop on Applications of Neural Networks to Telecommunications*, 1993, pp. 112–118.

[4] A. Varma and R. Antonucci, "A neural-network controller for scheduling packet transmissions in a crossbar switch," Technical Report, University of California, Santa Cruz, February 1994.

A Technique for Mapping Optimization Solutions into Hardware

Timothy X Brown
Bell Communications Research
Morristown, NJ 07960
`timxb@bellcore.com`

Abstract

Feedback neural architectures have addressed a myriad of optimization problems—almost exclusively in slow software simulation. Speed is promised only when implemented in high-speed recurrent hardware. Unfortunately, hardware's discrepancies from ideal conspire to produce poor solutions. This paper describes a recurrent hardware-in-the-loop learning procedure that maps idealized solutions of optimization tasks into non-ideal recurrent hardware. The technique is demonstrated on a general purpose analog VLSI chip with winner-takes-all and resource allocation problems. For all problem sizes that fit into the chip the network finds very good solutions with high probability. Theoretical analysis and empirical measurements indicate that the noise limits the hardware to 30 neurons, but this increases dramatically with reduced noise.

1 Background

Feedback neural networks have addressed optimization problems important to telecommunications such as resource allocation, traveling salesman, and scheduling. These have been adapted to a variety of applications in switching, control, and communications [1,2,3]. When designed correctly, they always satisfy the hard constraints defining valid solutions [1]. By applying annealing techniques, solution quality can be made near optimal [4]. The real advantage for these problems—demonstrated almost exclusively in software simulation—is promised when implemented in high-speed massively parallel neural hardware.

2 The Problem

The many feedback connections used in solutions to these optimization problems, plus the necessity of simultaneous settling of the many neurons required to guarantee stable valid solutions requires either time-consuming precise numerical calculations or, when speed and power consumption are critical, parallel analog hardware such as[5,6,7]. Unfortunately, when the carefully balanced optimization parameters are mapped to analog VLSI hardware; process and temperature variations, non-ideal components, and limited precision result in violated constraints, invalid solutions or significantly sub-optimal solutions. It is well known that the variety of ideal-software/hardware mismatch errors that are difficult to observe and/or quantify necessitate hardware-in-the-loop learning[6,7], i.e. ideal weights are first loaded, and then example patterns presented; performing learning whenever outputs are in error. For optimization, this is complicated by the need for a recurrent architecture learning algorithm; and recurrent hardware with adjustable synapses. Also, the complicated structures that arise, bring doubt on whether hardware-in-the-loop learning can be used with very large optimization problem sizes.

3 A Solution

3.1 The Training Procedure

The Boltzmann algorithm presents an input pattern, settling the network while the output units 'float', and then settling again with the output units 'clamped' to the correct values. Synapses use the difference in neuron activation to update their weights[8]. This local learning property is ideal for recurrent problems. An optimization problem typically has many equivalent good or optimal solutions which, while equivalent, typically have very different outputs. An arbitrary output solution chosen for learning will likely be different from the networks's solution and disrupt already learned weights. Successful training requires an algorithm that when given a potential network solution finds a nearby optimal solution.

3.2 The Hardware

Any analog hardware with suitable recurrent connections, software adjustable synapses, and a variable gain can with software implement the mean field Boltzmann algorithm. This paper was greatly facilitated by using an analog fully-interconnected 32-neuron neural network experimental prototype chip with on-chip Boltzmann learning[5]. This device performs the neural computations in parallel analog circuits while the weights are stored in 5 bit (–15 to 15) digital registers.

1	1	0	1	0
0	0	1	1	0
1	0	1	0	0
0	0	0	1	1
0	0	0	1	1

(a) Initial Matrix

0	1	0	0	0
0	0	1	0	0
1	0	0	0	0
0	0	0	1	0
0	0	0	0	1

(b) One Solution

0	1	0	0	0
0	0	1	0	0
1	0	0	0	0
0	0	0	0	1
0	0	0	1	0

(c) Alternate Solution

Figure 1: Example of the Binary Resource Allocation Problem. Initial matrix, (a), is reduced to no more than one 1 per row and one 1 per column (b). Equivalent alternate solutions can exist (c).

More importantly, the chips implement the Boltzmann and mean field learning algorithms on chip as part of each synapse. Currently the chips are mounted on a VME board under software control of an attached workstation.

4 Two Test Problems

The first is a stylized version of the winner-takes-all (WTA) problem that takes as input a vector of N 1's and 0's and chooses a single "winner" from any one of the 1's. The neural solution consists of N neurons with inhibitory connections between each pair of neurons. This WTA is a simple selector problem, but serves as a basis for the next problem.

RxC binary resource allocation (BRA) takes an RxC matrix of 1's and 0's and reduces 1's to 0's so that the matrix has at most one 1 in each row and one 1 in each column while maximizing the number of 1's in the final matrix. WTA is 1xN or Nx1 BRA. The neural solution consists of RxC neurons with inhibitory connections between pairs of neurons in the same row or column. The problem is significant since problems such as packet arbitration can be reduced to it [1].

BRA is equivalent to the cardinality graph matching problem which has a polynomial-time algorithm solution[9]. This polynomial algorithm allows us to compute the optimal target solutions readily. Figure 1 shows an example 5x5 problem and solution. As can be seen, the optimal solution is not always unique.

5 Experiments

We tested both the WTA and BRA problems using the 32 fully-connected neuron VLSI chip. The current hardware interface requires that each input uses one neuron requiring both an input and output matrix. Thus a 3x3 problem requires at least 18 neurons. The initial weights were prescribed by the theory in [1] and after scaling to the [+15,–15] range of the hardware are shown in Table 1. Due to fixed offsets in the neuron range, additional bias input units were needed to guarantee a large enough bias level. The number of bias units was found to be ~R+C,

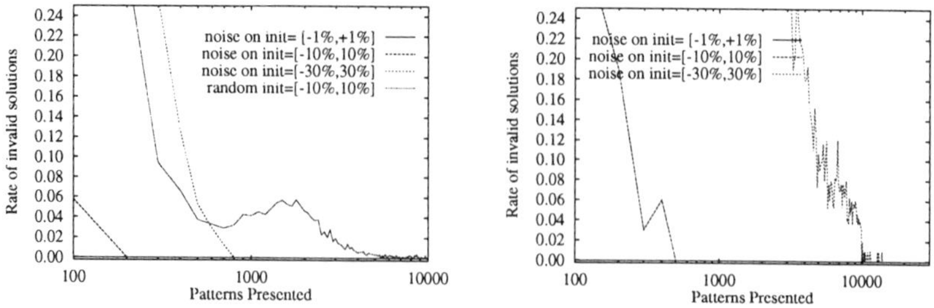

(a) (b)

Figure 2: Simulated Learning of 3x3 (a) and 8x8 (b) BRA with Different Levels of Initial Noise in the Weights. Shows rate of invalid solutions as a function of learning cycles.

although for larger sizes this was restricted by available neurons. To reduce the degrees of freedom we incorporated prior knowledge about the network connectivity. All extraneous synapses such as those between neurons not in the same row or column were permanently turned off and the one-to-one connections between input and output neurons were frozen at their initial values.

Random patterns were generated by setting elements in the initial matrix to 1 with probability 0.3. A learning cycle consisted of presenting a random pattern to the network, reading the output, calculating the nearest optimal pattern using the max-flow algorithm in [9] and using that as the target pattern for on-chip mean-field learning. An experimental run consisted of at least 4,000 (typically 10,000) learning cycles followed by 20,000 testing cycles. A solution during testing was judged by both its row/column constraint satisfaction, and whether it is an optimal solution. An experiment was the average of 20 runs.

5.1 Results: Software

To understand the algorithm's performance we first looked at a software simulation of the hardware. We ran a series of experiments where different noise levels were added to the prescribed initial weights. The noise was uniform zero mean with different maximum magnitude. Figure 2 shows learning curves for different perturbations on the initial ideal weights, and for completely random initial weights. Since the simulated network is ideal, with small noise, the network never makes a mistake. As noise is added, the network initially produces poor and even invalid solutions, but eventually delivers an error-free final set of

Table 1: Initial Weights.

output intra-row/column	input to output	total bias
–15	15	–8

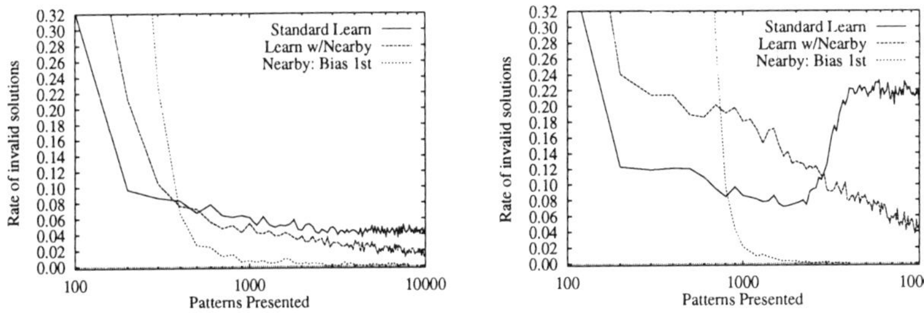

(a) (b)

Figure 3: Hardware Learning of 3x3 BRA (a) and 10 WTA (b) with: Arbitrary Optimal Patterns; Nearby Optimal Patterns; and Nearby Optimal Pattern and Bias First.

weights for both the 3x3 and the much larger 8x8 problem; suggesting that the hardware neural network can achieve good results even if weights are not initially ideal and it must learn them from example.

5.2 Results: Hardware

Figure 3 shows learning curves for 10-input WTA and 3x3 BRA problems. Despite the recurrent nature of the problem, the network learns to less than 5% errors within 10K iterations. Figure 3 also shows the erratic and poor performance when learning with an arbitrary optimal solution as opposed to the nearest optimal solution. Learning experiments showed that most significant weight changes were in the biases and that learning was greatly facilitated if the bias weights were first set to the correct value. Figure 3 shows the same learning when the first half of the learning cycles were used to find the bias, and the second half were used to learn all of the weights (the on-chip learning mechanism could be disabled for individual synapses under software control). In this case the network learns better solutions faster. Because of the improved performance, this learning first on bias and then on all weights was used for all of the later experiments.

To test the ability to learn problems in a variety of different dimensions, experiments were done for [2–15]x1, [2–7]x2, and [3–5]x3 size networks (all combinations that could fit onto the 32 neuron chip). A successfully trained network was defined as one that would find the optimal solution 99.9% of the time. Figure 4 shows the probability of a successful run as a function of the problem size. As can be seen problems can with high probability be learned for sizes up to 12x1, 6x2, and 4x3. The performance for the 5 larger sizes was limited due to lack of available bias neurons on the 32 neuron chip (e.g., the 15x1 and 5x3 problems had only 2 bias neurons available). This was confirmed by observing in these cases that at the end of learning the bias weights were always saturated.

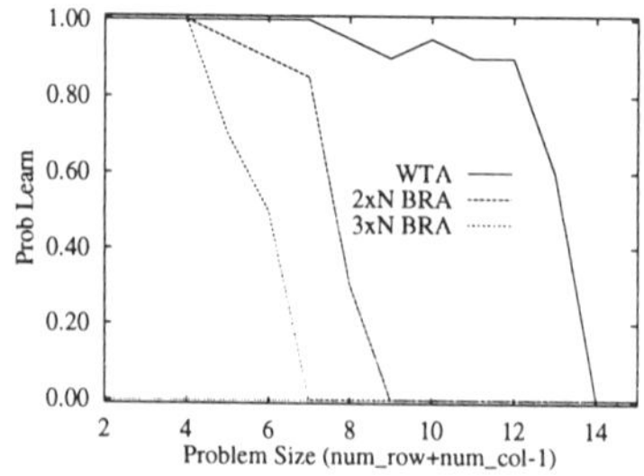

Figure 4: Probability of Learning to 99.9% Accuracy vs. Problem Size.

6 Extensibility to Larger Sizes

The previous section demonstrates in hardware the WTA/BRA for different sizes. We analyze the limits to the hardware from two perspectives—systematic weight variations and ambient system noise. For this analysis we focus on setting a bias weight, T_B, for each output neuron that works with all input combinations. With no noise, a range of bias weights will all work. As the noise grows, this range narrows; an effect that is amplified with more neurons. In this analysis we ignore learning and calculate the probabilities that any successful solution exists.

We look first at systematic weight variations with the following model. The connection between any two associated neurons (in same row or column) is nominally the maximum negative weight, $-T_{MAX}$. Each weight has a fixed additive error uniformly distributed in $[-\varepsilon/2,+\varepsilon/2]$. Since for large errors, we can adjust the weights to within the weight precision, we assume ε is less than the precision of the weights. The probability of there being a working bias is a function of the neuron on and off voltage, V_{ON} and V_{OFF}, and the size of the problem, $n = R + C - 1$. If all of a neuron's associates are off, then the input from the bias neuron should exceed the input from the associates so the neuron turns on:

$$V_{ON}T_B > N(-(n-1)V_{OFF}T_{MAX}, V_{OFF}\sqrt{n-1}\sigma_P), \tag{1}$$

where $N(\mu, \sigma)$ is a μ mean, σ standard deviation normal variable and $\sigma_P = \varepsilon/\sqrt{12}$ is the uniformly distributed precision's standard deviation. Similarly if one or more associated neurons are on then the neuron should turn off:

$$V_{ON}T_B < N\left(-(V_{ON} + (n-2)V_{OFF})T_{MAX}, \sqrt{V_{ON}^2 + V_{OFF}^2(n-2)}\,\sigma_P\right) \tag{2}$$

Normalizing by $V_{ON}T_{MAX}$ and taking the difference between (1) and (2) we get the normalized range of allowable values for T_B:

$$R = N(1 - V_{OFF}/V_{ON}, \sqrt{1 + (V_{OFF}/V_{ON})^2(2n-3)}\,(\sigma_P/T_{MAX})) \tag{3}$$

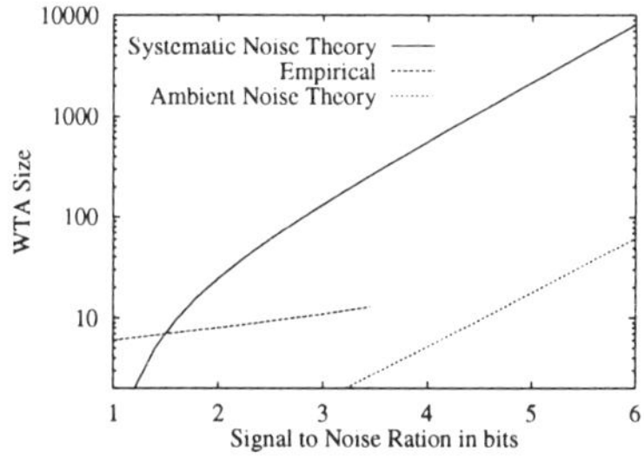

Figure 5: Precision Limitations on Addressable Problem Size.

We can find a valid bias if $R > 0$. In the hardware, V_{OFF} = 0.7V and V_{ON} = 4.3V. Conservatively, $V_{OFF}/V_{ON} = 1/4$. Using this value, Figure 5 plots the signal-to-noise ratio (SNR), $(1+T_{MAX})/\varepsilon$, such that a bias can be found for all n associated neurons with 99.9% probability as a function of n. Similar analysis on errors in the neuron output voltages yields:

$$R = N(1 - V_{OFF}/V_{ON}, \sqrt{2(n-1)}\,\sigma_N / V_{ON}) \tag{4}$$

where σ_N is the voltage variance. Figure 5 plots the SNR such that a bias can be found with 99.9% probability as a function of n. Comparing the two noises, clearly the network is not limited by systematic offsets—since, as the software results illustrate, these can be countered by adjusting the weights—and noise is a dominant factor. For the noisy VME system our measured SNR is about 6 bits indicating problems of up to size 60 can be tackled.

To get an idea for the largest size problems that we can solve in the hardware, we re-ran the simulations by permanently fixing T_{MAX} to smaller values, and observing the largest size WTA problem that could be learned with 99.9% accuracy. This is also plotted in Figure 5 (Note that it is not directly comparable with the other plots). The full 5 bits could not be used before the bias limits were reached. But extrapolating we see that problems of size 20–30 are possible, implying an ambient noise level closer to 5 bits. Steps are being taken to reduce the noise since according to Figure 5 every bit higher SNR doubles the feasible largest size. Using a more typical analog SNR of 8 bits, we see that problems up to size 200 are possible (e.g. 100x100 BRA with 10,000 neurons).

7 Discussion and Conclusions

We have shown in both software and analog hardware that the discrepancy between an ideal set of weights and a noisy weights can be corrected for with hardware-in-the-loop learning even in the difficult case of an optimization task with a recurrent architecture. This was demonstrated in hardware using an on-chip recurrent learning algorithm on a resource allocation task with excellent results on sizes up to 12x1, 6x2, and 4x3. Larger sizes were not always successful due to a limited number of bias units. Analysis of systematic offsets and noise on the neuron outputs suggested that systematic offsets will not limit significantly but ambient noise may limit the system to problems of size 200. Direct measurements in hardware suggested sizes of ~30, consistent with the noisy VME environment. New modified boards will both reduce noise and allow us to tackle problems with up to 64 neurons and probe these predictions. These encouraging results bring the promised speed of parallel analog hardware closer to significant optimization tasks.

Acknowledgments

I would like to thank Tony Jayakumar for help with the analog hardware and Josh Alspector for useful discussions.

References

[1] Brown, T.X, "Neural Networks For Switching," in *Neural Networks in Telecom.*, ed. Yuhas, B., Ansari, N., Kluwer, Boston, 1994, pp. 11–36.

[2] Kechriotis, G., Manolakos, E., "Implementing the Optimal CDMA Multiuser Detector with Hopfield Neural Networks," *Proc. of the Inter. Workshop on Appl. of Neural Networks to Telecom.*, ed. Alspector,J., et al., Erlbaum, Hillsdale, NJ, 1993, pp. 60–66.

[3] Barhen, J., Toomarian,N., Protopopescu,V., "Optimization of the Computational Load of a Hypercube Supercomputer Onboard a Mobile Robot," *Applied Optics*, Vol. 26, No. 23, pp. 5007–5014, 1987.

[4] Peterson, C., Anderson, J.R., "A Mean Field Learning Algorithm for Neural Networks", *Complex Systems*, Vol. 1, No. 5, pp. 995–1019, 1987.

[5] Alspector, J., Jayakumar, A., Luna, S., "Experimental Evaluation of Learning in a Neural Microsystem," in *Advances in Neural Info. Proc. Sys. 4.* ed. Moody,J.E., et al., Morgan Kaufmann, San Mateo, CA, 1992, pp. 871–878.

[6] Eberhardt, S.P., Duong, T., Thakoor, A.P., "Design of Parallel Hardware Neural Network Systems from Custom Analog VLSI "Building Block" Chips," *Proc. of the Inter. Joint Conf. on Neural Networks, 1989,* Vol. II, pp. 183–190.

[7] Holler, M., Tam, S., Castro, H., Benson, R., "An Electrically Trainable Analog Neural Network (ETANN) with 1024 'Floating Gate' Synapses," *Proc. of the Inter. Joint Conf. on Neural Networks*, 1989 Vol. II, pp. 191–196.

[8] Ackley, D.H., Hinton, G.E., Sejnowski, T.J., "A Learning Algorithm for Boltzmann Machines", in *Cognitive Science*, 1985, pp. 147–169.

[9] Lawler, E.L., *Combinatorial Optimization: Networks and Matroids*, Holt, Rinehart, and Winston, New York, 1976, p. 195

Decentralized Neural Dynamic Routing in Circuit-Switched Networks

W. K. Felix Lor and K. Y. Michael Wong

Department of Physics,
The Hong Kong University of Science and Technology
Clear Water Bay, Kowloon, Hong Kong
`phfelix@usthk.ust.hk, phkywong@usthk.ust.hk`

Abstract

We use a *Simplex* centralized algorithm to dynamically distribute telephonic traffic among alternate routes in circuit-switched networks according to the fluctuating number of free circuits and the evolving call attempts. It generates examples for training localized *Neural* controllers. Simulations shows that the decentralized *Neural* approach has a comparable performance in blocking probability with Maximum Free Circuit (*MFC*) and gives a surpassing performance in crankback.

1 Introduction

Dynamic routing in telecommunication networks is a complex problem of prediction and optimization, designed to cope with traffic fluctuations, network overload and unbalanced load due to partial network break down. In making a routing decision, one has to balance the gain and loss of the various components in the network, so that the overall time-dependent performance can be optimized. Often this can be done by a centralized controller which collects information of network status and makes a globally optimal decision [1]. However, this approach may involve heavy computational load for the controller, and is susceptible to network breakdown.

An alternative approach is to have decentralized controllers which make locally optimal routing decisions [2, 3]. This has the advantages of less computational load, and robustness against network breakdown. However, the local routing decisions may not be globally optimal.

Neural network is a candidate for combining the strengths of both approaches. Given a training set of globally optimal decisions, a local *Neural* controller learns to infer decisions based on locally available information. Here we propose a method to implement such decentralized *Neural* controllers.

Our work consists of two parts. In the first part, we simulate a circuit-switched network, and search for an intelligent centralized control algorithm which has a satisfactory performance. It can thus act as a "teacher" for supervised learning in the local *Neural* controllers. We then generate a training set consisting of numerous examples. Each example consists of network status information and associated routing decisions prescribed by the centralized controller. In the second part, we build a neural network for each node. Only local information of network status, together with some global background information, are available to a *Neural* controller. It is presented with the training set, and learns to make routing decisions.

2 Modeling Network Dynamics

Consider a hierarchical circuit-switched network consisting of a one-link or two-link paths. Telephone calls are randomly set up for a duration from different locations. When all the direct paths are exhausted, an alternate two-link path through an intermediate tandem is selected according to a routing scheme. If the first alternate path has no vacancy, the call will be cranked-back to search for a second priority path to connect, until it is blocked when all the possible paths are congested. Besides blocking, crankback is also an important performance measure, since it increases the load of the signaling network. Dynamic routing is implemented to minimize the blocking and crankback rates.

2.1 Traffic Profile

For simplicity, the signaling network is assumed to be highly efficient so that signals can be transmitted without negligible time delay among different nodes. Call attempts from node i to k originate randomly at the rate λ_{ik} per unit time, and are randomly terminated at a rate of μ per unit time independently. When the call attempt is blocked, no reattempt is considered.

2.2 Centralized Optimal Control

Various dynamic routing schemes have been proposed for circuit-switched networks. For example, the Maximum Free Circuit Routing (*MFC*) [3] is a simple but efficient algorithm which forms the basis for the Real Time Network Routing (*RTNR*) scheme implemented by AT&T in 1991 [4]. Recently, the Maximum Mean Time To Blocking Routing (*MTB*) has been suggested [5]. Simulations show that it performs better than *MFC*. However, the computation of "link first passage times" in *MTB* involves matrix inversion. Here we use a linearized version of *MTB* as the centralized controller. Moreover, whereas the original *MTB* neglects the effects of the routing decision itself on the "link first passage times", we predict their linearized limits *after* the routing decisions are made. This makes our scheme a genuinely global one. It turns out that the optimal scheme is equivalent to a *Simplex* algorithm [6], whose decision variables are the fraction of call attempts allocated along the alternative paths in the network.

Using *Simplex*, we determine the optimal fraction of traffic to be routed along each path, analogous to Load Sharing Dynamic Routing (*LSDR*) [7]. The routing fractions are updated at time intervals, typically of the order 5-10 seconds. Other routing actions, like routing in a fixed sequence of alternate paths, or routing in orders of priority according to a routing table, can be implemented after the routing ratios have been determined. Here we focus on *LSDR* for simulations, and monitor both the blocking probability and the crankback rate as performance measures.

The idea of *Simplex* is quite simple. For simplicity, assume that all paths between node pairs (i, k) are two-linked, bridged by intermediate tandems j. Suppose the vacancies of the trunk group from node i ingoing to tandem j is $n_{ij}^{(i)}$ and from tandem j outgoing to node k is $n_{jk}^{(o)}$. Let r_{ik}^{j} be the routing ratio originating from node i linking through tandem j and terminating at node k. We estimate the time to blocking $\tau_{ij}^{(i)}$, $\tau_{jk}^{(o)}$ respectively as

$$\tau_{ij}^{(i)} = \frac{n_{ij}^{(i)}}{\sum\limits_{k} \lambda_{ik} r_{ik}^{j}}, \qquad \tau_{jk}^{(o)} = \frac{n_{jk}^{(o)}}{\sum\limits_{i} \lambda_{ik} r_{ik}^{j}}. \tag{1}$$

The optimal routing ratios are determined by maximizing the minimum of the blocking times $\{\tau_{ij}^{(i)}\}$, $\{\tau_{jk}^{(o)}\}$, so that even the most congested link in the network is not very congested. This turns out to be a problem in linear programming which can be solved by the *Simplex* algorithm:

Mimimize b, the maximum consumption rate, subject to

$$
\begin{cases}
\sum_{k} \lambda_{ik} r_{ik}^{j} \leq bn_{ij}^{(i)}, \\
\sum_{i} \lambda_{ik} r_{ik}^{j} \leq bn_{jk}^{(o)}, \\
\sum_{j} r_{ik}^{j} = 1.
\end{cases}
\tag{2}
$$

The algorithm is divided into four stages to cope with the problem of degenerate optimal solutions. To provide good and unambiguous examples for supervised learning of the *Neural* controllers, this degeneracy has to be lifted. If the optimal solution for the minimum blocking time is degenerate, the next largest consumption rate is minimized in the original solution space and so on.

Let F be the set of "full" nodes, i.e. those with all of their outgoing or incoming paths blocked, and let NF be the set of "non-full" nodes. Optimization with respect to the "full" nodes has to be done separately, since they lead to infeasible solutions. The algorithm proceeds as follows:

- *Stage 1: Ignoring the "full" nodes, minimize the maximum consumption rate among the links to and from the "non-full" nodes. If the solution is degenerate, go to stage 2; otherwise go to stage 3.*

- *Stage 2: In the degenerate solution, ignore the links with maximum consumption rates (i.e. those which satisfy the hard constraints* $\sum_{\alpha \in l} \lambda_\alpha r_\alpha = bn_l$*) and minimize the maximum consumption rate in the rest of the network within the degenerate solution space. This leads to the optimization of the next largest consumption rate. Stage 2 is repeated until all degeneracies are lifted.*

- *Stage 3: For a link l to/from the "full" node, assign a virtual capacity* $n_l = N_l \mu t/2$*, which is the average number of call disconnections in an updating interval of duration t. Fixing r_α for the links to/from the "non-full" nodes, minimize the maximum consumption rate among the links to/from the "full" nodes. If the solution is degenerate, go to stage 4; otherwise exit.*

- *Stage 4: Same as stage 2 except that the links are to/from the "full" nodes.*

Comparing with *MFC*, *Simplex* has a comparable blocking probability, but a lower crankback rate. Hence, *Simplex* passes the test as a good teacher.

2.3 Decentralized Neural Control

Next we build a neural network for each originating node. Examples in the training set presented to this node consist of: the trunk status of the one-link and two-link paths from it, the call attempts from it (summed along all paths), and the total call attempts. Note that the total number of call attempts is the only global information fed into the *Neural* controller and provides measurement about the background traffic. In real networks, this can be broadcasted by the signaling network. The routing ratios determined by *Simplex* are supplied for training in each example.

For a single layered network, the inputs have to be normalized for convergence of the training stage. Hence the trunk status are normalized by the trunk capacity. Furthermore, to increase their weighting near congestion, nonlinear functions of their arguments, normalized by their averages, are used. Here we use a quadratic function. Similarly, the call attempts are normalized by their average. For an input vector $\vec{S}$, the local field for a path from i to k through j is given by:

$$h^j_{ik} = \sum_l J_{(ijk)l} S_l \tag{3}$$

where $J_{(ijk)l}$ is the synatic matrix to be determined by the training process.

The output for each path O^j_{ik} is the routing ratio computed by the *Neural* controller. It is normalized with a partition function so that the routing ratios through all tandems between a pair of nodes sum up to 1.

$$O^j_{ik} = \frac{e^{h^j_{ik}}}{\sum_{j'} e^{h^{j'}_{ik}}} \tag{4}$$

Learning is done by gradient descent of an energy function E with a learning rate η.

$$E = \frac{1}{2} \sum_{ijk} (r^j_{ik} - O^j_{ik})^2 \tag{5}$$

$$\Delta J_{(IJK)L} = \eta S_L O^J_{IK} [(r^J_{IK} - O^J_{IK}) - \sum_j (r^j_{IK} - O^j_{IK}) O^j_{IK}] \tag{6}$$

3 Results

We simulate a network with two inputs, two tandems and three outputs, shown in Figure 1, with N being the trunk group capacity. Traffic from inputs 3 & 4 to node 0 are each distributed by local *Neural* controllers through two alternative routes. The traffic to destination nodes 1 & 2 act

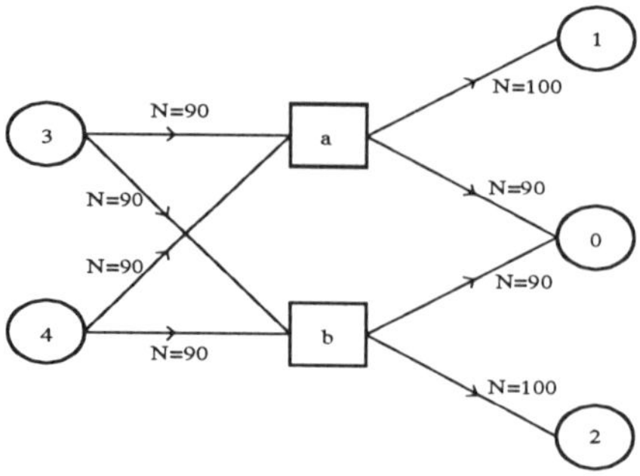

To \ From	3	4
1	0.15	0.15
0	1.00	1.00
2	0.35	0.35

Figure 1: Topology and traffic ratio between node pairs of a 2-2-3 network

as the background loading. We measure the overall blocking probability and crankback probability for a simulation period of 50,000 seconds. Figure 2 shows that *Simplex* and *Neural* have a comparable result in blocking probability with *MFC* routing. As shown in Figure 3, both *Simplex* and *Neural* have significant improvement in the crankback probability over *MFC*.

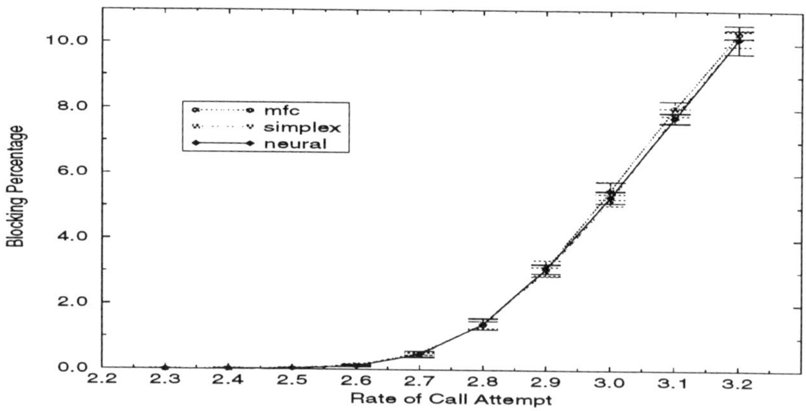

Figure 2: Blocking probability of a 2-2-3 network

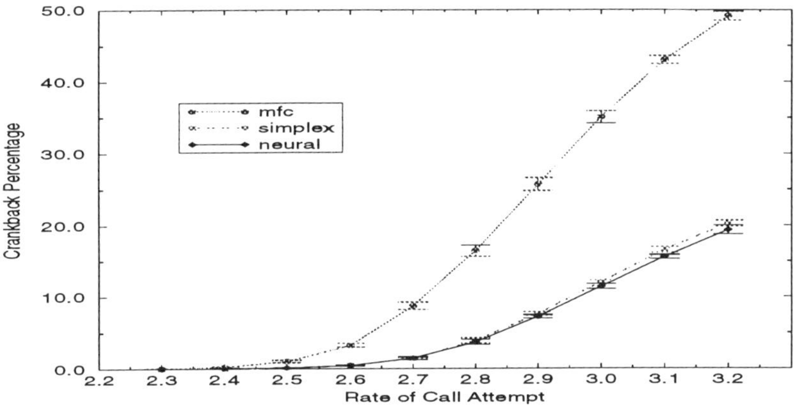

Figure 3: Crankback probability of a 2-2-3 network

4 Conclusions & Future Research Directions

In summary, we have found an optimal centralized control algorithm, and that neural networks can learn the controlling functions prescribed by the centralized controller. Despite the stochasticity of the call attempts and trunk status, learning is possible provided that sufficient examples are generated.

Since different traffic profiles result in disparate routing strategies, multi-layered perception control will function more superior than a single layered network in a real sophiscated telecommunication network. In our next step, we will build a two layered *Neural Network* in which the first layer performs the task of classifying input data, and the second layer implements the controlling action according to the weights determined by the classifier [8]. Such network architectures are believed to have fast response to evolving traffic profile, and able to generalize the control to more extreme situations. *Neural* controllers will be tested on more general network architectures, different topologies, and variable traffic profiles.

Acknowledgements

This work was supported by the Hong Kong Telecom Institute of Information Technology, HKUST.

References

[1] W.H. Cameron, J. Regnier, P. Galloy, and A.M. Savoie, "Dynamic

Routing for Intercity Telephone Networks", in *Proc. 10th Int. Teletraffic Cong.*, Montreal, Canada, June, 1983.

[2] R.J. Gibbens, "Dynamic Routing in Circuit-Switched Networks: the Dynamic Alternate Routing Strategy", Ph.D. Thesis, Univ. of Cambridge, 1988.

[3] E.W.M. Wong, T.S.P. Yum, and K.M. Chan, "Analysis of M and M^2 Routings in Circuit-Switched Networks", in *Proc. IEEE Globecom*, vol. 3, 1992, pp. 1487–1492.

[4] G.R. Ash, J.S. Chen, A.E. Frey, and B.D. Huang, "Real-Time Network Routing in a Dynamic Class-Of-Service Network", presented at the *13th Int. Teletraffic Cong.*, Copenhagen, Denmark, June 19–26, 1991.

[5] K.M. Chan and T.S.P. Yum, "The Maximum Mean Time to Blocking Routing in Circuit-Switched Networks", *IEEE JSAC*, vol. 12, no. 2, pp. 313–321, Feb. 1994.

[6] W.H. Press, B.P. Flannery, S.A. Teukolsky, and W.T. Vetterling, *Numerical Recipes in C: the Art of Scientific Computing*, Cambridge: Cambridge University Press, 1992, pp. 430–444.

[7] S. Avila and A. Guerrero, "Load Sharing Dynamic Routing in a Telephone Network", in *Proc. 11th Int. Teletraffic Cong.*, Kyoto, Japan, 1985, pp. 723–729.

[8] T. Fröhlinghaus, A. Wiechert, and P. Ruján, "Hierarchical Neural Networks for Time-Series Analysis and Control", *Network*, vol. 5, no. 1, pp. 101–116, Feb. 1994.

A Neural Network Algorithm to Solve the Routing Problem in Communication Networks

Michael W. Dixon, Matthew I. Bellgard, and Graeme R. Cole

Computer Science Programme
School of Mathematical & Physical Sciences
Murdoch University
Murdoch, Western Australia 6150
m_dixon@murdoch.edu.au

Abstract

This paper presents a Hopfield neural network that solves the routing problem in a communication network. It uses mean field annealing to eliminate the constraint terms in the energy function. Since there are no penalty parameters this approach should avoid the problems of scaling. Computer simulations of the neural network algorithm have shown that it can find optimal or near-optimal valid routes for all origin-destination pairs in a fourteen node communication network.

1 Introduction

The parallel nature of neural network models makes them suitable for simulation on parallel computer architectures. Neural networks are expected to be especially useful in the communications industry. Their promise as pattern recognizers and data correlators is naturally applicable to areas such as switching and queuing, transmission, error-correcting, data compression and routing of data communications traffic.

	1	2	3	4	5
1	1	1	0	0	0
2	0	0	0	0	0
3	0	0	1	1	1
4	0	0	0	0	0
5	0	0	0	0	0

Table 1: A neural representation of a route from node 1 to node 3.

Neural networks have often been formulated to solve difficult (for the most part, NP-complete) optimization problems. The routing problem may be viewed as an optimization problem where the objective is to minimize total cost when selecting a route for a given origin-destination pair. As a result there has been widespread research into the application of neural networks to the routing problem with varying degrees of success [2, 6, 7, 10, 12]. The major limitations that have been encountered are: i) some systems have failed to consistently converge to a valid route, ii) valid routes are not typically optimal, iii) most of these systems do not scale to larger sized communication networks, and iv) preprocessing is often needed to simplify the problem (e.g. calculating the minimum number of hops from origin to destination).

In this paper, we present a version of the Hopfield model that employs mean field annealing [4, 8, 13, 14] in an attempt to solve the routing problem. The Hopfield model has achieved prominence because of the relative ease of building it into hardware using currently available VLSI technology. Current simulations on a fourteen node communication network have shown that the system converges to a valid route for all origin-destination pairs. These routes are optimal or near optimal. Unlike other systems which use penalty constraint terms there is no need to tune parameters. This tuning has been found to be difficult and problem specific [1, 3]. Also, there is no need to pre-determine the minimum number of hops corresponding to the optimal route.

2 Mapping the Routing Problem Onto the Hopfield Model

NxN neurons will be used to represent a valid route from the source to destination node. The neurons are grouped into N groups of N neurons which are used to represent the path that the packet (message) will take. In a five node network which is fully interconnected (all the nodes have links to each other) twenty five neurons will be required to represent a route by

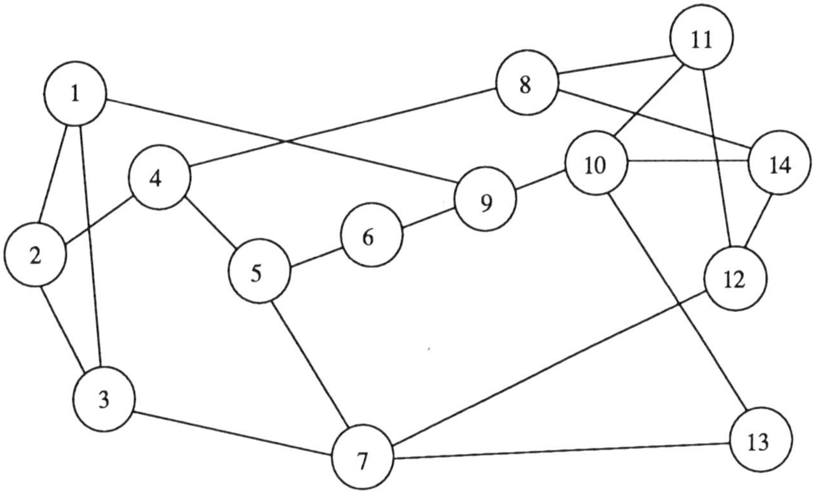

Figure 1: NFSNET-Backbone: 14 nodes & 21 bidirectional links.

an output matrix V.

The example in Table 1 illustrates a route the packet can take from node 1 to node 3 (e.g. where 1-1-3-3-3 is interpreted as 1-3). The same path can be represented in different ways, such as: 1-3-3-3-3, 1-1-1-3-3 and 1-1-1-1-3. For the purpose of demonstration the NFSNET-Backbone topology has been chosen (See Figure 1).

In many combinatorial optimization problems the neurons are grouped into vectors or clusters (each column will be considered a group). For all of these vectors a 1-out-of-N constraint is used, where only one from a group of N neurons can be equal to 1 and all other neurons in the group are to be equal to 0. Sometimes the optimization process ends up with a violation of these constraints, i.e. there are some vectors which have all neurons equal to 0 or more than one neuron is equal to 1. This follows from the fact that a neural network may converge to a local minima of the energy function [1, 3, 5]. A way to prevent this from happening is to use mean field annealing (MFA) to replace the classical sigmoid activation function of the form

$$V_i = g(U_i) = [1 + \tanh(U_i/T)]/2 \tag{1}$$

with a multi-dimensional activation function of the form shown in (2) below [4, 8, 13, 14].

$$V_{ia} = \frac{e^{u_{ia}}}{\sum_{b=1}^{N} e^{u_{ib}}} \tag{2}$$
$$(a = 1, 2, ..., N)(i = 1, 2, ..., N)$$

According to Cichocki and Unbehauen [4] the activation function generalizes the standard sigmoid function. By employing the multi-dimensional (generalized) activation function (2) the output variables V_{ia} of the neurons grouped in the vector V_i are not independent but always satisfy the constraint that the sum of the column must be equal to 1.

Our proposed energy equation is straightforward because it will only deal with the following distance term and an auxiliary term which adds local positive feedback around neurons which stabilize the neural network and eliminate chaotic behavior (parasitic oscillations) during the optimization process [9]:

$$E = \frac{1}{2}\sum_{x=1}^{N}\sum_{y=1}^{N}\sum_{i=2}^{N-1} GD_{xy}V_{xi}\left(V_{y,i+1} + V_{y,i-1}\right) + \frac{P_1}{2}\sum_{x=1}^{N}\sum_{i=1}^{N} V_{xi}^2 \tag{3}$$

where the subscripts x and y refer to the nodes while the subscript i refers to the position in the route. The GD_{xy} term is a function of both capacity and distance across a link. Decreasing distance and increasing capacity causes the generalized distance to decrease. For the energy function in (3) the generalized MFA equation will take the following form:

$$V_{xi} = \frac{e^{U_{xi}}}{\sum_{j=1}^{n} e^{U_{xj}}}$$

where

$$U_{xi} = -\frac{1}{T}\frac{\partial E(V)}{\partial V_{xi}} = -\sum_{y=1}^{N} GD_{xy}\left(V_{y,i+1} + V_{y,i-1}\right) + P_1 V_{xi}. \tag{4}$$

For each given Origin-Destination pair, the first and the last column in the neuron array are fixed (NxM matrix). The states of the neurons not in the first or last column are updated according to equation (4).

3 Simulation Results

We tested this model with the 14 node NFSNET-Backbone shown in Figure 1. The links, nodes and distances of the network can be represented by a

	1	**2**	**3**	**4**	**5**	**6**	**7**	**8**	**9**	**10**	**11**	**12**	**13**	**14**
1	0	9	13	L	L	L	L	L	20	L	L	L	L	L
2	9	0	7	9	L	L	L	L	L	L	L	L	L	L
3	13	7	0	L	L	L	15	L	L	L	L	L	L	L
4	L	9	L	0	7	L	L	16	L	L	L	L	L	L
5	L	L	L	7	0	7	11	L	L	L	L	L	L	L
6	L	L	L	L	7	0	L	L	7	L	L	L	L	L
7	L	L	15	L	11	L	0	L	L	L	L	14	9	L
8	L	L	L	16	L	L	L	0	L	L	7	L	L	8
9	20	L	L	L	L	7	L	L	0	7	L	L	L	L
10	L	L	L	L	L	L	L	L	7	0	5	L	8	5
11	L	L	L	L	L	L	L	5	L	5	0	7	L	L
12	L	L	L	L	L	L	14	L	L	L	L	0	L	4
13	L	L	L	L	L	L	9	L	L	8	L	L	0	L
14	L	L	L	L	L	L	L	8	L	5	L	4	L	0

Table 2: Generalized distance matrix.

14x14 distance matrix D with entries D_{ij} where each entry D_{ij} represents the distance between node i and node j. For nonexistent links a large value L was used to prevent the neural network from using the links. Also, a 14x14 capacity matrix C with entries C_{ij} was used to represent the capacity between the nodes. All the links were assumed to be bidirectional and to have identical capacity of 100. Using the distance and capacity matrices a (combined) generalized matrix (see Table 2) was calculated, which is represented by GD in equation (4).

For our simulations the following parameters were set: Starting Temperature of 1000, Ending Temperature of 200, the P_1 parameter was set equal to Temperature*2 and the Decay term to 0.995. Since the neural network should have no prior favor for a particular path, all the U_{xi} were set to 0. This corresponds to the initial state of the neural network being concentrated at the centre of the hypercube, with all neuron output voltages V_{xi} set to 0.5. Each simulation was stopped when the system reached the final temperature of 200 or if it settled into a stable final state. This was assumed to occur when the energy function changed by less than 0.0002 over the past 10 iterations.

In order to be valid a route need not be optimum but must not contain any nonexistent links. Between 10 and 200 iterations were required to reach the final solutions. Simulations were run for all origin-destination pairs to find valid routes and the summary of results are shown in Tables 3 & 4.

Hops	Number of optimal routes	Number of near optimal routes	Total number of routes
1	42	0	42
2	68	2	70
3	58	12	70

Table 3: Summary of simulation results for the 14 node NFSNET-Backbone network.

Source	Dest.	Simulation Route	Dist.	Optimal Route	Dist.
3	5	3–7-5	26	3-2-4-5	23
3	6	3-7-5-6	33	3-2-4-5-6	30
3	10	3-1-9-10	40	3-7-13-10	32
5	3	5-7-3	26	5-4-2-3	23
5	11	5-4-8-11	30	5-6-9-10-11	26
5	14	5-4-8-14	31	5-6-9-10-14	26
6	3	6-9-1-3	40	6-5-4-2-3	30
6	12	6-5-7-12	32	6-9-10-14-12	23
6	8	6-5-4-8	30	6-9-10-14-8	27
8	6	8-4-5-6	30	8-14-10-9-6	27
10	3	10-9-1-3	40	10-13-7-3	32
11	5	11-8-4-5	30	11-10-9-6-5	26
12	6	12-7-5-6	32	12-14-10-9-6	23
14	5	14-8-4-5	31	14-10-9-6-5	26

Table 4: Comparision of all non-optimal simulation routes with actual optimal routes.

4 Discussion

All of the source-destination paths selected were valid and 92.3% contained the optimum route. This rate of optimum route selection is higher than any currently published rate. The neural network was able to find optimal routes where the path was fairly short. The 7.7% of nonoptimal routes were mainly longer routes which contained many hops. As the data shows in Tables 3 & 4, the neural network had greater difficulty converging to a global minimum when the optimum route takes more hops to get from the source to destination than does another valid route. Usually, the neural network will favor a route which requires a smaller number of hops, even though the optimum route takes more hops. This can be attributed to how the updates are performed on the neural network. The updates are done left to right (from source to destination) and have introduced an unwanted bias toward selecting paths that have a smaller number of hops versus the shortest total distance. This should be easily corrected by updating the neural network randomly.

Currently, we are investigating ways of ensuring that all routes are optimal routes by using a hill climbing term such as that presented in [11] to prevent the neural network from getting trapped in local minima. Also, more tests on larger networks are being done to ensure that the neural network will always converge to valid routes no matter how large the communication network.

5 Conclusion

In this paper, we described a neural network routing methodology which does not use constraint terms. The two advantages of using mean field annealing are the ability to find optimal or near-optimal valid routes in a communication network, and the elimination of the constraint terms in the energy function which should help overcome the scaling problem.

References

[1] V.B. Aiyer, M. Niranjan & F. Fallside, "A Theoretical Investigation into the Performance of the Hopfield Model," *IEEE Transactions on Neural Networks*, Vol. 1, No. 2, pp. 204-215, 1990.

[2] J. Alspector, R. Goodman & T.X. Brown, *Proceedings of the International Workshop on Applications of Neural Networks to Telecommunications*, Lawrance Erlbaum Associates, Inc., Hillsdale, New Jersey, 1994.

[3] A.R. Bizzarrie, "Convergence Properties of a Modified Hopfield-Tank Model," *Biological Cybernetics*, Vol. 64, pp. 293-300, 1991.

[4] A. Cichocki & R. Unbehauen, *Neural Networks for Optimization and Signal Processing*, John Wiley & Sons, 1993.

[5] J.J. Hopfield & D.W. Tank, "Neural Computations of Decision in Optimization Problems," *Biological Cybernetics*, Vol. 52, pp. 141-152, 1986.

[6] F. Kamoun & M.K. Ali, "A Neural Network Shortest Path Algorithm for Optimum Routing in Packet-Switched Communications Networks," *in Proceedings of the Conference Globecomm,* Phoenix, Arizona, pp. 0120-0124, 1991.

[7] Y.C. Ouyang, *Development of a Neural Network Based Routing Algorithm for Communication Networks*, PhD Thesis, Memphis State University, USA, 1992.

[8] C. Peterson & B. Söderber, "A New Method for Mapping Optimization Problems onto Neural Networks," *International Journal of Neural Systems*, Vol. 1, pp. 3-22, 1989.

[9] C. Peterson, "Parallel Distributed Approached to Combinatorial Optimization: Benchmark on Travelling Salesman Problems," *Neural Computation*, Vol. 2, pp. 261-269, 1990.

[10] H.E. Rauch & T. Winarske, "Neural Networks for Routing Communication Traffic", *IEEE Control Systems Magazine*, Vol. 8, No. 2, pp. 26-31, April 1988.

[11] Y. Takefuji & K. Lee, "An Artificial Hysteresis Binary Neuron: A Model Suppressing the Oscillatory Behavior of Neural Dynamics," *Biological Cybernetics*, Vol. 64, pp. 353-356, 1991.

[12] S.C.A. Thomopoulos, L. Zhang & C.D. Wann, "Neural Network Implementation of the Shortest Path Algorithm for Traffic Routing in Communication Networks," *in Proceedings of the IEEE International Joint Conference on Neural Networks*, Singapore, pp.2693-2702, 1991.

[13] D.E. Van den Bout & T.K. Miller, "Improving the Performance of the Hopfield-Tank Neural Network Through Normalization and Annealing", *Biological Cybernetics*, Vol. 62, pp. 129-139, 1989.

[14] D.E. Van den Bout & T.KK. Miller, "Graph Partitionin Using Annealed Neural Networks", *Biological Cybernetics*, Vol. 62, pp. 129-139, 1990.

Classifying Loss Rates with Small Samples

Timothy X Brown
Bell Communications Research, Morristown, NJ 07960
timxb@bellcore.com

Abstract

In a modern telecommunication systems, loss rates are difficult to compute and, with sample sizes small compared to the loss rate, are difficult to measure empirically. When, based on such small samples, a classifier tries to identify what conditions result in loss rates that exceed a threshold, the decision boundary can miss the threshold by orders of magnitude. We develop two correction methods. The first weights samples to remove the bias. The second calculates confidence levels and combines these with sample aggregation to produce confident decisions. We analyze the methods and confirm the results using synthetic data.

1 Introduction

As telecommunication companies seek increasing flexibility to provide new services while customers develop innovative network uses, the tasks assigned to the network's elements become less constrained and frequently violate traditional engineering assumptions. For example, the advent of new services such as automatic redial, call forwarding, and beepers, as well as uses such as fax and modem has dramatically altered the call holding time distribution from the standard exponential [Bol94]. Computing which loads are overloads for either provisioning or control purposes becomes intractable in such complex environments.

The problem is shown in Figure 1. A network element (NE) accepts loads from different source types. The loads consist of discrete events such as call-setup requests if the NE is a switch call processor or bytes/packets if the NE is a data queueing system. Each item can either be acceptably serviced, or lost (rejected, blocked, overdue, etc. depending on the problem semantics). When the goal is to maintain the loss rate below a critical rate (CR), a key element is a classification device that identifies which system states will exceed the CR. When system com-

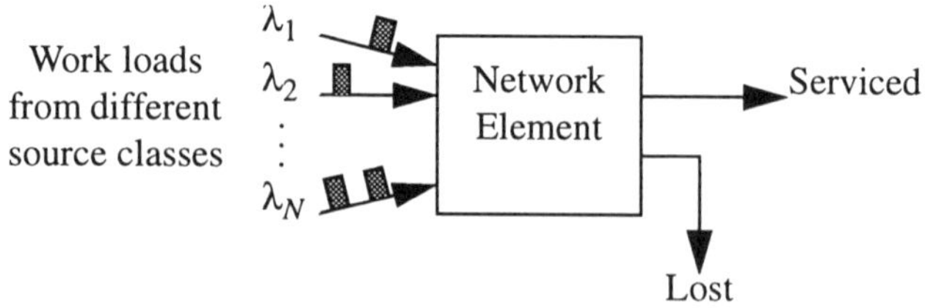

Figure 1: Work servicing model.

plexity precludes analytic decision boundaries, one solution is to base the decision on past samples of the observed loss rate as a function of the system state.

More formally, a NE traverses different states and either periodically, or at state transitions, the NE outputs a sample vector containing the number of events, T, the number of loss events, s, and the state feature vector, $\bar{\phi}$. The feature vector may correspond to the loads from the different sources and it may also include other information such as time of day. Each feature vector, $\bar{\phi}$, has an associated loss rate, $p(\bar{\phi})$, which can not be measured directly. The goal is for a given CR = $p*$ to have a classifier function, $C(\bar{\phi})$, such that:

$$C(\bar{\phi}) = \begin{cases} +1 & \text{if } p(\bar{\phi}) < p* \\ -1 & \text{if } p(\bar{\phi}) \geq p* \end{cases} \tag{1}$$

Depending on the properties of $p(\bar{\phi})$, the NE sample vectors could be used in a variety of statistical/neural classification techniques[Dud73][Ng91] to build an approximation to the classification function, $C(\bar{\phi})$. Unfortunately when the sample sizes are small relative to the CR a sample bias is introduced that leads to classifiers that can miss the $p(\bar{\phi}) = p*$ decision boundary by orders of magnitude. This phenomena can be observed in Figure 2. When the sample size is much smaller than $1/p$, the apparent boundary is biased towards $p >> p*$.

2 Solution Methods

2.1 Aggregate Method

In this section we describe a method of classifying a sample that assigns confidence intervals and aggregates a neighborhood of samples when necessary to achieve the desired confidence level. Given a sample of size T with s losses what is the probability that $p(\bar{\phi}) > p*$, the CR? In Appendix A we show:

$$P\{p(\bar{\phi}) > p*\} \cong e^{-Tp*} \sum_{i=0}^{s} \frac{(Tp*)^i}{i!} \tag{2}$$

From this we can set a confidence level, L, and evaluate whether the CR for this state will be met with probability L (i.e. $P\{p(\bar{\phi}) > p*\} < (1 - L)$). Unfortu-

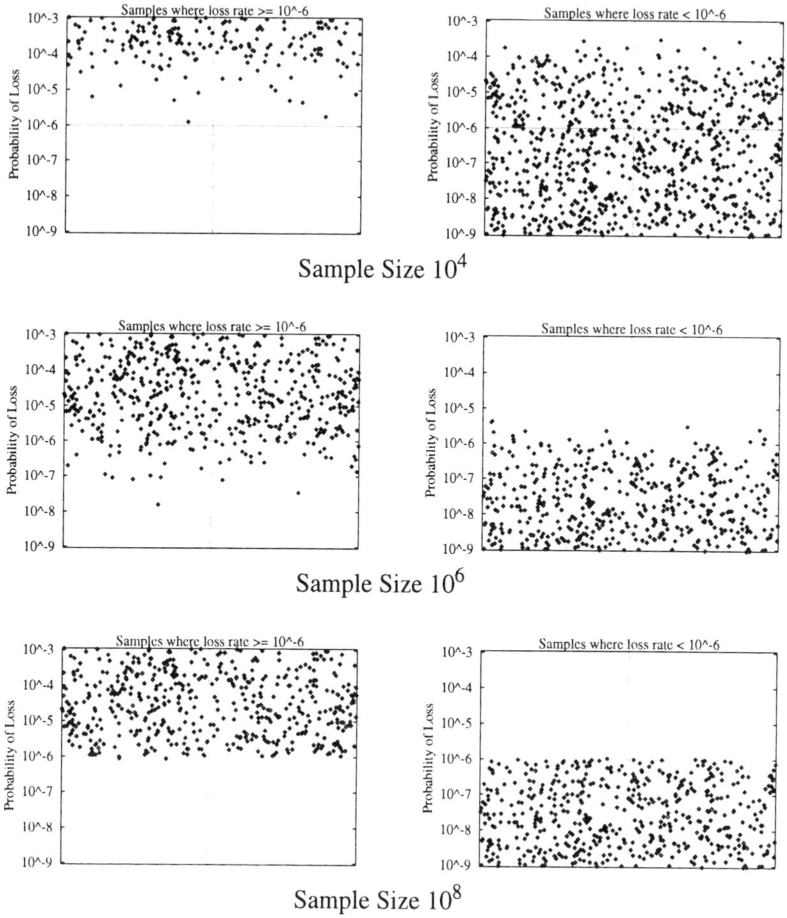

Figure 2: Small Sample Bias Demonstration. Experiment: 1000 samples drawn with loss rates uniformly chosen (on log scale) between 0.001 and 1000 times the critical rate, $p* = 10^{-6}$. After a sample size of Bernoulli trials, samples with observed loss rates below and above $p*$ are plotted separately left and right (the x-scale only provides visual separation). The experiment repeats for three different sample sizes.

nately, for a small T, even if $s = 0$, this level is not met. But, a neighborhood of samples with similar load combinations may all have $s = 0$ indicating that this sample can be classified as meeting the CR.

For each sample we choose its k-nearest neighbors (in state space), and classify the sample as meeting CR if, for the aggregate sample, $P\{p(\bar{\phi}) > p^*\} < (1 - L)$.

On one hand we want k larger for more robust estimates, on the other hand, we want it smaller for tighter boundaries. Using the above results, the neighborhood size is the smallest k such that the total sample T is greater than a critical size,

$$T^* = -\log(1 - L)/p^*. \quad (3)$$

From (2), (3) guarantees that if all items are serviced correctly in the aggregate sample we can classify it as meeting the CR within our confidence level.

For larger samples we use (2) to classify the samples. Figure 3 plots the acceptable number lost as a function of samples size and confidence level. This can be used in two ways. Given a sample of size T it shows the maximum acceptable number of loses. Conversely, given a number of losses, it shows how big the sample must be to be confident that the loss rate is below the CR.

2.2 Weighting Method

This method labels samples one at a time, and offsets the bias with weights on the positive and negative samples. We design a weighting, $w(s, T) \Rightarrow (0, \infty)$, and desired classification, $d(s, T) \Rightarrow \{-1, +1\}$, that minimizes the expected weighted mean square error, $E\left[w(s, T)\,(C(\phi) - d(s, T))^2\right]$ of the classifier in (1) (where the expectation is taken over s and T). Minimizing and solving:

$$C(\bar{\phi}) = \frac{E[w(s, T)d(s, T)]}{E[w(s, T)]} \quad (4)$$

which can be thresholded at zero. In Appendix B we derive solutions for d and w:

$$d(s, T) = \begin{cases} +1 & \text{if } s \leq \lfloor p^*T \rfloor \\ -1 & \text{if } s > \lfloor p^*T \rfloor \end{cases} \quad (5)$$

$$w(s, T) = \begin{cases} w^+ = \alpha T(1 - P(T)) & \text{if } s \leq \lfloor p^*T \rfloor \\ w^- = \alpha T P(T) & \text{if } s > \lfloor p^*T \rfloor \end{cases} \quad (6)$$

where $P(T) = \sum_{i \leq \lfloor p^*T \rfloor} e^{-p^*T}(p^*T)^i/i!$ and α is an arbitrary positive constant. The relative weighting of positive and negative samples is plotted in Figure 4.

3 Analysis of the Methods

We analyze the aggregate and weighting methods of Section 2 as well as the normal method that simply classifies each sample according to the sample loss rate (as in (5)), and weights all samples equally ($w^+ = w^- = 1$).

3.1 Expected Bias

In this section we analyze what decision boundaries will be chosen by the different methods. For a given p^* we choose the p s.t. the right side of (4) is zero where

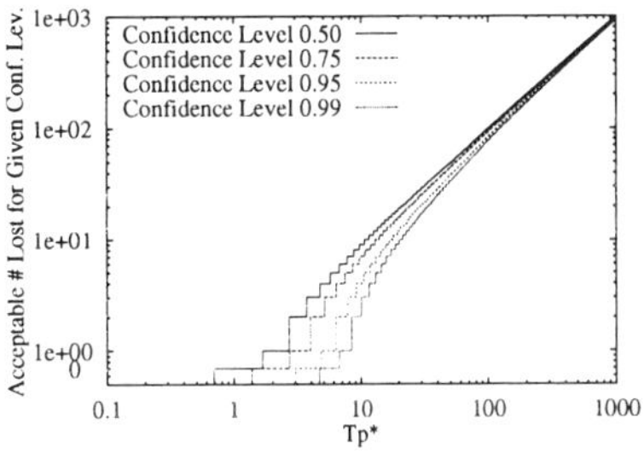

Figure 3: Acceptable Number of Losses for a Given Confidence Level vs. Sample Size Normalized by the Critical Loss Rate. Uses log scales, with 0 artificially added to the vertical axis.

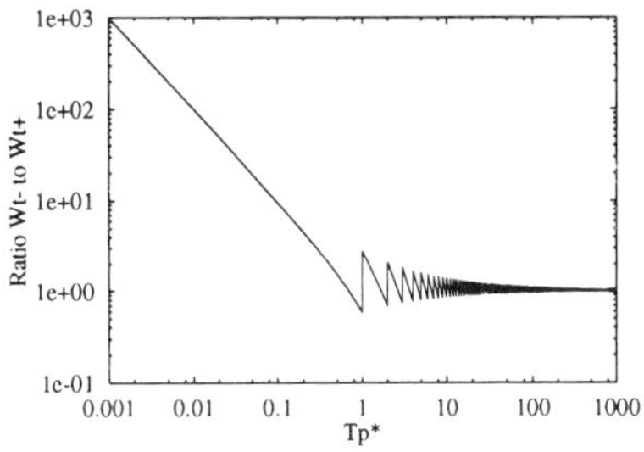

Figure 4: Plot of Relative Weighting of Samples with Losses Below and Above the Critical Loss Rate.

using the poisson approximation, the probability of s given p and T is $e^{-pT}(pT)^s/s!$. This is equivalent to finding the p such that:

$$\frac{w^-}{w^+ + w^-} = e^{-pT}\sum_{s \le c}\frac{(pT)^s}{s!} \tag{7}$$

Where the weights, $w^{+/-}$, and classification threshold, c, depend on the method used. For the normal method, $w^+ = w^- = 1$ and, according to (5), $c = \lfloor p^*T \rfloor$. For the aggregate method, $w^+ = w^- = 1$ and we choose a confidence level $L = 0.95$. If $T < T^*$ in (3) we choose a k s.t. $kT \ge T^*$ and solve (7) with $c = 0$ and kT instead of T. If $T \ge T^*$, we set c at the maximum acceptable number of losses from Figure 3, and solve (7). For the weighting method, we classify and weight points according to (5) and (6) so that w^+ is the top line in (6) and w^- the bottom. By definition $p = p^*$ is the solution to (7) in this case. These results are used in the next section.

3.2 Experiments with Simulated Data

To see these methods in a controlled environment, we performed the following experiments similar to those in Figure 2. We fixed CR = $p^* = 10^{-6}$. A sample consists of choosing a p uniformly from $0.001p^*$ to $1000p^*$ and then recording; the one-dimensional feature, $\phi = p/p^*$ in dB; the sample size, T; and the number of losses from T Bernoulli trials using p. An experimental run consists of generating a number of samples; applying one of the classification and weighting methods to the samples; then applying a linear classifier with a minimum mean squared error objective to the data and recording the decision boundary. An experiment consists of statistics from 100 runs.

The results for 1000 samples and $T = 10^4$, 10^6, and 10^8 are shown in Table 1. These can be interpreted using the expected boundaries in Section 3.1 and shown in Table 2. The normal case shows definite biases both negative and positive (a factor of 50 in the small sample case), the aggregate method is always biased down by differing safety margins, and the weighting scheme is close to the true boundary. The variance on these techniques is similar for a given sample size and number. Note the very large variance when the sample size is 10^4. This can be attributed to the fact that where as the top row has 10^{11} underlying Bernoulli trials, the bottom row has only 10^7. Table 3 has the same results using different sample sizes so that the total number of Bernoulli trials was fixed at 10^9. This produces similar results except that now the most bias is in the top row. This implies that many small samples are more informative than a few large samples. The observed down shifting from the expected values in some cases can be attributed (with analysis omitted for space reasons) to local sample noise.

Table 1: Boundary of Linear Threshold Classifier using Different Sample Weighting and Classification Methods. Results are average of 100 runs with the sample standard deviation. The true boundary is at zero.

#Trials, T	#Samples	Boundary for each technique ($p/p*$ in dB).		
		Normal	Aggregate	Weighting
10^8	10^3	0.0 ± 0.1	-0.8 ± 0.1	0.0 ± 0.1
10^6	10^3	-2.6 ± 0.4	-7.3 ± 0.7	-0.9 ± 0.4
10^4	10^3	17.2 ± 4	-10.6 ± 9	-0.4 ± 2

Table 2: Table 1 with Expected Boundaries.

#Trials, T	#Samples	Expected boundary for each technique ($p/p*$ in dB).		
		Normal	Aggregate	Weighting
10^8	∞	0.0	–0.8	0.0
10^6	∞	–1.6	–6.4	0.0
10^4	∞	18.4	–6.4	0.0

Table 3: Table 1 Normalized so #Trials·#Samples = 10^9.

#Trials, T	#Samples	Boundary for each technique ($p/p*$ in dB).		
		Normal	Aggregate	Weighting
10^8	10^1	0.0 ± 1	-0.2 ± 0.8	0.0 ± 1
10^6	10^3	-2.6 ± 0.4	-7.3 ± 0.7	-0.9 ± 0.4
10^4	10^5	17.5 ± 0.1	-7.4 ± 0.7	-1.3 ± 0.3

4 Conclusion

In a modern telecommunication systems, loss rates are difficult to compute and, with sample sizes small compared to the loss rate, are difficult to measure empirically. We analyzed three sample preprocessing methods for a statistical classifier that uses these samples to identify when the loss rate exceeds a threshold.

Building a classification model based on samples that are classified according to whether the sample itself meets the quality of service requirements (the normal method) can lead to classifying loads as having an acceptable loss rate, when in fact they have losses a factor of 50 or more greater. Using a technique that assigns confidence intervals and aggregates when necessary to achieve the desired confidence level (the aggregate method), prevents these mistakes at the expense of conservative estimates and greater computation if the samples are small. Classifying samples individually, but assigning weights to the samples to counter the bias (the weighting method) leads to unbiased estimates, although no guarantees can be given on the estimates. Artificially lowering the critical rate gains some confidence, but how much for a given confidence level is not clear.

All of the methods performed better when given many diverse, small samples, rather than a few large samples. The analysis in this paper only depends on the normalized sample size, $p*T$, and thus applicable to a variety of applications whether the critical loss rate is 10^{-3} or 10^{-12}.

Appendix A: A Confidence Level

We calculate $P(p > p^* | s)$. This is an ill-posed question without a prior distribution, g, on p, but we show that when T is large, then with mild constraints on g it is equivalent to assuming g is a uniform distribution in the interval 0 to 1. We note formally $P(p|s) = P(s|p)\,P(p)/P(s)$. So:

$$P(p > p^* | s) = \left[\int_{p^*}^{1} P(s|p)\,P(p)\,dp\right] \Big/ \left[\int_{0}^{1} P(s|p)\,P(p)\,dp\right] \tag{8}$$

For large T, $P(s|p)$ is well approximated by the Poisson distribution: $P(s|p) = e^{-Tp}(Tp)^s/s!$. For reference we note:

$$\int_a^b e^{-Tp}\frac{(Tp)^s}{s!}\,dp = \frac{e^{-Tp}}{T}\sum_{i=0}^{s}\frac{(Tp)^i}{i!}\Bigg|_{p=b}^{a} \tag{9}$$

Notation: $a = b \pm c \Rightarrow b - c \le a \le b + c$. Let f^* and f be the right side of (9) evaluated at p^* and 0 respectively.

Theorem: Choose ε, x, β, γ, and distribution g on $p \in [0, 1]$ with the properties: $\exists\alpha > 0$ s.t. $\alpha + \beta + \varepsilon\alpha\beta + 2\alpha\gamma < 1/3$; $1 > \varepsilon > (\alpha f^* e^{T - s\ln T})^{-1}$;

$1 \geq x > (s\log T - \log\delta)/T$ where $\delta = \alpha f^* \varepsilon$; $g(p) = g(0)(1 \pm \varepsilon\beta)$ when $p < x$ and $g(p) < \gamma g(0)$ when $p \geq x$. If T is large, then:

$$P(p > p^* | s) = (1 \pm \varepsilon) e^{-Tp^*} \sum_{i=0}^{s} \frac{(Tp^*)^i}{i!} \tag{10}$$

Proof: Noting $T \geq 1$ and $s \leq T$:

$$y = -\frac{1}{T}\log\left(\delta T / \sum_{i=0}^{s} \frac{T^i}{i!}\right) < \frac{1}{T}\log\left(\frac{T^s}{\delta}\right) = \frac{1}{T}(s\log T - \log\delta) < x .$$

Substituting this at one of the limits of (9):

$$\frac{e^{-Tp}}{T} \sum_{i=0}^{s} \frac{Tp^i}{i!} < \frac{e^{-Ty}}{T} \sum_{i=0}^{s} \frac{Tp^i}{i!} < \delta \sum_{i=0}^{s} \frac{Tp^i}{i!} \Big/ \sum_{i=0}^{s} \frac{T^i}{i!} \leq \delta \quad \text{for } (x \leq p \leq 1).$$

Thus the limits at $p = x$ and $p = 1$ in (9) are bounded by δ. We assume T is large so that we can use the Poisson approximations and (9) in (8). Noting that $f^* \leq f$ we get:

$$P(p > p^* | s) = \frac{\int_{p^*}^{x} P(s|p) P(p)\, dp + \int_{x}^{1} P(s|p) P(p)\, dp}{\int_{0}^{x} P(s|p) P(p)\, dp + \int_{x}^{1} P(s|p) P(p)\, dp}$$

$$= \frac{(f^* \pm \delta)\, g(0)\, (1 \pm \varepsilon\beta) \pm 2\gamma\delta g(0)}{(f \pm \delta)\, g(0)\, (1 \pm \varepsilon\beta) \pm 2\gamma\delta g(0)} = \frac{f^*(1 \pm \varepsilon/3)}{f(1 \pm \varepsilon/3)} = \frac{f^*}{f}(1 \pm \varepsilon)$$

which is the desired result. Q.E.D.

Appendix B: A Weighting and Classification

Given samples of size, T (from some unspecified distribution), and a number of successes, s (from a Binomial distribution with probability of success, p) we derive a weighting, $w(s, T)$, and a classification, $d(s, T)$, that provides maximum discrimination between success probabilities that are close to a threshold p^* in the following sense. If $F(p) = E[w(s, T) d(s, T) | p]$, where the expectation is taken over possible values of s and T:

$$F(p) = \sum_{T=1}^{\infty} \text{Prob}\{T\} \sum_{s=0}^{T} w(s, T) d(s, T) e^{-pT} \frac{(pT)^s}{s!} \Bigg|_p , \tag{11}$$

then we are looking for a weighting and classification such that $F(p) >,<,= 0$ if and only if $p <,>,= p^*$ with the maximum slope at $F(p^*)$. We remark that from (4) this implies that a statistical classifier trained on such weighted samples will choose a decision boundary at $p = p^*$. To make the result independent of T's distribution, we solve this for a fixed T. We contend that one solution is

$$w(s, T) = w^{+/-}(T) = \left[\sum_{s \in d^{+/-}(T)} e^{-p^* T} \frac{(p^* T)^s}{s!} \right]^{-1} \quad \text{if } s \in d^{+/-}(T) \tag{12}$$

where $d^+(T)$ and $d^-(T)$ are the set of s such that $d(s, T)$ is +,– respectively. That $F(p^*) = 0$ in this case is easily verified by substituting (12) into (11). Note that it can be multiplied by any positive constant, even by one that is a function of T, and still be a solution. One possibility suggested by (4) would be to normalize the weights so that they have expected values of 1. For small T this can lead to very large weights that may cause instabilities for some iterative classification algorithms. To avoid very large weights, while weighting small samples less than larger samples we choose this final solution:

$$w(s, T) = \frac{w^{+/-}(T)}{w^+(T) + w^-(T)} T\alpha \quad \text{if } s \in d^{+/-}(T) \tag{13}$$

where α is an arbitrary positive constant.

Interestingly, (12) and (13) are solutions with any classification scheme as long as both d^+ and d^- are non-empty. Logic and simplicity dictates a solution in the form $d^+(T) = \{0, \ldots, c\}$, $d^-(T) = \{c + 1, \ldots, T\}$ for some integer c. For maximum sensitivity, we choose the c that maximizes the expected change in (11) as p varies at $p = p^*$. Using solution (13) and skipping intermediate calculations, $c = \lfloor p^* T \rfloor$ where $\lfloor \ \rfloor$ is the floor operator. So, as our intuition would state, maximum fidelity is given when samples are classified as plus or minus by whether their rate is less than or greater than the target rate, and we weight the positive samples according to (13). Which if we assume that p^* is small enough that terms indexed by $i > T$ in $P(T)$ are negligible, reduces to (6).

References

[Bol94] Bolotin, V.A., "Modeling Call Holding Time Distributions for CCS Network Design and Performance Analysis," *IEEE Journal on Selected Areas in Communications,* Vol. 12, No. 3, pp. 433–438, Aug. 1994.

[Dud73] Duda, R.O., Hart, P.E., *Pattern Classification and Scene Analysis*, Wiley & Sons, New York, 1973.

[Ng91] Ng, K., Lippmann, R.P., "A Comparative Study of the Practical Characteristics of Neural Network and Conventional Pattern Classifiers," in *Neural Information Processing Systems 3*, ed. Lippmann, R.P., et al., Morgan Kauffman, San Mateo CA, 1991, pp. 970–976.

A Feature-Based Neural Network Movie Selection Approach

N. Karunanithi
2Q-265, Bellcore
445 South Street
Morristown, NJ 07960
karun@faline.bellcore.com

Joshua Alspector
2P-396, Bellcore
445 South Street
Morristown, NJ 07960
josh@bellcore.com

Abstract

The huge amount of information available in the currently evolving information infrastructure at any one time can easily overwhelm end-users. One way to tackle this situation is to use an "information filtering agent" which can select information according to the interest and/or need of an end-user. However, at present no information filtering agents exist for the evolving multimedia information infrastructure. In this study, we demonstrate the use of a feature-based neural network filtering agent for the video-on-demand application. We evaluate the neural network approach for 10 voluntary subjects who provided ratings for the movies. Our preliminary results suggest that the feature-based selection can be a useful tool to recommend movies according to the taste of the user and that the feature-based approach can be as effective as a movie rating expert. We also compare our feature-based approach with a clique-based approach.

1 Introduction

In recent years, computer-network-based information services have gained wide acceptance both within commercial and non-commercial sectors. The information content in such services is mostly textual. However, the currently evolving U. S. National Information Infrastructure (NII) is expected to support not only a variety of text-based information services but also various multimedia (hypertext, audio and video-based) information services. Some of the potential application domains in which the information infrastructure is likely to have impact are banking at home, access to electronic libraries, distance learning and laboratories, delivery of news and entertainment on demand, electronic shopping malls, law enforcement and security alertness, legal services, national health care and weather services, and telecommuting. Thus, the NII has the potential to change the way we work, communicate, travel, and generally access information.

The huge amount of information available in the information infrastructure at any one time can easily overwhelm end-users. Even within existing computer-network-based information services, providing information that is of interest to a particular end-user is not an easy task. For example, filtering relevant information in the "Internet" is not easy because a single message may be sent over a set of mailers, a message may consists of a "thread" (a sequence of "replies" to the original mail), or the header may not reflect what the actual content means. This situation is likely to worsen in the future multimedia information infrastructure unless the end-user has the ability to filter information based on what is relevant to him/her.

Several useful text-based tools exist for navigational purposes[5] on the "Internet". An example of a common-to-all user interface for the Internet is "Mosaic". Mosaic is a hyper-text based easy-to-use X-Window interface built on top of various Internet navigational, browsing tools such as "Gopher", WAIS, World Wide Web, Archie, FTP service etc. Even with such a common interface, these navigational aids require network support, and "active participation" of the end-user. However, at present no equivalent filtering systems exist for the evolving multimedia information infrastructure.

In this study, we evaluate the use of a feature-based neural network movie recommendation system for the evolving video-on-demand service. The initial study is based on a feature database of 1548 movies. We evaluate the approach for 10 voluntary subjects who provided ratings for the movies. Our preliminary results suggest that the feature-based selection can be a useful tool to recommend movies according to the taste of the user and that the feature-based approach can be as effective as a movie rating expert. We also compare our feature-based approach with a clique-based approach.

2 An Experiment on Movie Selection

In this study, we evaluate two fundamentally different approaches: a *clique-based* approach[1] and a *feature-based* approach. In the *clique-based* approach, movies are recommended using the ratings of a set of users who might have similar taste. In the *feature-based* approach, first a model is built using a set of important features of the movies that a user has seen and rated, and then that model is used to predict the ratings for movies that the user wants to see. It should be noted that a hybrid approach can also be developed by combining these two approaches. We present more details and a preliminary evaluation of these approaches in subsequent sections.

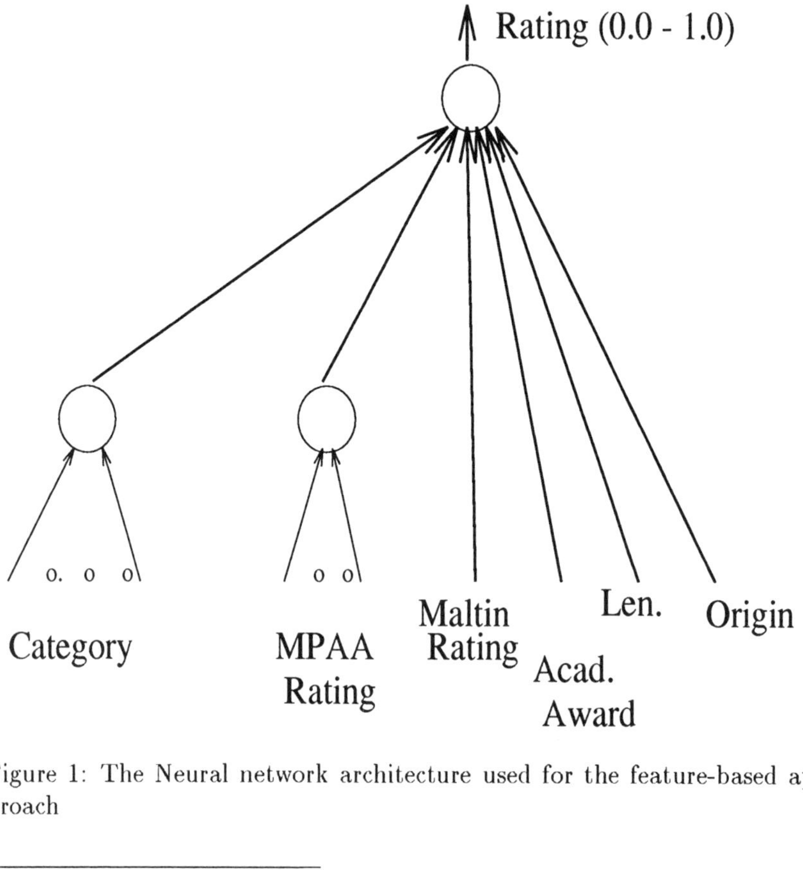

Figure 1: The Neural network architecture used for the feature-based approach

[1]The clique-based approach is a variant of the approach suggested by Will Hill of Bellcore.

2.1 A Feature-Based Approach

The feature-based approach is based on the notion that the features of the movies can be useful in recommending movies. Some of the features that can be used to recommend movies include MPAA ratings, expert critic ratings, movie category, name of the director, leading actor/actresses, and awards received. For example, a particular user may have a strong inclination to see only movies that are rated G, acted by Ben Kingsley, and belonging to the category Comedy or Drama. Thus, the feature-based approach exploits the bias of a user towards a set of important features of the movies.

The algorithm for the feature-based approach is as follows:

1. Extract relevant features from the movies that the user has rated.
2. Build a model for the user by associating selected features (as inputs) and the ratings (as output).
3. Estimate ratings for the movie that the target user wants to see by considering its features as new input to the model.

To implement this algorithm, one must select a proper set of features and a correct model to predict ratings. We already suggested several useful features for rating movies. However, in this study, we use only the six features that are shown in Figure 1.

The rationale for selecting a small set of features is that we wanted to evaluate whether we can build a reasonable model with as few features as possible. For step 2, we do not make any specific recommendation as to what modeling approach might be superior. The choice for modeling, for example, may include a linear or nonlinear regression model, an expert system, a neural network, or other approaches. In this study, we illustrate the use of the neural net.

The architecture of the neural network model is shown in Figure 1. The network is a simplified form of a modular network in that the inputs for features Category and MPAA Ratings are first fed to separate hidden units rather than directly into the network. The actual encodings used for these features are illustrated in Table 1.

2.2 A Clique-Based Approach

The clique-based approach is based on the hypothesis that the average rating of a *clique* of users is the best indicator of an individual's future rating. A set of users form a clique if their movie ratings are similar. A member of the clique for whom we want to predict a future rating is considered a *target user*. In order to identify a clique for a target user, we define two parameters: C_{min}, the *correlation threshold* and S_{min}, the *size threshold*. The parameter C_{min} defines the minimum correlation required for a user to become a member of the clique of a target user. Pearson's correlation

Feature	Type of Encoding	No. of Input Units
Category	1-of-N unary	25
MPAA Ratings	1-of-N unary	6
Maltin Ratings	Real value between 0.0 and 1.0	1
Academy Award	Real value between 0.0 and 1.0	1
Length (minutes)	Real value normalized by the mean	1
Origin	Real value between 0.0 and 1.0	1

Table 1: Encoding used for features

coefficient is computed between users using vectors of ratings for the movies that both users have seen and rated. Thus, the parameter C_{min} imposes a minimum limit on similarity of ratings required for a user to be a member of the clique of a target user. The second parameter, S_{min}, defines a lower limit on the number of movies that a user must have seen and rated in common in order to be a member of the clique. The parameter S_{min} is used to restrict users who have not seen the same movies that the target user has seen from entering the clique. Thus, a clique is defined for a target user by specifying suitable values for C_{min} and S_{min}.

Thus, the movie recommendation algorithm for the clique-based approach involves the following steps.

1. Initialize parameters C_{min} and S_{min}.
2. Identify a clique for the target user.
3. Estimate ratings for the movie that the target user wants to see by considering the ratings by the members of the clique who have already rated that movie.

In our preliminary implementation, we use a constant value of 10 for S_{min} and a positive variable value for C_{min} such that the number of users in a clique is held constant at 40. And in step 3 we use a simple arithmetic mean of the ratings of the members of the clique. Thus, the average ratings of the clique becomes the predicted ratings for the target user. The rationale behind computing the average is that a target user's taste may not be very different from that of his/her clique.

2.3 Data Collection

The feature database for the experiment was populated from the Microsoft Cinemania[2] CD-ROM for 1548 movies in Will Hill's survey. The features

[2]Cinemania is a registered trademark of Microsoft Inc.

collected include MPAA ratings, Maltin ratings, Category, Academy award, Length of the movie in minutes, the country of Origin, Director, leading Actor/Actresses, and a short review. As pointed out earlier, only the first six features are used in this study. The test-case users are 242 Internet subscribers who volunteered to rate the movies that he/she had seen. Each movie was rated on a scale of 0 to 10, with 0 being the worst, and 10 the best. The number of movies rated by the individual user varied from 0 to a maximum of 456 with the average being 177. For the purpose of evaluation, we selected 10 users who have rated approximately 350 or more movies as our *target users*. For convenience we label them as S3, S21, S39, S41, S46, S77, S111, S124, S145, and S178.

2.4 Results

The experiment for this study was conducted as follows: For each target user, we split the data set into a *training set* and a *test set*. The training set was used to build the model while the test set was used to validate the model. (In the clique-based approach, the training set was used to identify the clique for the target user.) The training set had 90% of the ratings and the test set had the remaining 10% of the data. In order to make a fair comparison, we cross-validated our model by splitting the data into 10 different mutually exclusive training and test sets. Thus, for each target user there are 10 experiments.

Target User Ids	No. of Movies Rated	Correlation Vs.		
		Maltin Ratings	Neural Net Ratings	Clique Averages
S3	364	0.23	0.29	0.51
S21	397	0.10	0.28	0.43
S39	436	0.43	0.45	0.67
S41	354	0.40	0.53	0.52
S46	456	0.38	0.48	0.57
S77	348	0.13	0.27	0.50
S111	374	0.58	0.58	0.64
S124	420	0.43	0.48	0.65
S145	367	0.41	0.39	0.64
S178	402	0.48	0.48	0.72

Table 2: A summary of ratings by different approaches.

Table 2 shows the results of our study. The results are presented in terms of the correlation coefficient between the actual ratings by the target users and the ratings by different movie recommendation approaches. For the neural network approach, the ratings represent the average over 25 experiments for each training set. The third column in Table 2 represents the correlation between target users and Maltin ratings. The fourth column represents the correlation between neural network predictions and actual ratings. The fifth column represents the correlation between ratings by the clique-based approach and target users. A higher correlation in Table 2 implies that the predicted ratings are close to the actual ratings. A comparison of results in column three and four suggests that the feature-based neural network ratings are better than Maltin (a movie rating expert) ratings for the target users S3, S21, S39, S41, S46, S77, and S124. For the remaining three users, Maltin ratings seems to be marginally better than the neural network ratings. The average difference between actual ratings and the neural network ratings for the target users is around 1.2 (on a scale of 10) with a standard deviation of 0.95. Thus, the neural network based movie recommendation approach may be used as an alternative to Maltin ratings. A comparison of the fifth column against other columns suggests that the clique-based approach is better than other two approaches for all target users except S41. Thus, for some users, the clique-baseed approach is a better movie recommendation system than our preliminary neural network implementation.

3 Conclusion

We developed a feature-based neural network rating system and compared against an expert and a clique-based approach. Our preliminary results, based only on a few important features, suggest that the feature-based neural network approach can be used as an alternative to an expert. The clique-based approach, on the other hand, may be used if there is a clique for the target user. The clique-based approach cannot be used for new movies and for users without a clique. So, the feature-based approach may provide an advantage if a new movie (or product) needs to be selectively targeted for the customers.

Acknowledgement

We are indebted to Will Hill of Bellcore for allowing us to use his movie list and ratings by Internet users. We are also thankful to Thomas Judd of Bellcore for populating the database with features from the Microsoft Cinemania CD-ROM.

References

[1] Dent, L., Boticario, J., McDermott, J., Mitchell, T. M., and Zabowski, D., "A Personal Learning Apprentice", In *Proc. of the National Conference on Artificial Intelligence (AAAI-92)*, August, 1992.

[2] Dumais, S. T., Furnas, G. W. and Landauer, T. K., "Using Latent Semantic Analysis to Improve Access to Textual Information", *Proc. of the 1988 ACM SIGCHI Conference*, pp. 281-295.

[3] Gifford, W. and Turock, D. "The Electronic Receptionist A Knowledge Based Approach to Personal Communications", *Proc. of the 1992 Computer Human Interactions Conference*, Monterey, CA, Sep. 1992.

[4] Harman, D. and Candela, G., "Retrieving Records from a Gigabyte of Text on a Minicomputer Using Statistical Ranking", *Journal of the ASIS*, Vol. 41, pp. 581-589, 1990.

[5] Obraczka, K., Danzig, P. and Li, S., "Internet Resource Discovery Services", *IEEE Computer*, Vol. 26, No. 9, pp. 8-22, Sep. 1993.

[6] Loeb, S., "Architecting Personalized Delivery of Multimedia Information", *Communications of the ACM*, Vol. 35, No. 12, pp. 39-48, Dec. 1992.

Text-independent speaker recognition using radial basis functions[1]

Steven E. Fredrickson and Lionel Tarassenko

Department of Engineering Science,
Oxford University, Oxford, England
sef@robots.oxford.ac.uk
lionel@robots.oxford.ac.uk

Abstract

Radial basis function (RBF) neural networks are applied to the telecommunications problem of identifying speakers from their voices. Both multiple-speaker networks and collections of single-speaker networks are investigated. Simulations conducted using the NTIMIT database indicate that RBF networks outperform multilayer perceptron networks for both methods of speaker grouping, and single-speaker networks offer advantages over multiple-speaker networks.

1 Introduction

'Biometric' recognition of telecommunication users can be beneficial to both service providers and their customers. For example, automatic speaker recognition can facilitate access to user-specific services and enhance system security. Unfortunately, there are relatively few current commercial applications of speaker recognition technology because speaker recognition to a high degree of accuracy remains a challenging problem. In this paper, we present recent findings in the application of neural networks to the task of speaker identification.

[1]This material is based upon work supported under a National Science Foundation Graduate Research Fellowship. Any opinions, findings, conclusions or recommendations expressed are those of the authors and do not necessarily reflect the views of the National Science Foundation.

2 Neural network theory

2.1 Multilayer perceptrons

A multilayer perceptron (MLP) neural network [4] has two (or more) layers of connection weights. It has been shown that an MLP with a single layer of hidden units is capable of universal approximation [1]. With such an architecture, the first layer maps an N_I dimensional input vector to N_J hidden layer 'neurons,' $(R^{N_I} \rightarrow R^{N_J})$, and the second layer maps the activations of the hidden neurons to N_K output neurons, $(R^{N_J} \rightarrow R^{N_K})$. For speaker identification, the input feature vector typically consists of cepstral coefficients and the output nodes employ a one-out-of-N_K encoding in which each speaker to be recognised is assigned an output class.

The activation V_k of a neuron k is a function of the weighted sum of its inputs:

$$V_k = f(\sum_j w_{jk} V_j) \tag{1}$$

where w_{jk} is the synaptic weight of the jth input to neuron k, V_j is the jth input, and $f()$ is the (sigmoid) activation function. Learning with a multilayer perceptron network consists of minimising the mean-square output error, usually with the error backpropagation algorithm [8], as successive input-output pairs are presented to the network.

2.2 Radial basis functions

A radial basis function (RBF) neural network [4] also has two layers. The architecture of an RBF is shown in figure 1. The first layer maps an N_I dimensional input vector to N_C hidden layer nodes or 'centres,' $(R^{N_I} \rightarrow R^{N_C})$, and the second layer maps the activations of these centres to N_K output nodes, $(R^{N_C} \rightarrow R^{N_K})$. As with an MLP, network outputs employ a one-out-of-N_K speaker encoding.

The basis functions are usually unnormalised Gaussian functions specified by centre locations and associated covariance matrices. The activation of an RBF centre is a function of its proximity in input space to the current input pattern.

$$\Phi_j(\vec{x}) = exp\left\{-\frac{1}{2}(\vec{x} - \vec{c}_j)^T \Sigma_j^{-1} (\vec{x} - \vec{c}_j)\right\} \tag{2}$$

where $\Phi_j(\vec{x})$ is the response of the jth RBF centre, $\vec{x}$ is the N_I dimensional input pattern, $\vec{c}_j$ is the jth N_I dimensional centre, and Σ_j is its covariance matrix. RBF centre positions and covariance matrices can be determined by unsupervised learning techniques, such as K-means clustering followed by the use of a width heuristic [6], in which case $\Sigma_j = \sigma_j I$.

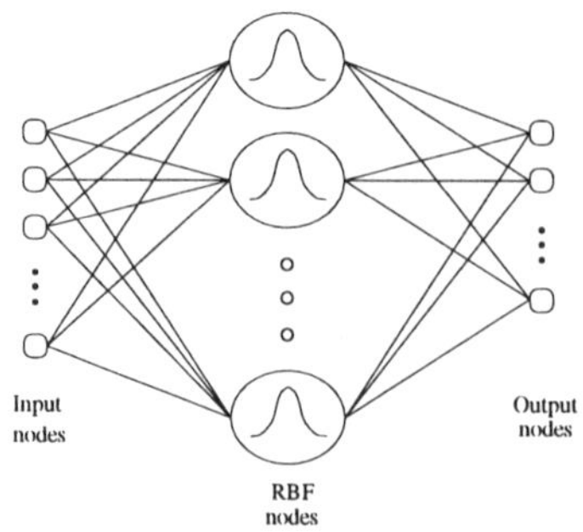

Figure 1: RBF Network Structure

The second layer of the network is a linear mapping from the RBF activations to the output nodes:

$$y_k = w_{0k} + \sum_{j=1}^{N_C} w_{jk}\Phi_j(\vec{x}) \tag{3}$$

where y_k is the kth output node, w_{jk} is the weight from the jth RBF centre to the kth output, and w_{0k} is the bias of the kth output. Output weights are computed via supervised learning techniques, such as least mean-square (LMS) gradient descent or matrix pseudo-inversion.

2.3 Class conditional probabilities

If MLP or RBF networks employ one-out-of-N_K output encoding with a squared error cost function (as specified above), network outputs are estimates of Bayesian *a posteriori* class probabilities [7]. Under such conditions, network outputs from different inputs can be combined probabilistically. For example, R speech vectors from the same speaker can be used collectively to make a speaker classification decision. Assuming the vectors form an independent sample set, the combined output probabilities can be computed by multiplying individual outputs:

$$[P_k]_{comb} = \prod_{m=1}^{R} [y_k]_m \tag{4}$$

where $[P_k]_{comb}$ is the kth combined output probability, and $[y_k]_m$ is the kth output from the mth observation.[2] The index $k \in \{1, 2, \ldots, N_K\}$ of the maximum combined probability, $\max\{[P_k]_{comb}\}$, is the index of the speaker. We have shown that the use of multiple independent observations reduces classification error rates significantly on a text-dependent speaker recognition problem [3].

[2]Network outputs should be thresholded or passed through a softmax function to ensure nonnegative probabilities.

3 Experimental procedure

3.1 Description of speech data

Simulations described in this paper were conducted using the NTIMIT [5] speech database. The NTIMIT database was produced by transmitting the DARPA TIMIT database across actual telephone channels. The original TIMIT utterances were recorded in a quiet recording booth using a high quality microphone. TIMIT speech was sampled at 20 kHz, and (following downsampling) stored as 16 kHz data. Due to network transmission of TIMIT, however, NIMIT utterances exhibit an authentic telephone bandwidth of 300–3400 Hz.

The full database contains 10 utterances (sentences) from each of 630 speakers. The 630 speakers are divided into 8 dialect regions of American English. For this paper we confine our testing to subsets of the 38 speakers (24 male, 14 female) in the training group of the New England (region 1) dialect. We use 7 sentences from each speaker for training and 3 for testing. Longer sentences are truncated so that the total duration of speech is roughly the same for all speakers. The resulting average utterance length is less than 2.5 seconds.

3.2 Data preprocessing

Converting from NTIMIT speech data to input feature vectors requires several steps. First, speech samples are segmented into $31ms$ frames every $15.5ms$. Next, an energy-based voice-activity detector is used to discard silent frames, and the remaining speech frames are pre-emphasised using a non-recursive filter with transfer function $H(z) = 1 - 0.95z^{-1}$. After applying a Hamming window, each frame is then processed with the fast-Fourier transform, and log energy values from 20 evenly spaced mel-frequency filter banks are used to generate mel-frequency cepstral coefficients [2]. We use 12 cepstral coefficients from each frame to form a feature vector labeled by speaker index, and we normalise each component to have zero mean and unit variance based on training set statistics.

3.3 Single-speaker networks and data sets

The probability of making a correct speaker identification with a trained neural network is maximised by using multiple input vectors as outlined in section 2.3. However, as the the number of speakers becomes large, the performance of a single '1-out-of-N' MLP or RBF classifier degrades. A common approach to handling large speaker sets is to replace one N-speaker network with N single-speaker networks. Single-speaker networks have two outputs each: One output estimates the probability $P(S_i|\vec{x})$ that the current input $\vec{x}$ corresponds to the associated speaker S_i, and the other estimates the probability that the input corresponds to any other speaker. If multiple input patterns are available, output probabilities can be combined probabilistically in the same way as with 1-out of-N networks. After computing output probabilities (single or combined), the final speaker classification is determined by the maximum of the 'speaker' probabilities as illustrated in figure 2. Note that the 'antispeaker' output probabilities are ignored.

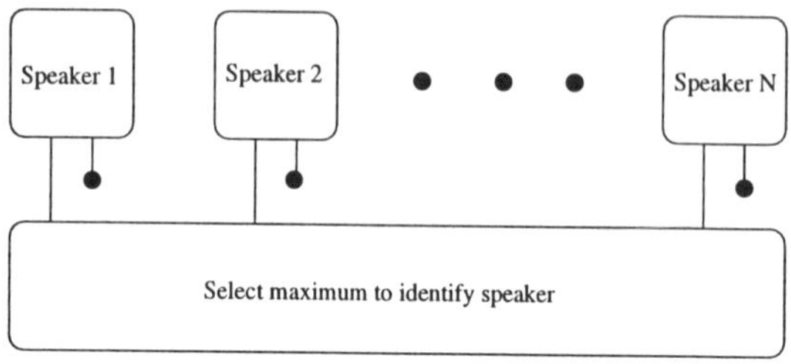

Figure 2: Classification using single-speaker networks

To use N single-speaker networks effectively, balanced training sets are needed so that meaningful comparisons can be made between the posterior probabilities from the N networks. Assuming there are $K = N \times M$ speech pattern vectors (M from each speaker), a *balanced* data set for a given speaker should have $2 \times M$ elements consisting of M 'speaker' patterns and M 'antispeaker' patterns. Our simulations compare three methods of generating M antispeaker vectors from the $K - M$ antispeaker candidates:

1. Select M patterns at random from the $K - M$ antispeaker candidates
2. Cluster the $K - M$ antispeaker vectors into M clusters and use the M centroids
3. Select the M antispeaker vectors nearest in input space to the M speaker vectors

4 Experimental results

The speaker identification experiments described in this section are 'closed-set' tests in which all test speakers are included in the training database. For the first two experiments, two 'conventional' classifications techniques are included for comparison with MLP and RBF networks. The least-mean-square (LMS) classifier classifies test patterns based on optimal (in a least-squares sense) *linear* decision boundaries computed from training data. The K-nearest neighbour (KNN) classifier classifies test speakers based on a 'vote' of the K nearest training patterns. The proportion of the votes is treated probabilistically for the purpose of combining decisions from multiple input patterns.

4.1 10 speaker test

The first experiment was conducted using 10 male speakers from the New England dialect group. The training database consisted of 672 patterns from each of the NTIMIT speakers. The simulation results are shown in table 1.

Table 1: Speaker identification results for 10 male speakers

Method	Test error rate (1 sentence)	Test error rate (3 sentences)
KNN (K = 9)	30.0%	0.0%
RBF (336 centres, K-means)	43.0%	20.0%
MLP (10 hidden units)	53.3%	30.0%
LMS	50.0%	50.0%

4.2 20 speaker test

For the second experiment, 10 additional male speakers from the New England dialect group were added to the database used in the first experiment, and the simulations were attempted again. Because the size of the RBF network exceeded the memory capacity of the computer workstation, single-speaker RBF networks were investigated instead. The RBF representation is 'localised' in input space, so the sub-division of the problem into local networks is a natural extension. The hyperplane decision regions created by MLP networks are less suited to the approach of single-speaker networks, but simulations are included for comparison purposes. The antispeaker patterns used by the single-speaker networks were selected at random. The results are shown in table 2.

Table 2: Speaker identification results for 20 male speakers

Method	Test error rate (1 sentence)	Test error rate (3 sentences)
KNN (K = 9)	38.3%	20.0%
RBF (336 centres, K-means)	—	—
MLP (15 hidden units)	76.7%	70.0%
LMS	73.3%	65.6%
20 Single-speaker RBFs (134 centres per net, K-means)	46.7%	40.0%
20 Single-speaker MLPs (10 hidden units)	68.3%	60.0%

4.3 Antispeaker generation test

The third experiment evaluated the three methods of generating 'antispeaker' patterns for the single-speaker networks. Except for varying the method of generating antispeakers, the database used was identical to that of the second experiment. The results for RBF networks with 134 centres per network are shown in table 3.

Table 3: Antispeaker generation comparison (20 male speakers)

Antispeaker generation method	Test error rate (1 sentence)	Test error rate (3 sentences)
Centroid	41.7%	35.0%
Random	46.7%	40.0%
Nearest	76.5%	69.4%

4.4 38 speaker test

The final experiment was conducted using all 38 speakers from the New England dialect group. Single-speaker RBF networks (with 134 centres per network, K-means optimisation, and centroid antispeakers) were compared against the K-nearest neighbour classifier. The simulation results are shown in table 4.

Table 4: Speaker identification results for 38 speakers

Method	Test error rate (1 sentence)	Test error rate (3 sentences)
KNN (K = 9)	44.7%	23.7%
38 Single-speaker RBFs	43.0%	34.2%

5 Conclusions

The 10 and 20 speaker experiments indicate that radial basis functions outperform multilayer perceptrons for both multiple-speaker and single-speaker configurations. The hypothesis that RBF kernels partition the feature space better than MLP hyperplane boundaries is supported by the fact that the K-nearest neighbour classifier provides the best overall results, whereas the linear classifier produces poor results.

The antispeaker generation experiment demonstrates that computing centroids is better than selecting antispeakers randomly or choosing nearest neighbours. Attempting to develop detailed boundaries between speaker and antispeakers by using nearest-neighbours proved entirely unsuccessful.

The 38 speaker experiment shows that relative to the K-nearest neighbour classifier, RBF performance improves as the number of speakers increases. (The 1 sentence results, which are statistically more reliable than the 3 sentence results, reveal a converging trend. With a larger, more diversified speaker set, the RBF performance may surpass that of the K-nearest neighbour.) Also, a key advantage of single-speaker RBF networks involves implementation. Unlike multiple-speaker networks that must be retrained every time a new speaker is added, single-speaker networks can be trained individually in a short period of time. K-nearest neighbour, in contrast, requires every training pattern from every speaker to classify each test pattern.

Overall, these preliminary experiments indicate that single-speaker networks are a promising technique, but absolute error rates are unacceptably high. The combination of speakers from the same dialect group and recordings over various telephone channels yields a difficult database. Further investigation is needed to assess whether compensating improvements in speech processing and/or the network training are possible.

References

[1] G. Cybenko, "Approximation by superposition of a sigmoidal function," *Mathematics of Control, Signals and Systems*, vol. 2, no. 4, pp. 303–314, 1989.

[2] S. B. Davis and P. Mermelstein, "Comparison of parametric representations for monosyllabic word recognition in continuously spoken sentences," *IEEE Transactions on Acoustics, Speech, and Signal Processing*, vol. ASSP-28, no. 4, pp. 357–366, August 1980.

[3] S. E. Fredrickson and L. Tarassenko, "Radial basis functions for speaker identification," in *Proceedings of the ESCA Workshop on Automatic Speaker Recognition, Identification, and Verification*, (Martigny, Switzerland), pp. 107–110, April 1994.

[4] D. R. Hush and B. G. Horne, "Progress in supervised neural networks: What's new since Lippmann?," *IEEE Signal Processing Magazine*, vol. 10, no. 1, pp. 8–39, August 1993.

[5] C. Jankowski, A. Kalyanswamy, S. Basson, and J. Spitz, "NTIMIT: A phonetically balanced, continuous speech, telephone bandwidth speech database," in *Proceedings of ICASSP'90*, pp. 109–112, April 1990.

[6] J. Moody and C. J. Darken, "Fast learning in networks of locally-tuned processing units," *Neural Computation*, vol. 1, pp. 281–294, 1989.

[7] M. D. Richard and R. P. Lippmann, "Neural network classifiers estimate Bayesian *a posteriori* probabilities," *Neural Computation*, vol. 3, no. 4, pp. 461–483, 1991.

[8] D. E. Rumelhart, G. E. Hinton, and R. J. Williams, "Learning representations by backpropagating errors," *Nature*, vol. 323, pp. 533–536, 1986.

Building Comprehensive AI and the Task of Speech Recognition

Nikola K Kasabov

Department of Information Science, University of Otago
P.O.Box 56, Dunedin, New Zealand, email: nkasabov@otago.ac.nz

Abstract

A new approach to building speech recognition systems has been suggested and experimented with. The approach is based on building *comprehensive AI* systems, i.e. AI systems which are built by using different standard and non-standard AI paradigms. It makes use of neural networks for low level pattern matching, of fuzzy systems for representing ambiguous and uncertain knowledge at a higher level and of symbolic production rule-based systems for high level control and decision making. Extracting fuzzy rules from speech data is considered to be important part of the process of building speech recognition systems. This approach has been implemented in a hybrid environment called FuzzyCOPE, which has been demonstrated on a few speech recognition tasks.

1. What is Comprehensive AI?

The area of integrating ordinary - symbolic artificial intelligence, neural networks, fuzzy systems and other AI techniques (genetic algorithms, evolutionary programming, etc) is called *"comprehensive artificial intelligence"* or *comprehensive AI* and is one of the long term objectives of many AI laboratories [1]. Comprehensive artificial intelligence deals with hierarchical, multi-paradigm systems. Building comprehensive AI is possible by integrating different AI methods [4]. Among the methods used for knowledge engineering so far, are the symbolic AI methods, the fuzzy logic methods and the methods of neural networks [1,2,3,4]. In order to better understand why these paradigms should be mixed when solving problems, we must look to their strengths and limitations. Each of the paradigms can contribute to the ultimate solution of a complex AI problem, such as speech recognition and language understanding, as follows:

• *Symbolic AI systems* can contribute with: rigid theory for symbol manipulation; theorem proving; rigorous exact reasoning mechanisms including chain reasoning and different reasoning strategies; universal computational models (for example - the production systems; the logic programming systems).
• *Fuzzy systems* can contribute with: well-developed fuzzy logic theory; human-like reasoning mechanisms; using linguistic terms; accommodating common sense knowledge, ambiguous knowledge, imprecise but rational knowledge; universal approximation techniques; robustness, fault tolerance; low cost for development and maintenance; low computational cost.
• *Neural networks* can contribute with: learning from data; modelling empirical behaviour of humans; universal approximation techniques; good generalisation; methods for extracting knowledge from data; methods for data analysis; associative memories and pattern matching techniques; massive parallelism; robustness.

The three paradigms are applicable to solving the same AI problem. They may provide similar or equivalent solutions, i.e. they 'overlap'. We can also view these three groups of methods as complementary ones at least for three reasons:

• They facilitate dealing with different kinds of knowledge representation, different inference, different accuracy and fault tolerance, etc.
• Each of them may be superior to the other when solving a concrete sub-task of a problem; for example, the symbolic approach and the connectionist approach have demonstrated strengths in solving different tasks. The former has been successful in high-level cognitive tasks, whereas the latter is more successful in low-level perceptual and learning tasks.
• One method may be applied to improve the performance of another, graphically shown in figure 1, for example: fuzzy rules can be used to 'pre-wire' a neural network structure for faster training and better generalisation; fuzzy rules can be used for initialisation of connection weights; neural networks can be used for learning fuzzy rules; neural networks can be used to refine fuzzy rules and membership functions; genetic algorithms can be used to learn fuzzy rules from data and membership functions or to adjust the connection weights during training; neural networks and fuzzy systems can be used as parts in a symbolic reasoning machine.

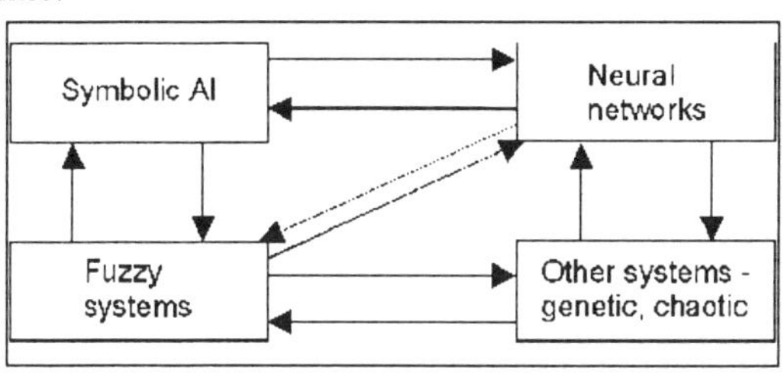

Figure 1. Different AI methods complement each other

2. The Model for Speech and Language Recognition

The model described here is based on the assumption that the whole process of speech recognition and language analysis is an integrated continuous process and the borders between the two are not well defined. Where does the speech recognition process end and where does the language analysis process really begin? Are they not two overlapping processes? This process can be viewed in two dimensions - time, and type of knowledge applied in order to recognise and understand a spoken sequence, where the speech recognition and language analysis phases overlap in both dimensions - figure 2. For example, correctly recognising a spoken phoneme may require understanding the context of the whole word, and also - some words pronounced before and after it. We adopt here the well accepted principle for continuous speech recognition, with unlimited vocabulary and speaker independence, which consists of the following steps: (i) identifying the sub-word units; (ii) grouping them into words; (iii) grouping the words into sentences. Different techniques can be used for the different phases of the whole task. The area of use of neural networks and fuzzy systems are shown in figure 2.

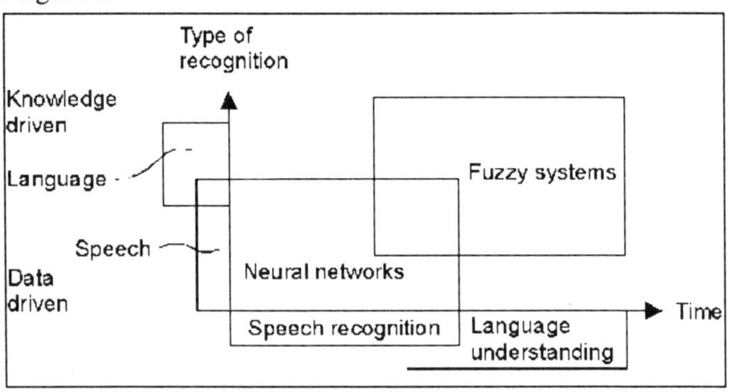

Figure 2. The sub-tasks of speech recognition and language understanding overlap over time and over the type of recognition

The speech and language knowledge is usually ambiguous and difficult to articulate and to process. It requires methods which reflect the ambiguity in the language. Higher level rule-based systems must have facilities to represent and process ambiguous, contradictory, incomplete, uncertain knowledge. Articulating rules for language processing, which are fuzzy by nature, is a difficult task for the language and the knowledge engineering experts. Such rules, used by humans, reflect their knowledge on the language which is not a simple 'flat' set of rules, but possibly - a hierarchical structure of many layers of such rules. "The more we learn and practice a language, the more layers we build in our brain, the more complex speech constructions we are able to recognise and comprehend" [6]. A comprehensive AI model for a speech recognition system is shown in figure 3. It comprises the following main modules:

• *Pre-processing module* - it does non-linear transformations over the raw speech signal. A module for fuzzifying the input signals could be used. Fuzzifying input signals into linguistically defined terms has the following aspects: it makes learning fuzzy rules, to be later on implemented at a higher level, easier; it is another non-linear transformation on the input speech signal which may improve the final recognition results.

• *A neural network module* for low level recognition. A neural networks can be trained by using a set of speech examples. The network is then used for a low level recognition and the results are passed to a rule-based higher level module. The module can be built as a single neural network, or a multi-network structure of many specialised neural networks for recognising patterns of sub-units, e.g. parts of phonemes or phonemes. Different types of networks can be used. This module is adaptable and trainable.

• *A higher level fuzzy rule-based system* for approximate reasoning over fuzzy rules for phoneme, words, sentence, concept and meaning recognition. The higher level is, in general, a multi-layer structure of fuzzy rules which represent different layers of speech and language knowledge. It can be a simple flat fuzzy rules system or quite a complicated system depending on the task.

• *Overall logic system for control* of the whole process and producing final results - this can be done by using symbolic AI system, for example - a production system (PS).

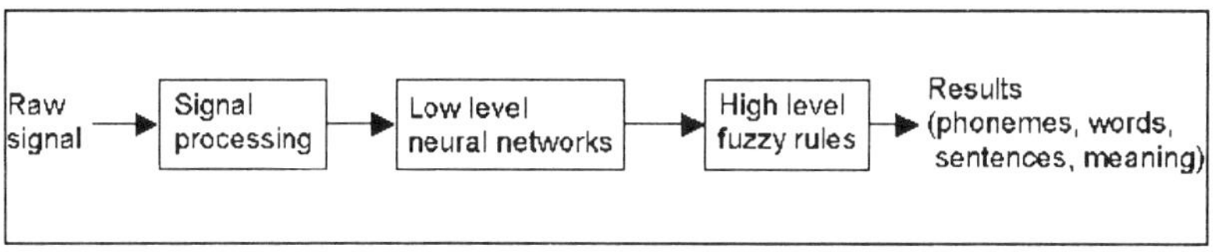

Figure 3. A comprehensive AI model for speech recognition

A module for extracting fuzzy rules from speech and language data (not shown in figure 3) can be an important part of the whole process of building a comprehensive AI system [3,4,5]. This module makes the environment adaptable, though it is not clear that complex speech and language recognition fuzzy rules can be captured in an automatic way to the same degree to which humans have learned them. The following principles can be applied when building a concrete speech recognition system based on the model of figure 3:

• The principle of *hierarchy,* i.e. the speech elements are recognised through their natural hierarchy of sub-units, units, strings, etc.

• The principle of a *delayed decision*, i.e. a lower level decision is temporarily accepted for further processing, but is subject to correction due to the higher level knowledge available and the interaction between the levels in the system.

• The principle of the *'fuzzy winner'*, i.e. at each level, not one, but many possible sub-units and units are kept for further processing and not just the one winning at that moment.

• The principle of *making use of both data and knowledge*. This principle means also applying both 'unconscious', 'blind', or 'stimulus- reaction' recognition, and 'conscious', rule-based or directed, recognition.

• The principle of *adaptability,* i.e. the system is adaptable to new variations of speech, new vocabularies, etc. The adaptability is achieved in three ways: additional training of the neural networks at the lower level with examples about new speech variations; adding or changing fuzzy rules in the higher level module for representing the new speech characteristics; learning fuzzy rules from new speech data and adding these rules to the already existing ones for a higher level processing.

• The principle of *biological, physiological and psychological plausibility*, i.e. at each level of recognition the system tries to imitate as much as possible that is known about the human's ear, brain and mind.

The higher level fuzzy processing, can be implemented by using: fuzzy inference methods, facilitated in different fuzzy engineering tools [1,3,4,5]; connectionist systems that process fuzzy rules, an example of such a system being NPS [6]. In order to represent in a computer system such a complex hierarchical model we need:

(1) an environment where we can represent such language structures and process them efficiently;

(2) methods and tools for learning fuzzy rules from huge speech and language data bases [3,7].

A software environment for building comprehensive AI systems is presented in the next section.

3. FuzzyCOPE

Hybrid fuzzy connectionist production systems environments are only one possible realisation of the methodology for building comprehensive AI systems [4,5]. They offer the following features:

• production rules and forward chaining reasoning;

• fuzzy rules and different methods of fuzzy reasoning, when fuzzy variables and fuzzy membership functions have to be defined in advance;

• different types of neural networks;

• different methods for rules extraction and methods for data analysis.

The idea of building such environments will be illustrated with a concrete environment called *FuzzyCOPE* [5]. FuzzyCOPE is a software environment which has five main modules - figure 4. Relevant modules are compatible from the point of view of linking output from one module as input to another. This is also shown on figure 4. The modules are explained below:

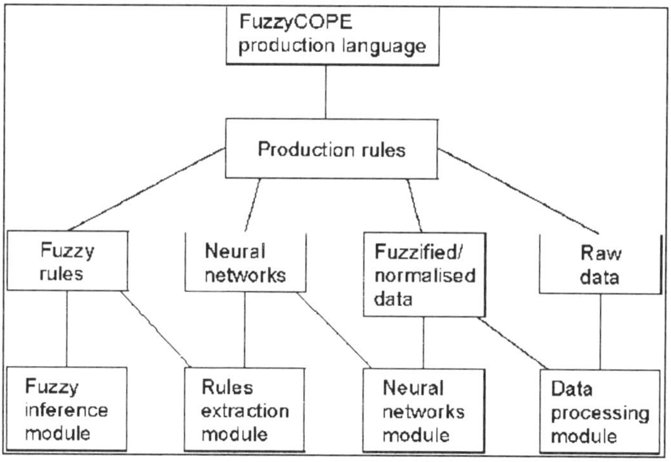

Figure 4. The architecture of FuzzyCOPE

• *FuzzyCOPE Production System Module* is an extension of FuzzyCLIPS (product of the NRC Canada) with new built-in functions to call fuzzy inference methods, to call rules extraction methods, to call neural networks and data processing methods, all of them being available for experimenting with in the corresponding modules of the environment. An user program can be written in FuzzyCOPE and compiled to executable code.

• *Fuzzy Inference Module* - offers different fuzzy inference methods for experimenting with, while tuning the fuzzy reasoning over a set fuzzy rules. The compositional inference methods, as well as the decompositional method for reasoning with fuzzy inputs and fuzzy outputs, are implemented [1,3,4,5]. This module processes a set of fuzzy rules either designed by the user or by the expert, or automatically extracted from raw or pre-processed data by the rules extraction module.

• *Neural Networks Simulation Module* - contains several neural network simulators which can be initialised, trained with raw or fuzzified data, and tested either as separate modules, or as a part of production rules written and executed at the higher level. The fuzzification of the raw data may be done in the data processing module. The output from this module can be used as input to the rules extraction module.

• *Data processing module* - does operations over raw data, like fuzzification, normalisation, speech data transformation. For example, a speech transformation function which transforms raw speech data into mel scale cepstrum coefficients is facilitated. The output of this module can be used as input to the neural network module for training or recall neural networks.

• *Rules Extraction Module* - includes several connectionist methods for rules extraction from raw or fuzzified data [7]. The neural network structures are initialised, trained and tested either in the neural network module, or directly in the production system at the higher level, where the neural network functions are called as actions in the production rules.

With the use of the above modules and functions, different ways of building a comprehensive AI system are possible.

4. Building Speech Recognition Systems in FuzzyCOPE

A small experimental system is presented here to illustrate the hybrid model from figure 2. The experiment is conducted on recognising eight speech units - seven phonemes and a silence: silence; /w/; /^/; /n/; /z/; /e/; /r/; /ou/. Eight neural networks are used, each for recognising segments of each of the units. A set of fuzzy rules is applied on each two consecutive segments recognised by the neural networks over the time scale, labelled by the label of the corresponding phoneme, in order to decide on the whole unit pronounced. The sequence of the recognised units is displayed and passed to the next unit which will create a string of continuously recognised phonemes. Here, details on the experimental system, are given.

• Speech data compilation, and pre-processing - digitized speech is recorded and a Fast Fourier Transformation (FFT) is performed over 11.6 ms segments of speech, with the use of a Hamming window and with overlapping between the segments of 50%. 256 point FFTs are calculated from the speech, sampled at 22,050 Hz. Mel-scaled cepstrum coefficients are then calculated from the FFTs. Mel-scale transformation is considered to be plausible in a sense with the way the inner ear works. 26 coefficients are obtained from each 256 point FFT for this experiment.

• Training neural networks at the low-level - to train the eight neural networks, examples of the allophonic realizations of the phonemes were extracted from the speech of selected group of 10 male and 10 female speakers. The extracted portions were taken from the stable parts of the phonemic realisation, therefore each of the segments taken carries substantial information about the pronounced phoneme. This meant the portions were very short in duration, but informative. Mel-scale cepstrum coefficients were calculated for these portions and used to train the neural networks.

• A fuzzy rule-based system for higher-level processing was developed. *General and specific fuzzy rules* are articulated for deciding on the spoken phoneme at a time period based on the segments recognised by the neural networks at shorter time moments t and (t+1). The conclusion in a rule is called "output". There are 58 (= 7 x 8 + 2) fuzzy rules articulated. The fuzzy inference is realised by using a min-correlation method [3,4]. The system was tested on newly spoken words which contain the above phonemes. The system performed well when recognising phonemes pronounced by 10 new speakers (5 male and 5 female). The system showed robustness which is an important characteristic when building speaker independent speech recognition systems.

5. Conclusions

A new approach to building speech recognition systems as comprehensive AI has been experimented and illustrated in the paper. It makes use of neural networks for low level pattern matching, a fuzzy rule-based system to represent ambiguous and uncertain knowledge at a higher level of the speech and language recognition and understanding task, and a production system for the overall control and final decision making. A hybrid environment for building comprehensive AI systems called FuzzyCOPE is presented and illustrated on a small scale phoneme recognition task.

Acknowledgement

The experiments presented in the last section of this paper was facilitated by Dr. Catherine Watson, Stephen Sinclair and Richard Kilgour from the Department of Information Science, at the University of Otago. The practical implementation of the hybrid system FuzzyCOPE was done mainly by Jay Garden, Paul Jones, Tim Albertson and Stephen Sinclair.

References

[1] T. Terano, "Trends in Fuzzy Engineering", *LIFE Technical News*, Vol.3, No.1 LIFE, Japan, 1992.

[2] D. Morgan and C. Scofield, "Neural Networks and Speech Processing", Kluwer Academic Publishers, 1991.

[3] B. Kosko, "Neural Networks and Fuzzy Systems: A Dynamical Approach to Machine Intelligence", Prentice Hall, 1992.

[4] N. Kasabov, "Neural Networks, Fuzzy Systems and Knowledge Engineering", MIT Press, in print, 1995.

[5] N. Kasabov, "Building Hybrid Fuzzy Connectionist Rule-based Systems. *Automation and Soft Computing,* in print, 1995.

[6] N. Kasabov "Towards Using Hybrid Connectionist Fuzzy Production Systems for Speech Recognition". *Lecture Notes in Computer Science/Artificial Intelligence*, in print, 1995.

[7] N. Kasabov, "Learning Fuzzy Production Rules for Approximate Reasoning with Connectionist Production Systems", in: S.Gielen and B. Kappen (Eds) *Proceedings of the International Conference on Artificial Neural Networks ICANN'93*, Amsterdam, 13-16 September), pp.337-345, Springer Verlag, 1993.

Real-world speech recognition with neural networks

Etienne Barnard, Ronald Cole, Mark Fanty and Pieter Vermeulen

Center for Spoken Language Understanding
Oregon Graduate Institute
P.O.Box 91000, Portland OR 97291-1000
barnard,cole,fanty,pieter@cse.ogi.edu

Abstract

We describe a system based on neural networks that is designed to recognize speech transmitted through the telephone network. Context-dependent phonetic modeling is studied as a method of improving recognition accuracy. Preliminary results from a realistic test of the system (a field trial for the U S Census bureau) are reported.

1 State of the art

Current speech-recognition algorithms are still substantially inferior to human capabilities, in almost every important way. Humans employ semantic and syntactic information to understand speech despite an astounding variety of degradations, they perform noise suppression, channel normalization, and speaker separation to gain considerable robustness, and they are excellent at recognizing short stretches of speech even when contextual information or sources of normalization are absent.

Despite lagging way behind in all these respects, automatic speech recognition is establishing several important niches for itself. It is becoming clear that human-like performance is often not necessary: if the user's goal is to obtain information over the telephone, or communicate data to a computer when keyboard use is impractical, a much lower level of performance will suffice. While it remains a fascinating challenge to improve automatic speech recognition, current algorithms can be adapted for use in a variety of circumstances.

In this paper we describe some of the steps we have taken to make practical speech-recognition systems. Our approach is based on neural-network classifiers which operate at a fixed frame rate; these classifiers estimate the (posterior) probability that each phoneme occurred at each time frame. A Viterbi search is then performed to match these phoneme probabilities to specified word models, and various optional post-processing steps are performed to decide which word actually was spoken. This architecture is reviewed in Section 2. In Section 3 an example of such an application is described, namely the automation of the collection of the U S Census, and Section 4 contains initial evaluation results with this system.

2 Architecture

Conceptually, our speech-recognition system consists of a number of stages as shown in Fig. 1. The incoming speech (in our case, the mu-law encoded digital samples, with sampling rate 8kHz, which are transmitted over a telephone line) is first converted to a representation which is more suitable for recognition. We use Perceptual Linear Predictive (PLP) analysis[4]; this is a modification of linear predictive coding which takes into account some of the properties of human hearing. We compute seventh-order PLP coefficients in a 10 msec window, and advance the window 6 msec at a time. We thus obtain seven PLP coefficients (and the energy) every 6 msec – we refer to this as a "frame" of speech.

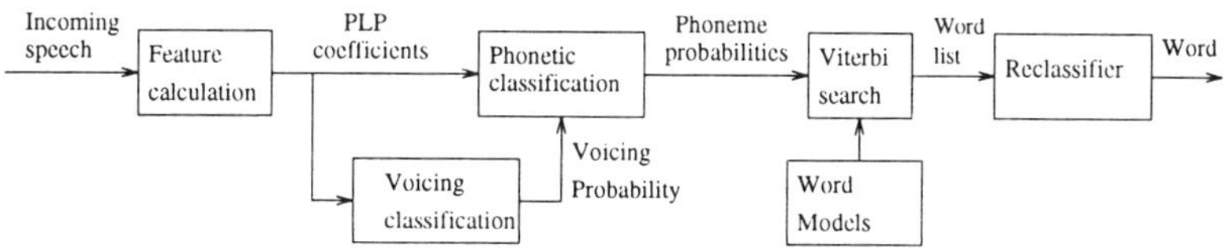

Figure 1: General architecture

These PLP coefficients are used in two ways. Firstly, they are used to classify each frame as being voiced or unvoiced. The distinction between voiced and unvoiced speech is useful for phoneme recognition, and we have trained a neural-network classifier to estimate the probability that each frame is voiced based on the values of the PLP coefficients in that frame, and also the PLP values in several frames surrounding the frame of interest. The computed PLP coefficients are combined with the voicing probability and used as input to a further neural network which performs phonetic classification. This network thus has 70 inputs: ten feature values (seven PLP coefficients + energy + two voicing values) at each of the same seven sampling offsets used for voicing classification. We represent all English words in terms of 39 phonemes, and this classifier has 39 outputs corresponding to the probability that each of these phonemes is present.

The role of the phoneme recognizer can be described as mapping from a series of acoustic descriptors to a series of phoneme occurrences. The system described above attempts to do so directly: the recognizer tries to determine which phoneme was intended at each time frame of the input signal, based on the acoustic evidence in a window surrounding that time frame. Unfortunately, this places unreasonable requirements on the phoneme classifier, given the substantial variability of different acoustic realizations corresponding to the same phoneme. We have found that large improvements in performance can be obtained by using context-dependent modeling[2]. In our initial system we used the same architecture as Cohen et al.

Each phoneme is modeled in three parts: left, middle and right. The left and right part of each phoneme is further modeled by six different output nodes depending on the class of phoneme which is adjacent. (The six broad classes we use are "silence", "fricative", "stop", vowel", "semivowel", and "nasal"). The middle part of each phoneme is modeled by a single context-independent output node. Thus, whereas a phoneme is represented by a single output node in our baseline system, it is now represented by 13 output nodes. Although this increases the recognition-time computational complexity of the neural network by more than a order of magnitude, we have found the performance gains to be sufficient to warrant this cost – on several problems we have found a decrease in error rate ranging between 20% and 50%. Since a fast DSP can still do the computation in real time, context-dependent networks are used throughout our current real-time system.

As a result of phonetic classification we have an estimate of the probability that each phoneme is present at each 6 msec time frame.

The Viterbi search uses this matrix of classification scores to decide which word (or sequence of words) was spoken. Thus, each word is expressed in terms of the sequence of phonemes that is expected when that word is uttered. To compute the likelihood that a word was spoken, one assumes that the acoustic vectors in different time frames are independent, so that the likelihood of a phoneme occurring in a sequence of time frames is the product of the likelihoods in each of the individual time frames. The Viterbi search thus computes likelihoods for each of the possible target words, finding the optimal alignment of the word in time in the process.

Our reclassifier processes these likelihoods to make a final decision on word identity. The values of p_i are only known for those words actually in the vocabulary. Simply choosing the largest as the recognized word will not detect out-of-vocabulary utterances. This typically involves a decision on whether the largest p_i is too small compared to some threshold – indicating that even that word is unlikely to have occurred. The threshold needs to be set dynamically, since effects like speaker variations, background noise, and channel distortions affect the overall scale of the estimated probabilities. If fine word distinctions are required, a further processing step which recalculates word likelihoods based on the hypothesized boundaries found during the Viterbi search.

3 Collecting U S Census data

We have conducted a study to determine the feasibility of using an automated spoken questionnaire to collect information for the Year 2000 Census in the United States of America. The goal of the study was to develop and evaluate a telephone questionnaire that automatically captures and recognizes the following information: (1) full name, (2) sex, (3) birth date, (4) marital status (now married, widowed, divorced, separated, never married—choose one), (5) Hispanic origin (yes or no); if Hispanic: Mexican, Mexican-American, Chicano, Puerto Rican, Cuban or other (specify), (6) race: White, Black or Negro, American Indian (specify tribe), Eskimo, Aleut, Chinese, Japanese, Filipino, Asian Indian, Hawaiian, Samoan, Korean, Guamanian, Vietnamese or other (specify).

After preliminary rounds of data collection to refine the selection and wording of the system prompts, a large, regionally diverse data collection effort resulted in approximately 4000 calls. We then used a semi-automatic procedure that involved hand transcription at the word level of about a quarter of the corpus and automatic generation of "forced" phonetic alignment of these transcriptions using a classifier trained on a different task. A new classifier was then trained on the automatically aligned census data and used to realign it. The process was repeated a couple of times until performance asymptoted.

Word Spotting. While the vast majority of the responses in the 4000-call corpus are succinct, there are enough responses with "extraneous" speech that it needed to be dealt with. In addition, there was a great deal of background noise in many of the calls.

Because the majority of responses are succinct, our initial system has a simple word spotting approach in which all words and sounds not in the target set match a single garbage model. We use the approach described in [1], in which the output score for the garbage word is computed as the median value of the top N phoneme scores for each frame, where N varies with the task and is set empirically. The grammar for the responses (except numbers, which are more complex) is [garbage/silence] [target word] [garbage/silence].

Confidence. The recognizer always returns a target response. This means that if we are trying to recognize "male" or "female" and the caller just coughs or asks "Do I press a button or what?" then either "male" or "female" will be recognized. We assign a confidence score to these matches so they can be rejected. ** sentence deleted

An ideal confidence score will be low for all incorrect responses and high for all correct responses. So far, the best confidence score we have found is to take the difference of the average frame score for the top-scoring target word and the average frame score for the second highest scoring target word.

Dialogue. Our efforts have focused on designing a system for cooperative users. Even with cooperative users, unexpected dialogue situations may arise. Some responses will inevitably fall outside the preferred response set. The ability to cope with unexpected dialogue situations is essential to achieving good system robustness. The current system includes three aspects to dialogue repair:

1. Detecting breakdowns. The system is capable of identifying certain difficulties when they arise (e.g., low confidence for a recognized response).

2. Recovering from breakdowns. Repair strategies currently supported by the system include repeating the question if confidence is low; confirming the response if confidence is medium; and taking the best guess and continuing with the next question if the system fails to recognize a response on a second attempt.

3. Review, followed by confirmation or identification of errors. The dialogue concludes with a summary of the information recognized by the system. The caller is asked if the system's information is correct; if not, the caller is asked to identify the incorrect categories. A human operator may be able to resolve errors either during or following the call, as described in the next section.

Operator Monitoring The system includes a graphical user interface that enables a human operator to monitor active calls or review any call at a later time. The caller's responses can be played at any time. Each recognition response is displayed as text, color coded to indicate the system's confidence. The operator can change the record to correct recognition errors.

At the end of each call, the system reviews each recognized response using pre-recorded speech; the caller's first and last name are spelled. After the review, the caller is asked if the information is correct. If not, the system signals the operator that the call has an error and asks the caller which of the questions have incorrect answers. The operator can then listen to this information, and correct errors in the appropriate fields of the record without reviewing the whole call.

4 Evaluation

Our analyses of the protocols focused on three questions[3]: (1) What percentage of callers completed the protocol? (2) What percentage of responses contained the desired information? (3) What percentage of responses could be recognized with a small vocabulary key-word spotter?

Completion. We eliminated spurious data such as wrong number and crank calls and then totaled the number of callers who completed the protocol. Only 2.2 percent failed to complete the protocol. (Note that the respondents were Census employees or friends and relatives of Census employees.)

Desired Information. Next we analyzed how many responses provided the requested information. The percentage of informative responses ranged from 99.1 to 99.9 percent.

Conciseness. A detailed analysis was made of the distribution of informative responses. It was found that about 97 percent of the responses contained the desired word, either by itself ("Male") or in a common phrase ("I'm male"). About 3 percent of the informative responses did not contain the exact word or phrase, but did provide the desired information. Subsequent analysis of this category revealed that, in fact, well over half of the responses were concise, and could be recognized without natural language processing. For example, instead of the target word "White" the caller may have said "Caucasian".

We have also evaluated the ability of our recognizers to cope with the speech produced by these prompts. Below we describe, as an example, our word-recognition performance on some sub-tasks.

Word Recognition. We have developed task-dependent recognizers for each question. They were trained on the hand-transcribed portion of the corpus, using automatically located phoneme boundaries. As examples of system performance, tables 1 and 2 show confusion matrices obtained for the "gender" and "race group" questions that form part of the questionnaire. Results were obtained on an independent test set, and only responses containing a target word were tested. In each case, rows correspond to the word of interest that appeared in the test utterance, and columns correspond to the word detected by the system.

Our conclusion is that the context-dependent networks seem to be able to deliver satisfactory performance, even though the data collected for this task is noisier than other corpora we have collected, and is also regionally

	male	female	operator	Percentage correct
male	229	6	2	96.62%
female	5	558	0	99.11%
operator	0	2	341	99.42%

Table 1: System performance: "gender" question

	white	black	negro	other	operator	correct
white	685	7	1	1	0	98.70%
black	1	35	0	0	0	97.22%
Negro	0	1	6	0	0	87.50%
other	0	2	0	21	0	91.30%
operator	1	1	1	2	338	98.54%

Table 2: System performance: "race group" question

very diverse.

Discussion Our initial results show that careful structuring of the task dialogue can produce concise and informative responses on a task such as this one. Also, acceptable recognition performance is obtained with even a simple recognition system. We have implemented this system on the Linkon FC-3000 processor board, including using context-dependent networks for classification. Other extensions have been added to our protocol for this field test: the system now loops over all family members in a household, asks questions about coverage, provides operator hand-off, etc. This system, which can handle 20 incoming calls, is currently (Feb. 1995) being prepared for deployment in a field test managed by the U S Census Bureau.

5 Conclusion

We have shown that current speech-recognition technology based on neural networks can be applied to real-world tasks. With these adaptations it is possible to implement systems which are very useful in specific niche areas, such as the form-filling task described in Section 3. Such applications are characterized by limited recognition vocabularies and either isolated word recognition or (at worst) word spotting.

There is considerable divergence in the community on where this technology is headed. One group of researchers believes that incremental improvements to current methods will continue to deliver improved performance[5]. In this case, the abovementioned limitations will be overcome by simply using more training data, more powerful processors and more processor memory – and the availability of all three of these components is clearly on the horizon. Another school of thought holds, however, that we have reached the phase of diminishing returns with current techniques[6]: new ideas need to be generated to overcome the restrictions placed by our HMM-based algorithms.

While these issues will probably take a while to resolve, there is no doubt that the range of application for speech recognition will continue to expand rapidly. Although more competent recognizers are certainly desirable, current technology offers much scope for application.

Acknowledgements We thank the U.S. Bureau of the Census, U.S. Office of Naval Research, the National Science Foundation, Digital Equipment Corporation, and U S WEST for their research support.

References

[1] J. Boite, H. Bourlard, B. D'hoore, and M. Haesen, "A new approach towards keyword spotting," in *Proceedings of the 3rd European Conference on Speech Communication and Technology*, (Berlin), pp. 1273–1276, Sept. 1993.

[2] M. Cohen, H. Franco, N. Morgan, D. Rumelhart, and V. Abrash, "Context-dependent multiple distribution phonetic modeling with MLPs," in *Adavances in Neural Information Processing Systems, 5* (S. J. Hanson, J. D. Cowan, and L. C. Giles, eds.), (San Mateo, Calif.), pp. 649–657, Morgan Kaufmann, 1993.

[3] R. A. Cole, D. G. Novick, M. Fanty, S. Sutton, B. Hansen, and D. Burnett, "Rapid prototyping of spoken language systems: The year 2000 census project," in *Proceedings of the International Symposium on Spoken Dialogue*, (Tokyo, Japan), Nov 1993.

[4] H. Hermansky, "Perceptual linear predictive PLP analysis for speech," *Journal of the Acoustical Society of America*, vol. 87, no. 4, pp. 1738–1752, 1990.

[5] J.Makhoul and R.Schwartz, "State of the art in continuous speech recognition," in *Voice communication between humans and machines*, (Washington D.C.), pp. 165–198, National Academy Press, 1994.

[6] S. E. Levinson, "Speech recognition technology: a critique," in *Voice communication between humans and machines*, (Washington D.C.), pp. 159–164, National Academy Press, 1994.

Experiments with Statistical Connectionist Methods and Hidden Markov Models for Recognition of Text in Telephone Company Drawings

Atul K. Chhabra[1] and Vishal Misra[1,2]

[1] *NYNEX Science & Technology*
500 Westchester Avenue, White Plains, NY 10604 USA
atul@nynexst.com
[2] *Department of Electrical and Computer Engineering*
University of Massachusetts at Amherst
Amherst, MA 01003 USA

Abstract

At the last IWANNT workshop, we presented the framework for statistics-driven high-order feature selection for neural network based printed character recognition in the context of telephone company engineering drawing conversion. Here, we discuss the results of various experiments designed to improve the accuracy of text recognition in drawings. First, we study the effect of limiting the maximum number of higher order feature pairs that are selected for improving the discrimination of any given class pair. Classifiers obtained with different limits are found to make different mistakes. These classifiers could potentially be used together with a voting scheme to further reduce the classification error. Next, we attempt to model the limited-vocabulary language as a single hidden Markov model of the first order. This helps in improving segmentation in the case of touching characters and helps in correcting some errors made by the isolated character recognition system.

1 Introduction

Telephone companies have a huge number of records stored in the form of tabular or quasi-tabular data on engineering drawings. Assignment drawings, wiring lists, front equipment drawings, and distributing frame drawings are a few examples of such structured drawings (see figure 1 for examples of some of these drawings). The ASCII data in these drawings has traditionally been stored on paper or microfilm because some of the data dates back to the days when computers were not commonly used in telephone company operations. Conversion of the drawings to computer data is crucial for efficient operation and for utilizing emerging technologies such as automated mapping and facilities management. Conversion using manual data entry is very tedious and labor intensive.

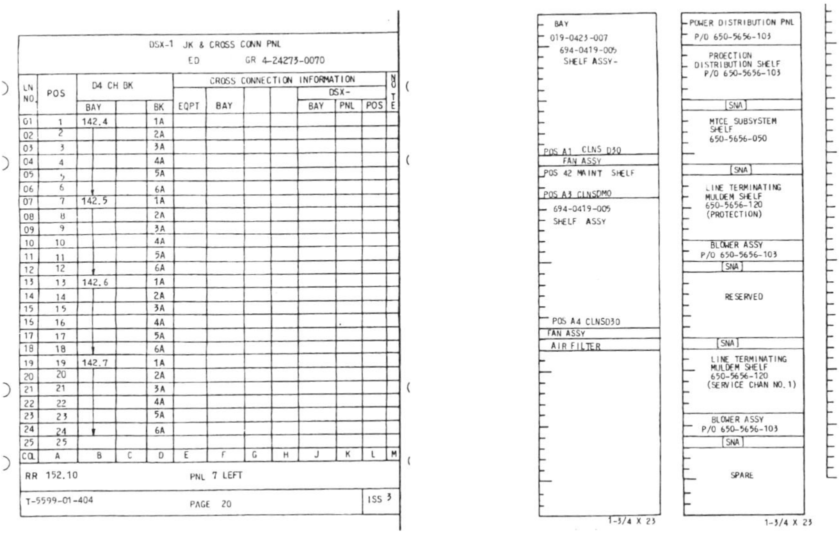

(a) (b)

Figure 1: Small extracts from E-size telephone company drawings: (a) DSX Assignment Drawing and (b) Front Equipment Drawing.

At the last IWANNT workshop, we presented early work on a semi-automatic system to convert the telephone company drawings using line drawing interpretation and neural networks based text recognition [1]. In this paper, we describe progress made since then on improving the character recognition performance. We present experiments with statistical high-order feature selection used in conjunction with a feed-forward neural network for isolated character recognition, and experiments with hidden

Markov models for text string recognition. First, we study the effect of limiting the maximum number of higher order feature pairs that are selected for improving the discrimination of any given class pair. By imposing differing limits, it is found that the resulting classifiers have a significant non-overlap in their classification errors. Fusion of the results of several such classifiers can be used to reduce the overall classification error. Next, we attempt to model the lexicon as a single first order hidden Markov model. This helps in improving segmentation in the case of touching characters and helps in correcting some errors made by the isolated character recognition system.

2 The Nature of Characters in the Drawings

The telephone company drawings contain text that is machine-printed (typewritten), drafted, and/or hand-printed. Even though most of the text is typewritten or drafted, the recognition of this text is much harder than conventional OCR of printed characters. This is due to several reasons listed below.

- The drawings are drawn/typed on a medium known as mylar. The E-size mylars undergo severe warping, shearing, and other distortion due improper handling over several years or decades, and due to hundreds of passes through roller fed copying machines. This causes the characters and the line work to deform severely.
- With age, patches of ink begin to fall off from the mylars because of wear and tear. This results in fragmented characters.
- Different characters (even characters belonging to the same character string) may have been typed/printed at different times and with different pitch and font, resulting in misalignment, touching characters, uneven/unpredictable spacing, etc.

Due to the above reasons, it is not possible to use an off-the-shelf OCR package. We need a system that can be trained with real image data. We described such a system in [1, 2]. Here, we present enhancements to that system.

3 Statistical High-Order Feature Selection

Our intelligent character recognition (ICR) system consists of two major components – the single character recognizer (SCR) and the text string segmenter/interpreter. The SCR uses statistically derived high-order combinations of raw geometric features and a feed-forward neural network classifier [1, 2].

	Error on Training Data	Error on Test Data
NN1	1.09% (491 errors)	4.11% (491 errors)
NN2	0.84% (379 errors)	4.16% (497 errors)

Table 1: Performance of the two classifiers at 0% rejection rate. NN1 used high-order features computed as described in [2]. For NN2, a limit of four high-order features per class pair was imposed.

Most neural network based character recognition systems treat neural networks as a 'black box'. We showed [1, 2] that significant improvement in classification accuracy and speed can be obtained by intelligent selection of inputs to a neural network classifier. Specifically, we showed that high order combination of geometric features leads to better accuracy and speed than using pixel data or raw features as input to the neural network. In essence, the high-order combinations of raw features provide higher discriminating ability than the raw features themselves. The combinations are derived by a statistical discriminant analysis through an iterative optimization process. This process tends to allocate more high-order features for class pairs that are difficult to distinguish (such as the hand-printed letters O and D). This results in too little attention being given to the class pairs that are relatively easier to distinguish. To avoid this problem, we experimented with the idea of limiting the maximum number of high-order features devoted to a particular class pair. This results in a more even allocation of high-order features across the class pairs. Although this causes the discrimination of classes such as hand-printed O and D to deteriorate, such problems can easily be fixed with the help of context.

For our experiment, we used hand-printed characters obtained from the NIST special database 3 (training data) and NIST test data 1 [3]. We used only the upper case alphabets in our experiment. The training set consisted of 44,951 character images and the test data contained 11,941 character samples. We trained two feed-forward neural networks, NN1 and NN2, using error backpropagation. The inputs to the two networks were two different sets of high-order features computed from the same training data. The first set of features was computed by imposing no restriction on the feature selection algorithm of [2]. For computing the second set of features, we limited the number of high-order features that could be selected for any class pair to four. The performance of the two networks is shown in table 1.

Although there appears to be no significant difference in the overall performance of the two networks, we observed that they differ in the kinds

	D-O	H-M	V-Y	K-X	U-V	L-C	C-I
NN1	40	13	12	14	47	19	9
NN2	48	19	17	18	36	11	2

Table 2: Number of errors made by classifiers NN1 and NN2 in distinguishing between similar looking characters. Errors for several confusing class pairs are shown.

of mistakes they make. The difference in errors is highlighted in table 2. We observed that approximately one third of the errors made by the network NN1 are not repeated by NN2. The converse was also found to be true, i.e., about one third of the errors made by the network NN2 are not repeated by NN1. Therefore, the two networks can potentially be used together with a voting scheme to further reduce the error rate.

As an aside, note that the performance of both the networks is better than that reported earlier in [3]. In the NIST test [3], the NYNEX recognizer stood second out of 37 participating upper case alpha recognizers. The error rate of the NYNEX recognizer was 4.9%±0.5% as compared with 3.7%±0.5% for the top scoring recognizer. With the new improved networks, our error rate of 4.1% is within the statistical error bounds of the top scoring network.

4 Segmentation and Hidden Markov Model based Recognition of Text Strings

Once a text string has been extracted from the image of a drawing, and prior to applying the SCR, it is necessary to segment the text string into isolated characters. When neighboring characters are not touching, and when the characters are well formed, this is a trivial task. However, in practice, text strings frequently contain fragmented characters and touching characters. We address this problem by initially oversplitting the image of a text string into a large number of segments. The segmentation is done by computing the upper and lower profiles of the touching characters and by interpreting them along with their first and second derivatives. The profiles and the associated derivatives yield a set of likely 'pinch-points'. We then calculate the best split path which passes through a pair of these pinch-points.

Following this, we try and recognize the initial segments and valid groups of the initial segments as single characters. In order to better guide the segmentation and grouping process and the overall string interpretation process, we can pre-define grammars or hidden Markov models (HMM's) of the lexicon for text strings.

A grammar defines the structure of a text string. Text entries in tabulated data are often very structured. For instance, text entries in a column of data are likely to conform to similar structure. When a grammar is specified for a text string, the string interpretation procedure simultaneously performs splitting-regrouping and string-grammar matching. The grammar based system was referred to in [1] and is described in detail in [4]. Here, we discuss the HMM based system for text string interpretation.

For entries in the drawings which have no associated grammar, we use HMM's as an aid to recognition. HMM's have been used in speech recognition for several years. Recently, they have been applied to the recognition of cursive handwriting [5]. We have improved upon these ideas while applying them to printed character recognition. As stated above, our printed character recognition task is much harder than that can be handled by conventional OCR systems.

We treat each character as a state in the Markov Model and treat the corresponding image segment grouping as the symbol. We use a single path discriminant HMM for the lexicon. The HMM is built by first collecting all the ASCII strings that are expected in the drawings. Once the dictionary is ready, we calculate the transition probabilities between letters and the probability of occurrence of each letter as the first character in a word. In a first order HMM we use transition probabilities only from one character to the next. The performance of the system is expected to improve with the use of a second or higher order model, i.e., if we consider transition probabilities across two or more characters. However the growth in computational complexity for higher order models makes it infeasible for us to implement them for the real-time interactive system that we are building.

We have tried both the stationary and non-stationary versions of the HMM. In a non-stationary model the transition probabilities are calculated separately for each position in a string. In a stationary model the location of the transition in the string is ignored. If the dictionary is completely known then the non-stationary model performs better. However, the non-stationary model becomes too rigid and the chance of it recognizing a new word is considerably lower than that of a stationary model.

Conventionally the Viterbi algorithm is used to estimate the most likely state sequence which resulted in a given string image. Experiments showed that retaining only the best transitions at each stage of the trellis frequently missed out the correct answer. We therefore use a modified version of the Viterbi algorithm in which, at each stage, we retain two to three best transitions to a state from the previous state. We pass each segment grouping through the modified Viterbi algorithm to come up with the most likely string interpretations. We also make sure that there is no string length bias

Strings recognized correctly as the top choice of the HMM	93.97 %
Strings recognized correctly in the top two choices of the HMM	97.41 %
Strings recognized correctly in the top 10 choices of the HMM	98.27 %
Strings not recognized correctly in the top 10 choices of the HMM	1.73 %

Table 3: Performance of the HMM and Viterbi algorithm on test data extracted from drawings

in the recognizer. Use of the HMM significantly improves the recognition of badly degraded and/or touching strings in the drawings. We ran experiments on a few hundred random samples collected from the drawings. We used a lexicon of about 200 words and used a first order stationary HMM. The results are presented in table 3 and a sample run is shown in figure 2. The picture in the top of the figure shows the scanned image of the deformed word 'ASSIGNMENT.' Also shown is the initial segmentation of the image into 'atomic' symbols which are then regrouped in different ways to find a good match (using the Viterbi algorithm) with the HMM of the lexicon. The bottom part of the figure shows the top ten choices of text strings reported by the Viterbi algorithm. Next to the reported answers, you can see the confidence level for each character and an overall score from the Viterbi algorithm. Although the confidence levels for 'A' and 'N' are almost zero, the system is able to recognize the word correctly due to the HMM of the lexicon.

5 Conclusion

In this paper, we presented techniques for improving the performance of single character recognition and word recognition. This work is part of a larger effort towards building an interactive system for converting telephone company drawings from paper into CAD and database form.

References

[1] A. Chhabra, S. Chandran, and R. Kasturi, "Table structure interpretation and neural network based text recognition for conversion of telephone company tabular drawings," in *Proc. IWANNT'93*, Princeton, NJ, pp. 92–98, October 1993.

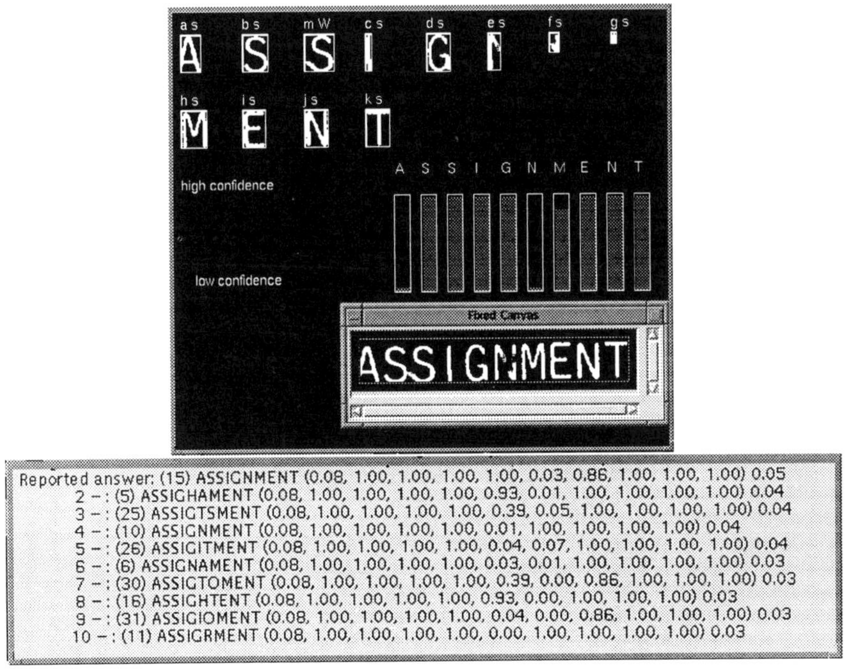

Figure 2: Recognition of a text string with touching and fragmented characters.

[2] A. Chhabra *et al.*, "High-order statistically derived combinations of geometric features for handprinted character recognition," in *Proc. IAPR International Conference on Document Analysis and Recognition*, Tsukuba Science City, Japan, pp. 397–401, October 1993.

[3] R. Wilkinson *et al.*, "The first census optical character recognition system conference," Tech. Rep. NISTIR 4912, National Institute of Standards and Technology, Gaithersburg, Maryland, USA, May 1992.

[4] A. Chhabra, "Anatomy of a hand-filled form reader," in *Proc. IEEE Workshop on Applications of Computer Vision*, Sarasota, FL, December 1994.

[5] M. Chen, A. Kundu, and J. Zhou, "Off-line handwritten word recognition using a hidden markov model type stochastic network," *IEEE Transactions on PAMI*, vol. 16, pp. 481–496, May 1994.

Learning Synonyms and Related Concepts in Document Collections

R.A. Bustos and T.D. Gedeon

School of Computer Science Engineering
The University of New South Wales
Sydney NSW 2052, Australia
{tom | robertb}@cse.unsw.edu.au

Abstract

For very large document collections or high volume streams of documents, finding relevant documents is a major information filtering problem. Traditional full text retrieval methods can not locate documents which use specialised synonyms or related concepts to the formal query. We use a neural network approach to learn synonyms and related clusters of words defining similar concepts from a sample document set. The task is then to filter the document collection to find more of the same.

1 The Information Filtering Problem

A *high-volume information source* is one where the rate of document arrival makes it infeasible for an individual to examine and assess the importance of every document. Each *document* is a (possibly structured) piece of text which is concerned with a set of topics. Users of the information source are interested in a particular set of *topics*. The *information filtering problem* is to select, from the incoming document stream, *all* documents which are closely related to the user's interests, and to select *only* those documents [1].

A number of researchers [2-4] have developed systems to assist with information filtering. A local example of an information filtering system is the grapeVINE system developed at the University of New South Wales [5]. The well-known SMART information retrieval system [6], has also been applied to the task of information filtering on Internet news. All of these systems, however, require significant assistance from the user/information-provider in creating and/or maintaining the filters, specifying categories, building synonym lists, and so on. Hence we are interested in automatic indexing as applied to information filtering [7].

2 Automatic Indexing

The aim of indexing text items is to (implicitly) summarise their content. The possible approaches to this problem can be categorised according to whether they are syntactically-based or semantically-based. One extreme, the semantic, natural language understanding approach would construct a deep representation of the semantic content of documents. This is an extremely difficult, complex and as yet unresolved problem. At the other extreme, the information retrieval approach extracts a list of key terms from the document via simple syntactic processing, and devises a document signature based on the following significant measures [8]:

Frequency-keyword approach
: Take the complete text, remove the "stop" words, then sort all the remaining distinct words (keywords). Count the frequency of each keyword. Assign significance to keywords according to their frequency.

Title-keyword Approach
: Compile a list of keywords from the title, subtitle and headings of the document on the basis that the main concepts of the document are likely to be mentioned there. Higher significance can be assigned to the keywords from the main title, and so on.

The location method
: A keyword occurring at the introduction and/or conclusion of a paragraph is likely to be the most central to the theme of the text.

The Cue method
: Is based on the notion that certain of the words, which are not keywords, nonetheless increase or decrease the score of certain keywords, for example by the use of "significant" versus "impossible".

The Indicator-Phrase method
: Phrases about the topic of the text, for example "the purpose of this work," "the main aim of this paper" lend extra significance to following keywords.

Structure of the document

> This includes header information such as the title of this section, markup languages [9] and meta-language constructs used by the source community. Examples of these deriving from the Internet, for example, are: ":-)" to indicate something is being said in jest, use of repeated "!!!" and ALL CAPITALS for emphasis, and so on. Such information affects the weight attributed to key phrases in its vicinity.

The above significance measures form the indexing parameters for an automatic indexer. Current automatic indexing mechanisms assume that there is a best way for combining the significant measures to arrive at a particular signature for a document. Yet, the criteria for arriving at a particular combination is based on intuition rather than active observation of user behaviour. For example, the linear combination of such diverse indications of significance would require the assumption that each of them is providing independent evidence, which can not be readily justified to be correct. The problem of determining the most appropriate way of combining these indexing parameters appears to be amenable to solution using a neural network approach which can learn an appropriate composition function. This involves initial training of the neural network with the inputs which are the individual indicators of significance, and the desired output is the level of relevance of a document as judged by the user reader based on current interests. The hidden neurons in the network could learn the composition function required to best match the inputs (the significance measures) and the outputs (the relevance of the examples).

The first significance measure listed is a global measure, but has been used most extensively, and successfully in the past [10]. We use this method to find consistent indices in full text as follows.

The set of documents with keywords already known is used as a training set in supervised (Hebbian) learning, and their *L* link weights updated. We retain from prior work the concept of *Textual-Associative* (*T-A*) extra links between words and documents to indicate the statistical relationship between each document and the words it contains. This involves the size of the document, frequency of word occurence, and the overall rarity or commonness of the word. These factors are aggregated to produce the linking *T-A* weight.

Weight correlations in the multi-dimensional space of input patterns have been shown to be a good indicator of the difference of functionality of neurons [11]. The most consistent index generation will take place when the weights linking *word* and *document* neurons are most different, as the words then can best distinguish between different documents, via the generalisation of the keywords it has learnt. To discover the keywords for a new document, we dynamically add an extra neuron to the trained network. The extra *document* neuron has weights derived from its similarity (correlation) to the training documents (see Figure 1).

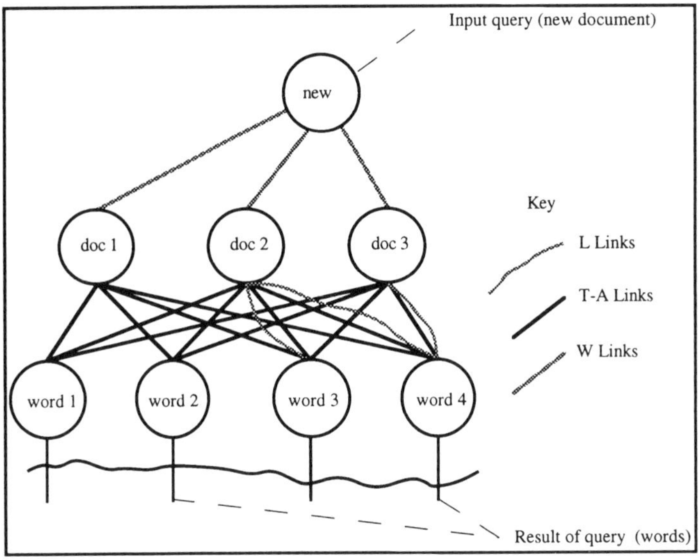

Figure 2. Dynamic (temporary) addition of a new document

This extended neural network is iterated (cycled) until convergence, when the *word* neurons which are active indicate the keywords generated for the new document. These keywords for the new document are used to determine whether it is to be filtered or retained.

3 Description of study

We wish to train a neural network to reproduce the word frequency measure component of a retrieval index.

A collection of 306 documents (being sections of a legal textbook) comprising of some 1,901 different words was chosen. By removal of stop words and stemming, the number of words was reduced. All words which occurred only once or twice in the whole document collection were removed, further reducing the number of words to some 831 words. This number of words was still too large to input into a supervised training neural network, so some other means of reducing the number of words needed to be found. All words occurring only once or in more than half the documents were removed leaving 200 words. From these, 75 words were selected manually looking for words with a moderate frequency and moderate document occurrence.

The network topology is shown in Figure 2. Note that not all connections are shown. The network was then trained normally using error back-propagation.

When being used for retrieval, it is necessary to create a representation of each document in the collection for comparison to the query vector. For these experiments we have used the [12] technique of inverse document-term frequency weighting (IDTW) to produce these vectors.

This technique (for reducing the size of vectors to be used in the comparison) seeks to identify the most important words in the collection by examining the cumulative total of word weights for each document. This was done by totalling the word-document significance values for each word. This resulted in a second short list of 75 words which were likely to be useful for distinguishing documents in the given collection.

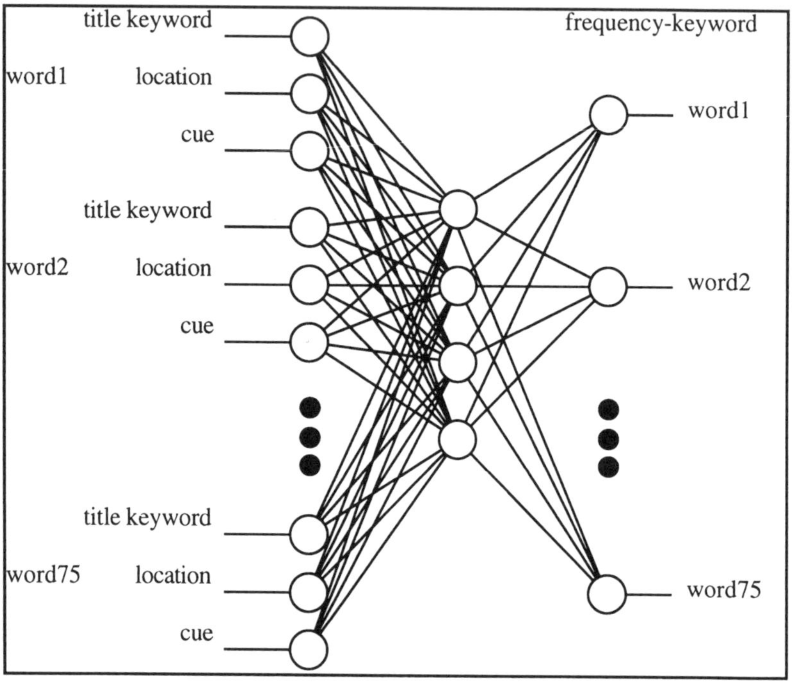

Figure 3. Neural network topology

The experiment is to use the title-keyword measure, the location measure and the cue measure as training input to a network, and attempt to predict the frequency-keyword measure. The network will learn the clusters of words forming related concepts, and will produce significant output activations for words which do not occur in documents if enough of the rest of the cluster is present. In effect, the network will be learning to perform a word cluster completion task.

The neural network topology we have chosen has three inputs for each of the

75 words. The network is using a relatively small proportion of the document texts as its input. Further, all of these measures can be calculated locally and do not require global document collection information as is required for the frequency-keyword calculations. This may provide efficiency gains in the future in highly parallel implementations.

4 Results

There are two ways we can test the network produced using comparing lists of documents retrieved.

The more difficult test is to use each document as a query to find similar documents. We compare its index (or vector) in the 75 dimensional space formed by the frequency-keyword values of the words to the vector of all of the other documents to determine which of these are similar. The top ten similar documents using the actual frequencies to form the vector (the traditional method) are compared to the top ten retrieved using the network's outputs as the components of the index vector. The average overlap is 60.9%. That is, six of the top ten documents in both lists are identical.

The other comparison that can be made is using the original queries which are relevant to the short list of 75 words that were used. The queries are converted into a 75 component vector, and compared to the vectors for the 306 documents, using both the frequency-keyword values and network output values as before.

The average overlap in this case is 100%. That is, we have shown that the neural network can reproduce the behaviour of the frequency-keyword vector retrieval using only the title, location and cue information, in a specialised subset of the document domain. The previous result of 61% demonstrates that some generalisation to the overall document domain was also taking place.

We have also applied a simple analysis technique to the weigth matrix of the neural network to determine the relative importance [13] of the three input measures, as shown in equation (1):

$$C_{title} = \frac{\sum_{i=1}^{75}\sum_{j=1}^{4}\left|w_{title_{i\_j}}\right|}{\sum_{h=title}^{cue}\sum_{i=1}^{75}\sum_{j=1}^{4}\left|w_{h_{i\_j}}\right|}.100\,\% \tag{1}$$

This is calculated by summing the absolute magnitude of the weights from a particular measure to the hidden units, and dividing by the sum of all the weights connecting to the hidden units.

Table 1 shows the results for the relative importance of the three kinds of syntactic processing derived significance information represented by the inputs, as produced for the network before and after normal back-propagation training.

Table 1: Relative importance of inputs

	Title	Location	Cue
before training	32.4%	33.5%	34.2%
trained, simple choice	14.0%	40.7%	45.3%
trained, IDTW choice	22.8%	37.4%	39.8%

Before training the relative significance of the three measures is the same, as expected due to the random initialisation of network weights.

After training, the Cue method is shown to be the most important, closely followed by the location method. The title method in this domain is relatively unimportant. This accords with observations regarding the quality of the titles of the documents used, which all seem to have fairly similar words in their titles. This clearly demonstrates our contention that the relative importance of various measures will differ across domains. In many domains it is accepted that title keyword measures are important.

The 'better' choice of words using the cumulative inverse document term weighting scheme produced slightly higher contribution for the Title method, however the overall greater significance of the Location and Cue methods remains, as well as the slightly greater relevance of the Cue method.

5 Conclusion

For very large document collections or high volume streams of documents, finding relevant documents is a major information filtering problem. We use a neural network approach to learn synonyms and related clusters of words defining similar concepts from a sample document set. The task is then to filter the document collection to find more of the same.

The network will learn the clusters of words forming related concepts, and will produce significant output activations for words which do not occur in documents if enough of the rest of the cluster is present.

We contend that the difference in the predicted and actual frequency keyword values are the most significant measure of the usefulness of any potential cluster member words found.

References

[1] Ngu, AHH, Gedeon, TD and Shepherd, J "Discovering Indexing Parameters for Information Filtering,"*Proceedings 2nd International Conference on Intelligent Systems*, 6 pages, Singapore, 1994.

[2] Fischer G and Stevens C, "Information access in complex, poorly structured information spaces", *CHI'91 Conference Proceedings*, 1991.

[3] Foltz, PW and Dumais, ST, "Personalized information delivery: an analysis of information filtering methods", *CACM*, vol. 35, no. 12, pp.51-60, 1992.

[4] Goldberg, D, Nichols, D, Oki, BM and Terry, D, "Using collaborative filtering to weave an information tapestry", *CACM*, vol. 35, no. 12, pp.61-70, 1992.

[5] Brookes, C, *grapeVINE: Concepts and Applications*, Office Express Pty. Ltd., 1991.

[6] Salton, G. (1971) *The SMART Retrieval System - Experiment in Automatic Document Processing*, Englewood Cliffs, Prentice-Hall.

[7] Gedeon, TD and Ngu, AHH "Index Generation is better than Extraction," *Proc. International Conf. on Non-Linear Theory*, pp. 771-774, Hawaii, 1993.

[8] Paice, CD "Constructing Literature Abstracts by Computer: Techniques and Prospects," *Info. Proc. and Management*, vol. 26, no. 1, pp. 171-186, 1990.

[9] Botham, A, Fuller, M, Mackie, E, Sacks-Davis, R, Wilkinson, R "An SGML-based Hypertext Information Retrieval System," *Australian Computer Science Communications*, vol. 15, no. 1, pp. 111-123, 1992.

[10] Gedeon, TD and Mital, V "Information Retrieval in Law using a Neural Network Integrated with Hypertext," *Proceedings International Joint Conference on Neural Networks*, pp. 1819-1824, Singapore, 1991.

[11] Good, RP and Gedeon, TD "Network Analysis Techniques as Visualisation Tools," *Proceedings International Conference on Non-Linear Theory (NOLTA)*, pp. 945-952, Hawaii, 1993.

[12] Rose, D and Belew, "A Connectionist and Symbolic Hybrid for Improving Legal Research," *International Journal of Man Machine Studies*, vol. 35, 1991.

[13] Wong, PW, Gedeon, TD and Taggart, IJ "An Improved Technique in Porosity Prediction: A Neural Network Approach," *IEEE Transactions on Geoscience and Remote Sensing*, (in press) 1994.

Neural Compression: an Integrated Application to EEG Signals

R.Battiti*, A.Sartori, G.Tecchiolli, P.Tonella and A.Zorat*

Istituto per la Ricerca Scientifica e Tecnologica
Trento 38100 ITALY

-

** Universita' di Trento*
Trento 38100 ITALY

Abstract

In this paper a lossless and lossy Neural Compressor for electroencephalographic (EEG) signals is presented and compared with statistical and spectral techniques. The system is integrated on a board comprising a dedicated Neural Chip and is trained using a novel VLSI-friendly algorithm (RTS). The compression factors obtainable (50%-80%) make the on-line transmission of real EEG signals possible.

1 Introduction

Electroencephalographic signals (EEG) are an example of a context where data compression techniques specialized for this particular kind of signal can give very good results. EEG signals (20 tracks acquired in parallel at 128 Hz with 8 bits per sample) are recorded during a session that typically lasts 45 minutes. The record is shorter because the signal is not acquired for the entire duration of the test. The resulting track usually lasts for about 10 minutes and requires about 1.5 MByte of memory space.

In several cases EEG records stored at one hospital are needed at a different site. In addition, an EEG test might be administered outside the hospital (e.g. EEG are recorded in out-patients departments, first-aid posts, or patients' homes), while the analysis of the data and the subsequent diagnosis are performed at the hospital. In these cases, the data must be transmitted from one site to the other over a telephone line by means of modems [8]. Storage requirements and transmission needs for such signals demand that data compression techniques be used.

An application of Neural Networks for lossless and lossy compression for EEG signals is presented here. These signals are usually slowly varying signals, having a first derivative of limited value. Low amplitude and high frequency noise is superimposed on the signals.

The learning paradigm used to train the lossless and the lossy compression Neural Network is the Reactive Tabu Search (RTS). A hardware architecture was designed and TOTEM, a full-custom VLSI chip implementing it, was realized to build Neural Networks that work with RTS.

While the primary motivation for the work is the compression specifically of EEG signals, results reported here can be easily extended to other signal families having these characteristics.

In Section 2 the Reactive Tabu Search learning technique is briefly described, Section 3 is devoted to the features of TOTEM, in Section 4 and 5 lossless and lossy compression methods are presented and compared.

2 Neural Networks and RTS

While derivative-based methods for neural net training have been used with success in many contexts, they are applicable only to differentiable performance functions and are not appropriate in the presence of local minima. In addition, the random initialization problem is not trivial [4] and the calculation of derivatives is expensive and error-prone, especially if special-purpose VLSI hardware is used [6].

To overcome these problems, the task can be transformed into a *combinatorial optimization* problem, which is then solved by means of a heuristic method: the *Reactive Tabu Search* (RTS) [2]. The solution thus found provides the solution to the original problem as well. This method is based on the construction of a *search trajectory*, in which the neighborhood of a point consists of the strings differing by a single bit and the selected move is one that causes the largest decrease in the error function f, *among those that have not been already executed in the most recent part of the search.* The prohibition period is adjusted by a cycle-detection and reaction mechanism based on the previous history of the search.

RTS escapes rapidly from local minima, is applicable to non-differentiable and even discontinuous functions, and is very robust with respect to the choice of the initial configuration.

3 The Totem Chip

The digital data stream SIMD computational structure was used as the paradigm for the development of an architecture specially tailored for Neural Network and the RTS algorithm. The digital realization is particularly attractive because RTS allows Neural Networks to operate with limited word widths of weights and inputs without performance degradation [2].

Architectural simplicity was achieved by optimizing the execution of multiply-and-accumulate on long data streams, the most compute-intensive operation for the forward pass of the Multi Layer Perceptron (MLP). Ancillary operations such as the calculation of the changes in the weights during the learning phase are delegated to other circuits such as standard microprocessors. The fixed-point, short-word digital realization made possible by RTS is intrinsically more economical than its floating-point, long-word format counterparts in terms of silicon estate, power dissipation and speed.

The data stream approach proposed uses a *broadcast* scheme to feed the features to an array of N_{neu} "neurons" realized with multipliers and accumulators (MACs) (see Figure 1), thus providing the desired full connectivity of the MLP network while requiring the lowest possible bandwidth at the input port of one I/O transfer per accumulation cycle. Access to the output of individually addressable neurons is provided.

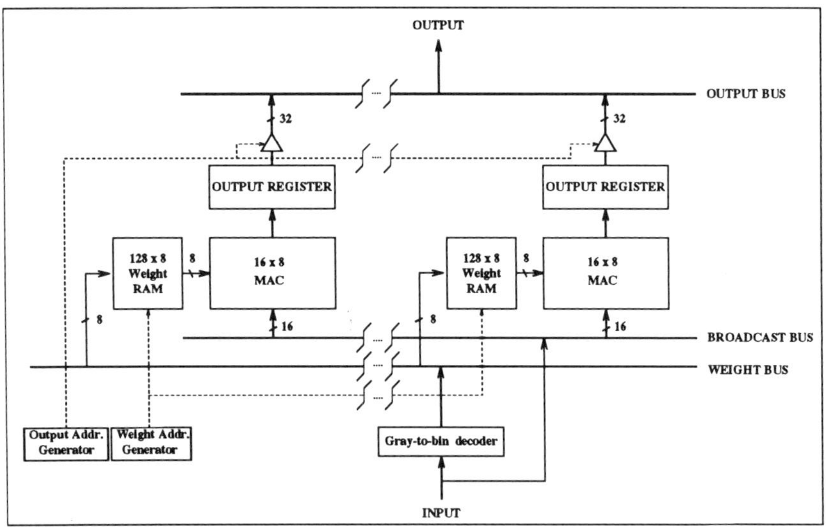

Figure 1: Simplified block diagram of the TOTEM chip.

The TOTEM chip comprises an array of 32 parallel processors with associated on-chip weight memory and control logic with broadcast and output buses. The key design constraints of very compact memory and multiplier cells were satisfied by full-custom design of these elements and by careful physical matching of the form factors of the memory and MAC blocks.

The chip was fabricated in a conventional 1.2 μm CMOS process. By partitioning the internal memory into smaller modules, placed in close physical proximity to their corresponding MACs, an access-time in the range of 10 ns has been achieved. Cycle-time is under 30 ns, for a sustained performance of 10^9 multiply-accumulate operations/s and a 4-cycle latency.

4 Tracking Electroencephalographic Signals

Lossless compression through Neural Network can be achieved by using the net as a predictor and coding the prediction error. The predictive Neural Network uses a certain number of past samples to predict one future event trying to *track* the signal. If the algorithm used for the prediction is good, many of the differences between actual and prediction values will be zero or close to zero. A variable length coding (*e.g.*, Huffman) of the differences will assign short code words (few bits) to these frequently occurring patterns, thus resulting in high degree of compression, as only the *coding of the differences* need to be transmitted (see Figure 3). At the receiver's end, the same prediction algorithm is executed to generate a sequence of values which are then corrected by adding the differences obtained by decoding the received sequence (see Figure 2).

To implement the tracking neural network we chose a 4-6-1 architecture of a multilayer perceptron, with feedback from both hidden and output units (*i.e.*, a Recursive Neural Network (RNN)) so that the symbols to be encoded are given by (1):

$$d(t) = x^{(t)} - f(x^{(t-1)}, \ldots, x^{(t-4)}, h_1^{(t-1)}, \ldots, h_6^{(t-1)}, o_1^{(t-1)}) \qquad (1)$$

where $x^{(t)}$ represents the actual value of the input signal at time t, $d(t)$ the difference between actual and predicted value, and $f(\ldots)$ the output of the neural network. h and o are the values produced by the hidden and output layers, respectively.

A conventional, statistically sound approach to signal prediction is based on the theory of linear digital filters.

Consider the linear filter (2):

$$y_t = b_1 x_{t-1} + b_2 x_{t-2} + \ldots + b_k x_{t-k} \qquad (2)$$

The coefficient vector $\bar{b} = (b_1 \ldots b_k)^T$ that minimizes the prediction error

$e_t = x_t - y_t$ is given by the solution $\overline{b}$ of the linear system:

$$\overline{b}^T \cdot \overline{R}_x = E[x_n \overline{x}] \tag{3}$$

where R_x is the correlation matrix of process x_n, and $\overline{x} = (x_{t-1} \ldots x_{t-k})^T$ (for details see [5]).
The linear predictor can be regarded as a a single layer perceptron consisting of one output neuron and with a linear thresholding function. For this reason, TOTEM can be used to provide a very fast and economical hardware implementation of the linear predictor.

High degree ($k = 20$) digital filters have been designed and applied to some examples taken from the EEG database, without a substantial improvement of the results obtained with a much simpler second order linear predictor. Thus the result presented here are related to the case $k = 2$ over the whole database.

The results of data compression are shown in Table 1, comparing some popular file compressors (*gzip* and *lharc*) to the linear predictor (LP) and the recursive neural network (RNN) described here.

The results shown are obtained by processing a database of over a million samples corresponding to several minutes of $10 + 10$ EEG tracks. Of these patterns, half were used to train the network and the remaining half were used for testing.

Table 1: **Neural Network Compression (Percentage Values)**

Data Set	Gzip	Lharc	LP	RNN
Train	42.1	38.6	51.0	51.4
Test	37.3	34.6	43.8	45.0

The designed transmission system consists of a compression board (Figure 3) and a decompression board (Figure 2), both equipped with a TOTEM chip and a modem for connection over a telephone line.

The compression board contains an A/D converter to acquire the signal, a TOTEM chip to implement the Neural Network tracking the signal, a lookup table to encode the error symbols with a variable length alphabet, and a modem that transmits the bits of code. The lookup table mapping the error symbols into variable length bit-streams can be computed from an estimate of the probability distribution of the error symbols.

The decompression board comprises the modem to receive the bit strings, an associative memory to restore the error symbols, and a TOTEM chip to implement the Neural Network tracking the EEG signal.

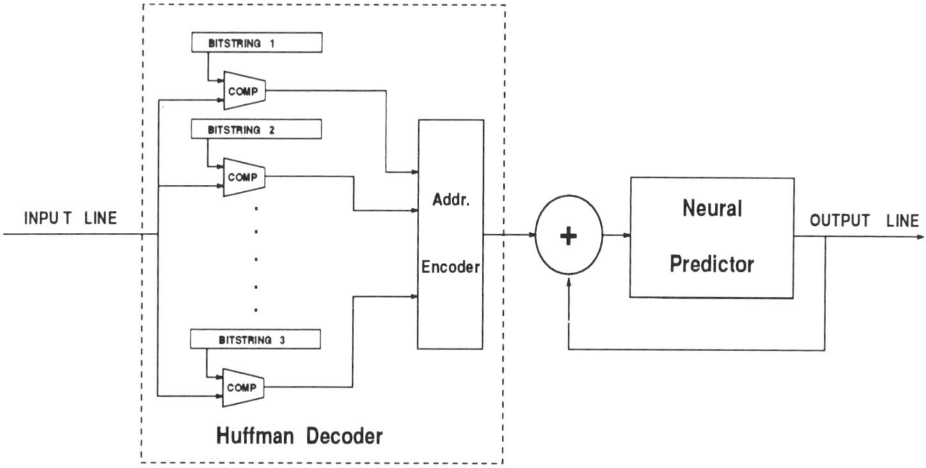

Figure 2: Schematic diagram of the decompression stage.

5 Lossy Compression

Lossy compression through Neural Networks can be achieved by using the activations of a hidden layer as encoding, after training the Neural Network as an Auto-Associative Memory.

As a first test of this neural compressor, a 10-6-2-6-10 architecture was chosen, resulting in an 80% compression. During the training phase, the entire network was used: it was fed with the sequence $(x_t, x_{t-1}, \ldots, x_{t-9})$, and its target was to produce an output equal to the input, i.e. $(y_1, \ldots, y_{10}) = (x_t, x_{t-1}, \ldots, x_{t-9})$. After the training steps, the network was split into two sub-networks: one with architecture 10-6-2, that receives $(x_t, x_{t-1}, \ldots, x_{t-9})$ as inputs and produces (h_1, h_2) as outputs, and a network with architecture 2-6-10 that receives (h_1, h_2) as inputs and produces $(y_1, \ldots, y_{10})$ as outputs. The first network is used to compress the signal; (h_1, h_2) represent the compressed signal, and $(y_1, \ldots, y_{10})$ are the result of the de-compression process.

As $(y_1, \ldots, y_{10})$ is generally different from $(x_t, x_{t-1}, \ldots, x_{t-9})$, the compression is lossy. Preliminary experiments indicate that the quality of the reconstructed signal is acceptable.

A spectral lossy compressor has been developed as well. As EEG signals have a limited spectral range (4 to 14 Hz), while the 128 Hz sampling rate allows high frequency noise to be recorded, it is possible to re-build the plot of the signal, without loosing any diagnostic information, using only

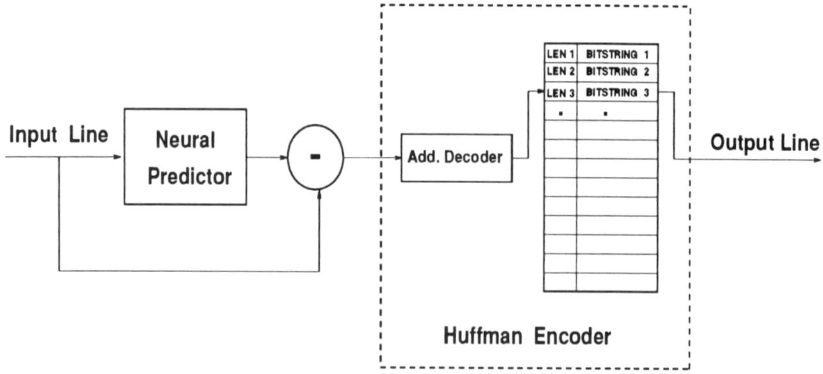

Figure 3: Schematic diagram of the compression stage.

the lowest frequencies as a code for the original data. The Fourier analysis required by this process can be implemented by the TOTEM chip.

In Table 2 the mean squared error between the original and reconstructed signals is shown for the auto-associative neural compressor described here, for the Fourier-based spectral compressor and for the JPEG compressor. The spectral and JPEG compressors are tuned to obtain the same compression factor (80%) as the neural compressor. The two rows present result obtained on two different data.

To interpret correctly the result shown, it is important to note that the electrical noise of the signal has a magnitude of approximately two bits, which corresponds to an equivalent mean squared error of about 10. While the JPEG compressor shows the best performance, the computational overhead of the algorithm implementing it makes it too slow for this application. The compression board equipped with the TOTEM chip can compress signals at a rate up to 10^6 input samples per second, thus it is usable even for high-performance application, *e.g.* for high-speed data storage.

Table 2: **Mean Squared Error of Lossy Compressors**

Data Set	AutoAssNN	Spectral (DFT)	JPEG
Train	15.8	28.5	10.5
Test	26.5	33.9	16.5

6 Conclusions

A compression system based on the TOTEM chip and the RTS algorithm gives good performance in terms of speed, compression ratios, memory requirements and interface with telecommunication packages for applications in the context considered here. Current activities are focussed on the tuning of the system parameters (neural network structure, sampling window, ...) and the accompanying software.

Acknowledgements

Giuliano Antoniol for his ideas and suggestions in the selection of techniques adequate to track the EEG signal. S.Chiara Hospital for system validation experiments. This research was partially founded by INFN and EU under project MInOSS.

References

[1] R. Battiti, P. Lee, A. Sartori, G. Tecchiolli, "TOTEM: a Digital Processor For Neural Networks and Reactive Tabu Search," *MICRONEURO 94,* Turin, September 1994.

[2] R. Battiti and G. Tecchiolli. (1993) "Training Neural Nets with the Reactive Tabu Search." Tech. Rep. UTM 421, Dip. di Matematica, Univ. di Trento - Italy. to appear in *IEEE Transactions on Neural Networks.*

[3] D. Hammerstrom (1990), "A VLSI architecture for high-performance, low-cost, on-chip learning." *Proc. International Joint Conference on Neural Networks 90, S.Diego,* **2**:537–543.

[4] J. F. Kolen and J. B. Pollack. (1990) Backpropagation is Sensitive to Initial Conditions. *Complex Systems* **3**(4):269–280.

[5] Allen Gersho, Robert M. Gray *Vector Quantization and Signal Compression* 1992, Kluwer Academic Publishers.

[6] T. Nordström and B. Svensson. (1992) "Using and Designing Massively Parallel Computers for Artificial Neural Networks." *Journal of Parallel and Distributed Computing* **14**(3):260–285.

[7] F. Vaz, O. Pacheco, A. Martins da Silva. (1994) "Long distance EEG transmission using the public telephone network". *Bolet. Epileps.* 1(3):35–39.

Neural Networks for Efficient Adaptive Vector Quantization of Signals

Fisseha Mekuria * , *Tore Fjällbrant*
Dept. of Electrical Engineering, University of Linköping, S-58183 Linköping, Sweden
fisseham@isy.liu.se
* **Now at Ericsson Mobile Communications AB, Lund, Sweden**

Abstract

In this paper an adaptive vector quantization (VQ) algorithm is described based on a computation and memory efficient neural network model. The adaptive neural VQ (ANVQ) algorithm is based on the full search VQ, and utilizes the modified frequency sensitive competitive learning algorithm for training the developed neural network. The ANVQ algorithm improves the codebook search speed and utilization and gives better output SNR compared to conventionally designed VQ algorithms, and for signals not included in the codebook training-sequence during actual coding. Furthermore, the ANVQ model architecture is suitable for VLSI implementation. A C++ interface to the DSP32C signal processor is used for experimental simulation of the ANVQ algorithm.

1.0 Introduction

Telecommunication systems require efficient coding and signal enhancement schemes for voice and image signal processing. Vector quantization,VQ, has recently been an area of active research to meet these demands. In particular VQ is being used in the popular CELP coders for mobile radio [2,8], in video signal coding for applications in video teleconferencing [2], for combined source and channel coding to get improved performance during poor channel conditions and in speech recognition applications for feature extraction and pattern matching[2,8]. One important drawback of VQ based systems is the fact that the VQ operation consumes a large amount of computational power. A number of algorithms for reducing the required amount of computational complexity have been suggested[2-5]. In a previous publication, [3,5] a fast neural net model for searching tree structured VQ (TSVQ) codebooks have been presented. Using such a model it was possible to reduce the computation and memory requirements of tree structured VQ codebooks by an order of 2 but with the same output SNR as that of the traditional tree searched VQ. In this paper a general type neural network models for fast adaptive vector quantization of signals is described based on popular distortion measures frequently used in speech and image signal processing for telecommunication applications.

2.0 Vector Quantization

Vector quantization, VQ is a block coding scheme and has demonstrated a superior performance over scalar coding systems. VQ has been applied successfully in speech coding for mobile communications, as front-end in speech recognition and image compression algorithms [1,2,4,8]. A simplified view of the operation of a VQ coder decoder system is shown in figure 1.

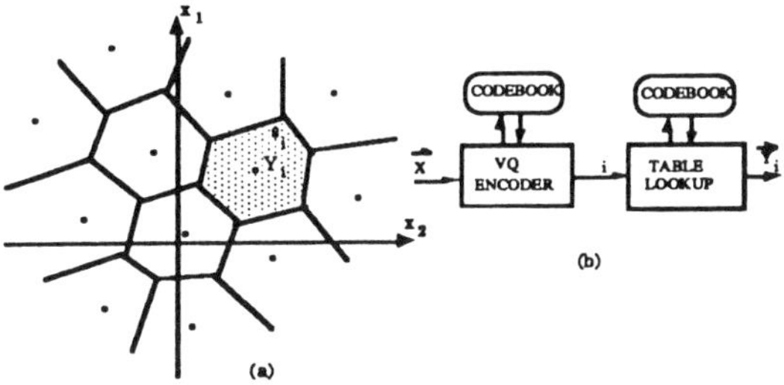

Figure.1 The VQ paradigm

Assume that the input analog signal is sampled at a particular frequency and a sequence of speech samples x(n) are to be vector quantized. Group k samples of x(n) into a k-dimensional input vector $X = \{ x_1, x_2, x_3, \dots, x_k\}$. In a vector quantizer each input vector X is mapped onto another k-dimensional vector Y called the reconstruction or codebook vector. The reconstruction vector Y is chosen from a codebook C of size N containing $\{Y_j, 1 < i < N\}$, where each $Y_j = \{ y_{j1}, y_{j2}, y_{j3}, \dots, y_{jk}\}$. The output of the vector quantizer is actually the index i of the minimum distortion d(X,Y) reconstruction vector Y_i according to:

$$\text{index}=i \text{ iff } d(X,Y_i) \leq d(X,Y_j) \quad \text{for all } j = 1,2,\dots N \text{ and } j\neq i \tag{1}$$

This is the full search VQ (FSVQ) algorithm. For each vector X, $k*2^R$ ($R=\log_2 N$), multiply/ accumulate (MAC) operations and 2^R comparisons are required for the FSVQ algorithm, where R is the rate of the VQ in bits/vector. This mapping of a vector of samples onto a single R-bit number acounts for the high data compression achievable by vector quantization.

The fixed codebook C is trained first using a long sequence of input vectors and appropriate iterative algorithms to produce a codebook of locally optimum signal shapes called codewords. The extraction of robust and invariant feature vectors from the input signal is also a necessary step for a good signal compression system based on VQ. Such an example of a high performance speech coding system is the CELP coder. Although the VQ codebook performs well for signals within the training data, the performance of VQ, for outside of training data is dependent on how much the codewords Y_j, are globally representative. One solution can be increasing the training sequence length to include all types of expected codeword shapes, which is a tediuos and practically impossible method. A more interesting approach is to adapt the codebook following the statistics of the input signal. The latter method with increased codebook utilization during the initial codebook design is described in this paper.

3.0 The Neural VQ Model

Distortion measures are the usual tools used to asses performance and indicate differences in performance between similar systems. Some criteria for a useful distortion measure are: computational simplicity and subjective relevance i.e., differences in distortion values should indicate similar differences in subjective signal quality. The computation of distortion measures accounts for the most part of the computation requirement in many speech processing algorithms. Hence the adaptive neural VQ model to be developed here aims to reduce the computational complexity of vector quantization, utilizing the structure of the distortion measure and the speed and adaptivity of neural network models. Since the main computation load of VQ lies in calculating the distortion measure d(X,Y), the ANVQ algorithm is developed by first simplifying d(X,Y) into a neural model equation for fast computation. In the next sections an attempt is made to develop the ANVQ neural model equation for three important distortion measures frequently used in speech processing.

3.1 Squared Error Distortion Measure

Let us compare each input vector X, of dimension k, with all the codebook vectors Y_j , where j=1,2,..L, based on the squared error distortion measure (SEDM) given by:

$$d(X,Y_j) = \|X - Y_j\|^2 = \sum_{i=0}^{k-1}(x_i - y_{ij})^2 \qquad (2)$$

It is possible to compute equation (2), directly by performing squaring operations. An alternative and computation efficient form can be obtained by expanding (2) as follows,

$$d(X,Y_j) = \| X \|^2 - 2(X \cdot Y_j + Q_j) \qquad (3)$$

where

$$Q_j = -1/2 \| Y_j \|^2 = -1/2 = \sum_{i=0}^{k-1} y^2_{ij}, \quad \text{and} \quad X \cdot Y_j = \sum_{i=0}^{k-1} x_i\, y_{ij}.$$

for all j codebook vectors. From equation (3) it clear that the minimization of the distortion d(•) for each input vector X is equivalent to maximizing the dot product X•Yj or an activation function of the form,

$$F(j) = f\left(\sum_{i=1}^{k+1} C_{ji}\, x_i\right) \qquad (4)$$

for each input vector X and neural unit j. The Q_j term represents the autocorrelation sum of the components of each codebook vector and has been included as the k+1st element in Y_j, and an additional element of unity is inserted into the input vector, i.e., $\underline{X}=\{x_{j1},x_{j2},x_{j3},...,x_{jk},x_{jk+1}=1\}$. Cjs are weighting vectors where $C_j=\{y_{j1},y_{j2},y_{j3},..., y_{jk},y_{jk+1}=Q_j\}$. The algorithm for the ANVQ uses this model together with a modified version of the frequency sensitive competitive learning (FSCL) network given in figure 4, to increase the search speed of the VQ. Equation (4) has similar form as the single-layer perceptron neural-net equation given in [7]. Once the autocorrelation sums Q_j of the codebook vectors and augmenting C_j and the input vector X is done, computation of equation (4) for the design of the VQ codebook and finding the maximum out of all N competing units can easily be done by a neural parallel VLSI computing architecture of type given in [5]. The pipelined architecture of the DSP32C signal processor is used for high speed computation of the simple neural sum of products equation given in (4). The reduction in computational complexity due to the ANVQ model compared to FSVQ , TSVQ and the TSNVQ algorithm is shown in figure 2, based on SEDM. Note this reduction in complexity is achieved if the implementation of the ANVQ algorithm is done on a parallel processing machine based on the network in figure 4.

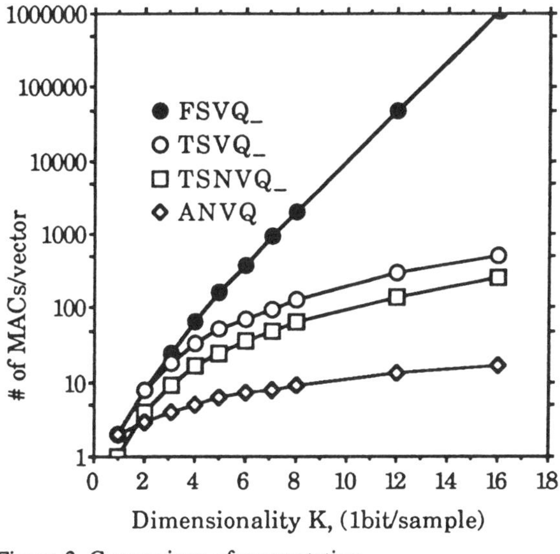

Figure 2. Comparison of computation

3.2 Weighted Squared Error

In the weighted squared error, WSE, distortion measure unequal weights are introduced to render certain contributions to the distortion more important than others. A general WSE distortion measure is defined by,

$$d_w(X,Y_j) = (X - Y_j)^T W(X - Y_j) \tag{5}$$

where W is a fixed k x k symmetric positive definite matrix and the vectors X and Y_j are regarded as column matrices of of order k. Equation (4) can be expanded in a similar manner as for for the SEDM and simplified to yield,

$$d_W(X,Y_j) = X^T W X - 2X \cdot \underline{Y}_j + Y_j \cdot \underline{Y}_j \qquad (6)$$

Thus the minimization of d_W for fixed W and X, over the codevectors (weights) C_j can be achieved by maximizing a function of the form given in equation (4) with the only exception here being that, $\underline{Y}_j = W\ Y_j$ so that, $C_{ji} = \underline{Y}_{ji}$, and $q_j = -1/2\ Y_j \underline{Y}_j$.

3.3 Itakura-Saito Distortion Measure

The Itakura Saito distortion measure (ISDM) evolved from studies of spectral matching effects of linear prediction. The ISDM does fulfil the demands imposed on metrics used as distortion measures, such as analytic tractability and subjective relevance. ISDM is defined in terms of the power spectral density of windowed speech samples and the magnitude squared of a conventional LPC all-pole filter model, G(z), used to model the windowed and z-transformed speech samples x(z) such that [1],

$$d_{IS} = (|X|^2 ; |G|^2)$$

$$= \int_{-\pi}^{+\pi} [|X/G|^2 - \ln(|X/G|^2) - 1] d\theta / 2\pi \qquad (7)$$

where $G(z) = \sigma / A(z)$

and the inverse filter polynomial,

$$A(z) = \sum_{n=0}^{M} a_n Z^{-n}, \qquad (8)$$

with $a_0 = 1$,

the polynomial A(z) is used to minimize the prediction residual energy a, which arizes due to modelling x(z) by G(z). The predictor coefficients a_k are obtained as a result of minimizing the prediction residual over a frame of speech samples. Thus to minimize the ISDM, one has to minimize the residual energy given by the expression [1],

$$a = R_a(0) R_x(0) + 2 \sum_{n=1}^{M} R_a(n) R_x(n), \qquad (9)$$

where R_a and R_x are autocorrelations of the predictor coefficients for the LPC model of interest, and the input speech frame respectively. Thus for a fixed frame of speech X it is possible to find one out of a number of representative model parameters of G(z) by minimizing a function of the form

$$F(j) = f\left(\sum_{n=0}^{M} C_{jn} u_n\right) \qquad (10)$$

where a term 1/2 Ra(0)Rx(0) is inserted as the last component in $C_j = R_a(n)$ and $u_n = R_x(n)$. It can be seen that equations (4) and (10) have similar structure and can be implemented to find the minimum distortion codeword out of stored representative vectors by using the ANVQ algorithm given in figure 3. Note that in equations (4) and (10) it is the maximum and the minimum of F(j) respectively that is sought in the modified FSCL algorithm shown in figure 3. The minimax problem can also be solved using the binary search neural nets given in [7].

4.0 The ANVQ Algorithm

The flowdiagram of the ANVQ algorithm is given in figure 3. The algorithm for the ANVQ uses the activation function in (4) and network training using modified version of the frequency sensitive competitive learning (FSCL) to increase the search speed of the VQ as shown in figure 2. In figure 3 the algorithm selects the neural unit with a maximum of actvation,

$$D_j = F(j)/h(u_j) \tag{11}$$

for the SEDM measure. The term $h(u_j)$ is a frequency weighting function and u_j is a counter of the number of times the neural unit j won competition. This function is used to improve the utilization of the initial codebook by spreading the probability of winning to all neural units[5]. This is achieved as follows: if a neural unit wins competition many times its count u_j and hence its frequency weighted activity D_j decreases, increasing the chance of other units to win competition. For our purpose $h(u_j)=1.45u_j$ has been used.

Improved coding performance is achievable if a vector quantizer can adapt to follow some local statistical characteristics of the input vectors. The ANVQ algorithm describes a forward adaptive VQ based on a model algorithm similar to that in [4,6]. The neural weights C_j are updated for each winner unit by adding the difference vector between the input and codebook vector multiplied by an error function e(m). The error term, $\delta = \varepsilon(m)[X(m) - C_j^W(m)]$, is separately quantized and transmitted as a side information to the decoder in the simulated speech coder.

5.0 Simulation Results

A speech coder based on the ANVQ algorithm is simulated to assess the performance of the algorithm. The programming environment used is a C++ interface for the DSP32C signal processor. The DSP32C development card (Intelligent Instr., USA) is plugged on HP-486-AT machine running at 33 MHz. The ANVQ program code is written in what we call DSP-C for getting as optimal code as possible for the DSP32C signal processor, the only exception is that, the Q_j term is computed by the control C program in the AT while the DSP32C processes the next input vector. A speech data base of six 64k lång segments, composed of both male and female speakers, sampled at 8kHz and DCT transformed, were used as the source. Using the C++ interface and the analog interface capability of the DSP32C card both subjective and objective tests can be done. The codeword index is quantized at 1 bit per sample, giving a total rate of 12kbps, with a segmental SNR of 18.4 dB, which is comparable to the 20dB SNR [4] obtained at 16 Kbps.

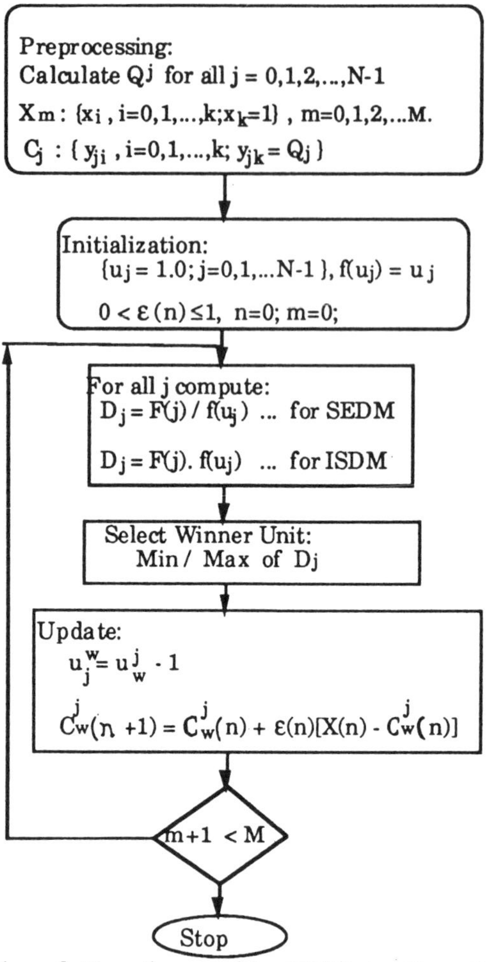

Figure 3, Flow diagram of the ANVQ algorithm with FSCL training

6.0 Conclusion

In this paper we have presented the development and simulation of a reduced complexity adaptive VQ algorithm based on a neural network formulation. The main theme of the paper was the reduction of the computaional complexity of the VQ algorithm. The algorithm also achieved high performance by utilizing the neural network efficiently, which is due to the modified FSCL training. As such, an extensive test of the system with a large speech corpora should be done to assess the ANVQ performance for different dimensions and transmission bit rates. The preprocessing delay can in the future be addressed by using multi-signal-processor environment. However, for realtime VQ applications an Applications Specific VLSI chip having a parallel processing unit architecture is the best solution.

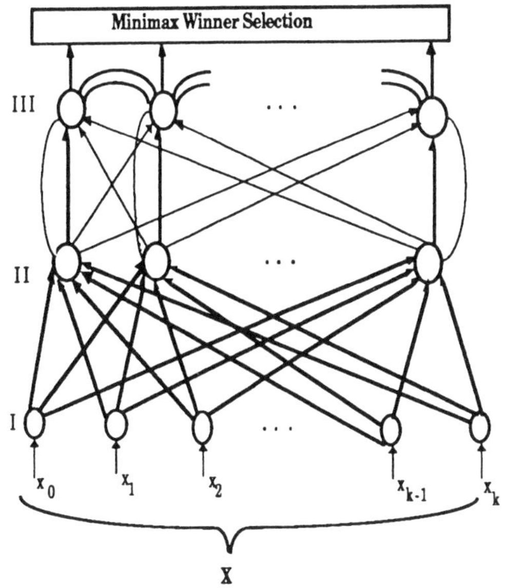

Figure 4. The Modified FSCL neural network

Acknowledgements

This work is supported by the Swedish national board for industrial & technical development, NUTEK under project No. 9302772.

References

[1] Buzo A., et al, " Speech Coding based UponVector Quantization ",IEEE Transactions on ASSP, Vol. ASSP 28, pp. 562-574, Oct. 1980.
[2] Abut H.(1990)."Vector Quantization". IEEE Reprint Collection, IEEE Press,Piscataway, New Jersey.
[3] F. Mekuria, T. Fjällbrant, (1994) "An adaptive Neural Network VQ Algorithm and its Implementation on the DSP32C Signal Processor." in Proceedings of IEEE ISCAS '94, May 30-June 2, London, pp. 6.443-6.448.
[4] T. Kwang, J. Foster, S. Ardalan,(1992) "Adaptive Vector Quantization for Waveform Coding." ICASSP '92,pp. I-101-104.
[5] Mekuria F.(1993) "On Reduced Complexity Vector Quantization and Perceptual Processing of Speech Signals with DSP implementation." PhD.Thesis,Dept. of Electrical Engineering, University of Linöping, Sweden.
[6] Kohonen T. (1988), " Learning Vector Quantization." Neural Networks, Vol. 1, suppl. 1, pp 303-310.
[7] Lippmann R. P.,(1989) " Pattern Classification Using Neural Networks," IEEE Comm. Magazine, Nov., 1989.
[8] Furui S., Sondhi M.M., (1992) " Advances in Speech Signal Processing," Marcel Dekker, New York., U.S.A.

A Complexity Analysis of Telecommunications Software Using Neural Networks

Simon Field†*, Neil Davey*, Ray Frank*

†BNR Europe Limited,
London Road, Harlow, Essex, UK CM17 9NA
E-mail: S.D.H.Field@bnr.co.uk
**University of Hertfordshire,*
Hatfield, Hertfordshire, UK. AL10 9AB
E-mail: N.Davey@herts.ac.uk, R.J.Frank@herts.ac.uk

Abstract

Units of software are represented as points in a multidimensional space, by calculating 12 measures of software complexity for each unit. Two large sets of commercial software are thereby represented as 2236 and 4456 12-ary vectors respectively. These two sets of vectors are then clustered by a variety of competitive neural networks. It is found that the software does not fall into any simple set of clusters, but that a complex pattern of clustering emerges. These clusters give a view of the structural similarity of units of code in the data sets.

1. Introduction

In this paper we examine whether units of software fall into natural clusters when represented by a set of complexity measures. The work extends an analysis [1] in which C source code was mapped onto vectors in a 12 dimensional space, and then clustered using a simple competitive neural net. They found that the vectors were represented by 3 classes, which were then labelled as identifying standard, marginal and non-standard code. If legitimate, such a classification has obvious value as a tool for software maintenance and system engineering. Our work extends this analysis in two ways:

- The source code used is two large sets of commercial telecommunications software, giving the opportunity to evaluate the stability of the results with respect to the selection of software.
- The clustering methods used make use of several different types of self organising neural network, to give results less dependent on the limitations of one particular algorithm.

1.1 The Data

The first data set is a set of 2,236 procedures drawn from a single software product written in the proprietary BNR language PROTEL[1], a block structured language designed for the control of telecommunication systems. The second set is of 4,456 PROTEL procedures drawn from a variety of software products.

1.2 Networks Used

We use four self organising networks with a variety of parameter settings. These are: a simple competitive net (SCN); a competitive net with a conscience mechanism — Frequency Sensitive Competitive Net (FSCN); a self organising map (SOM); and Fuzzy ART. The motivation for this choice is:

- SCN: To attempt to replicate the original results of [1].
- FSCN: To examine if the small number of nodes used in the SCN classification and concomitant dead units could be overcome.
- SOM: To investigate if global order was present in the representation space.
- Fuzzy ART: To avoid prescribing the number of categorising units used.

2. Background

2.1 Metrics Used

Software measurement is a valuable technique for understanding, guiding, controlling and improving software development [2], and many metrics have been proposed, including those based on complexity used here.

Twelve metrics were used in [1], and thus the same twelve were used for this work. The measures include McCabe's Cyclomatic Complexity (V_g) [3], Halstead's Software Science measures [4,5], and a number of others detailed in Table 1.

Notice that the last five measures are defined from the numbers of operand/operators in the code. We would not, however, expect a self organising neural network to find the same clusters without this information as with it, since all such networks use a first order measure of similarity [6].

---

[1] As in [1], the method and metrics used are language independent.

Table 1: Complexity measures used

LOC	number of lines of code
V_g	McCabe's cylomatic complexity measure
D	V_g:LOC
n_1	number of unique operators
n_2	number of unique operands
N_1	total number of operators
N_2	total number of operands
n	$n_1 + n_2$, the vocabulary of the procedure
N	$N_1 + N_2$, the length of the procedure
V	N(log n), the volume of the procedure
$\hat{L}$	$\frac{2}{n_1} \times \frac{n_2}{N_2}$ program level
E	$\frac{V}{\hat{L}}$ effort

2.2 Sheppard and Simpson's work

Here the three categories identified were Standard (95.1%), Marginal (4.75%), and Non-Standard (0.15%). Standard code was based on the value of D, and in fact classified code with a high control density which was contrary to what was expected. Marginal code was based on the values of n_1 and n, meaning that the vocabulary was excessive. Non-Standard code had a high value for LOC and V_g.

3. Networks Used

3.1 SCN and FSCN

Initially these nets consist of a set of nodes with small random initial weights. When an input is presented the winning node is selected as that with greatest similarity to the input. The winner is then modified to become more similar to the input. We use Euclidean distance as the measure of similarity and therefore simply move the weight vector of the winner towards the input in Euclidean space. The size of the update is determined by a learning rate.

This type of network is prone to poor classifications, due partly to the problem of dead units – units that accidentally never win in the competition and thus never move towards the inputs. In order to overcome the above problem one mechanism [7], is to make it harder for frequently winning units to win again. This is done by subtracting a bias term from each unit, that is proportional to the number of times the unit has one.

A potential problem with this type of network is that the biasing mechanism, whilst producing reasonable code vectors for the data set, may obscure natural clustering in the data.

3.2 Self Organising Map

In a SOM [8] the output units are arranged in a fixed topology, usually a two dimensional grid. We use a grid, wrapped around at the boundaries. Winning units pull their neighbours in the grid with them, towards the input. The neighbourhood of a unit should initially be set a large fraction of the output space and decrease over time. A useful heuristic for setting the neighbourhood is that it should start at roughly half the output space and decrease to unity, to ensure global order. We use a ten by ten SOM and therefore initialise the neighbourhood to five. We also include a conscience mechanism as in 3.1. A successfully trained SOM provides more information than the other networks used here. The SOM not only captures clustering in the input space but also the relative position of these clusters, which can be located from their position in the output space.

3.3 Fuzzy ART

Fuzzy ART [9] is an extension of the basic ART algorithm to allow for continuous, rather than binary, data. In the original ART1 network, proximity is measured by the dot product, or equivalently by mapping the Boolean AND over both vectors and by taking Norm 1 of the result. In Fuzzy ART the fuzzy logic AND connective, min, is used to extend the method to real values.

The behaviour of the net is determined by the value of the vigilance parameter and the number of nodes in the initial disabled set. In our experiments we used a pool of nodes greater than required, and a variety of vigilance parameters.

Fuzzy ART is appealing as the clusters are formed dynamically but the algorithm has some specific problems. As noted by [6] the weight vectors drift towards the origin. This is due to the update rule: components of weight vectors are monotonically decreasing. As described by [10] some of the difficulties of this can be overcome by complementary coding; but this has the unfortunate side effect of doubling the length of the input vectors.

For further details of the Fuzzy ART algorithm see [11].

4. Results and Discussion

The results obtained from the simple competitive net were radically different from the other three nets; the SCN found only 4 clusters where the others found many. This is unsurprising since all the other nets have mechanisms for dealing with dead units. The FSCN and the SOM produced very similar clusters. Fuzzy ART produced many more clusters than the other nets, and clustered the data in a different way from the FSCN/SOM. The results are now given in more detail.

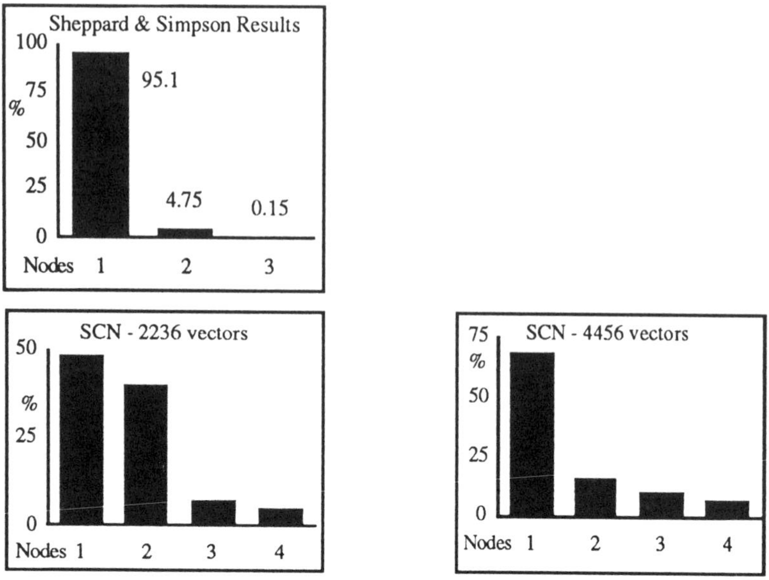

Figure 1: The number of clusters by three SCNs

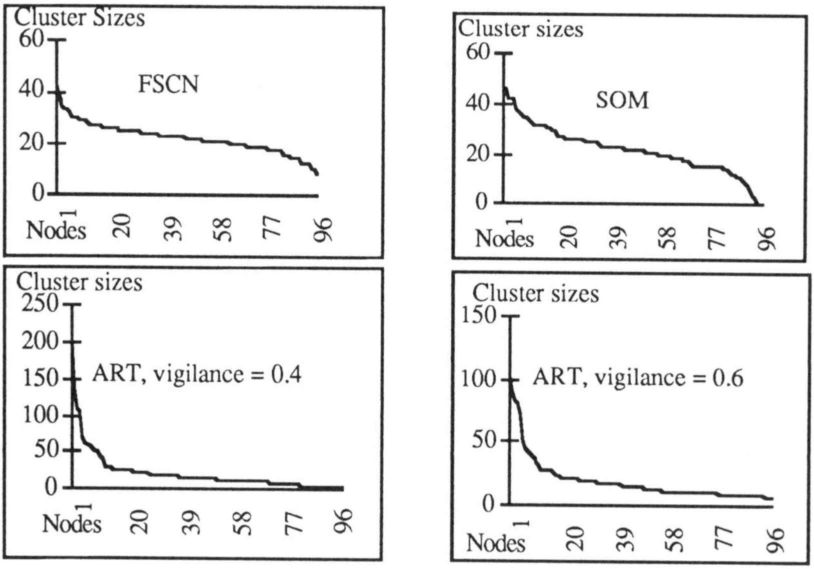

Figure 2: 2236 Vectors set, classified by four networks

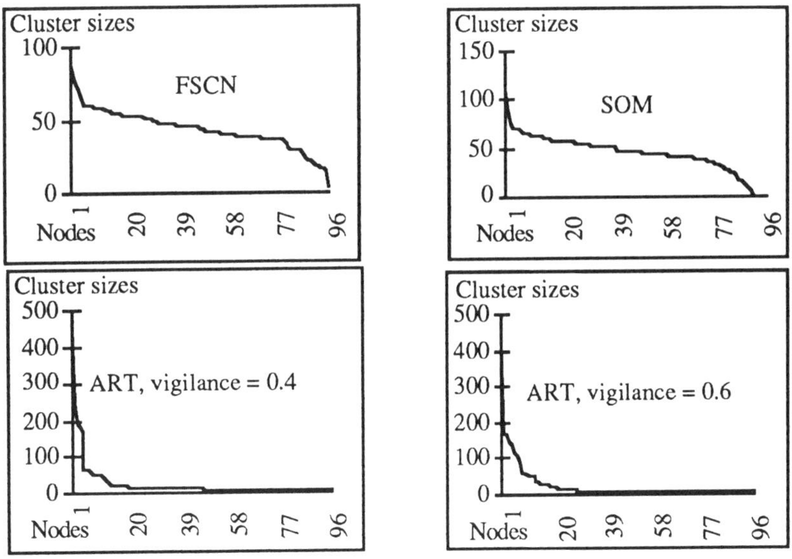

Figure 3: 4456 vectors classified by four networks

4.1 SCN

This experiment was an attempt to replicate the results obtained in [1] and so the net was initialised for both data sets with 12 units. In both cases only four of the twelve units were used. The clusters were identified as follows: Node 1 clustered "average" procedures, i.e. those with no particularly special features; Node 2 clustered procedures with a large language content compared to their size; Node 3 clustered very small procedures; and Node 4 clustered very large procedures.

4.2 FSCN

For both data sets the net was initialised with 100 units, all of which were used after training. For the smaller data set the resulting clusters ranged in size from 43 vectors to 8, with a uniform spread.

4.3 SOM

Again 100 units were used, this time arranged in a ten by ten grid. The initial neighbourhood was set to five and decreased over the training time to one. After training two units of both nets did not participate in the classification. The classification produced was quite similar to the FSCN. However with the SOM additional proximity information could be inferred from the position of the classifying nodes in the SOM grid. For example two vectors grouped by Fuzzy ART were found to be classified by adjacent nodes in the SOM.

4.4 Fuzzy ART

The net was trained over both data sets with two settings of the vigilance parameter: 0.4 and 0.6, and a pool of 2000 uncommitted nodes. As expected the net produced more clusters with the higher vigilance than lower.

Table 2: Nodes used by Fuzzy ART

Data set	2236 vectors		4456 vectors	
Vigilance	0.4	0.6	0.4	0.6
Total Number of nodes	155	227	619	930
Number of Dead Units	17	17	70	252

It is noticeable that the number of clusters found is not linear in the size of the data set; the net may be identifying increased variation in the larger data set, due to its multi-product source.

A problem with the Fuzzy ART network becomes apparent when the weight vectors of the classifying units are examined – they are generally very small. The mean weight vector for the smaller data set and vigilance 0.4 is: (0.026, 0.003, 0.120, 0.062, 0.031, 0.036, 0.079, 0.040, 0.024, 0.031, 0.031, 0.001) with length 0.178. This compares with the mean vector of the raw data: (0.076, 0.034, 0.245, 0.101, 0.075, 0.071, 0.126, 0.079, 0.057, 0.150, 0.140, 0.013) with length 0.394.

Another major difference between the earlier nets and Fuzzy ART is that the former are sensitive to frequency densities in the input space: if an area of the input space is heavily populated then many units will move there. This is not the case with Fuzzy ART, since a single unit will classify an area if all the vectors within it are sufficiently close and satisfy the vigilance test.

5. Conclusions

This work highlights the care with which unsupervised clustering must be undertaken. It is clear that resultant clusters are highly dependant on the architecture and parameter settings used. In particular SCNs may give unrepresentative results due to the problems of dead units.

Some of the limitations of using the ART type architecture are described above and it is difficult to make a direct comparison due to the different distance metric used. The FSCN and the SOM gave similar results, with the SOM providing additional information: the topological map.

These methods do not provide a simple way of partitioning software units on their complexity, however the pattern of similarity could be potentially useful in locating procedures of related complexity. A similar application of neural nets to the location of software clones is described by [12].

Finally, it should be noted that the major activity of work such as this is the pre-processing of data.

Acknowledgements

We would like to acknowledge the financial assistance of the Department of Trade and Industry of the UK Government for partly funding the placement of Simon Field on a Teaching Company Scheme at BNR Europe Limited, through the Teaching Company Directorate organisation, under grant no. TCS-1326.

References

[1] J.W. Sheppard and W.R. Simpson, "Using a competitive learning neural network to evaluate software complexity," *ACM*, pp262-267, 1990.

[2] C.M. Lott and H.D. Rombach, "Measurement-based guidance of software projects using explicit project plans," *Info. and Soft. Tech.* Vol. 35 No. 6/7 pp407-419, 1993.

[3] T.J. McCabe, "A Complexity Measure," *IEEE Transactions on Software Engineering*, Vol. 2 No. 4, pp3-15, 1976.

[4] M.H. Halstead, *Elements of Software Science*, Elsevier Science, 1977.

[5] P.G. Hamer and G.D. Frewin, "M.H. Halstead's Software Science - A Critical Examination," in *Proc. 6th Intl. Conf. Soft. Eng.* pp197-206, 1982

[6] B. Moore, "ART1 and Pattern Clustering," in *Proc. 1988 Connectionist Models Summer School*, San Mateo:Morgan Kauffman, 1988, pp174-185.

[7] S. Grossberg, "Adaptive Pattern Classification and Universal Recoding: I. Parallel Development and Coding of Neural Feature Detectors," *Biological Cybernetics* Vol. 23, pp121-134, 1976.

[8] T. Kohonen, "Self-Organised Formation of Topologically Correct Feature Maps," *Biological Cybernetics* Vol. 43, pp56-69, 1982.

[9] G.A. Carpenter, S. Grossberg and D.B. Rosen, "Fuzzy ART: Fast Stable Learning and Categorisation of Analog Patterns by an Adaptive Resonance System," *Neural Networks* Vol. 4, pp759-771, 1991.

[10] G.A. Carpenter, and S. Grossberg, "A Massively Parallel Architecture for a Self-Organising Neural Pattern Recognition Machine," *Computer Vision, Graphics and Image Processing* Vol. 37, pp54-115, 1987.

[11] S.D.H. Field, N. Davey, and R.J. Frank, "Using neural networks to analyse software complexity," *University of Hertfordshire Technical Report.*

[12] S. Carter, R.J. Frank, and D.S.W. Tansley, "Clone Detection in Telecommunication Software Systems: A Neural Net Approach," in *Proc. Intl. Workshop Application Neural Networks to Telecoms* pp273-280, 1993

Dynamic Competitive Learning Applied to the Clone Detection Problem

P. Barson†, N. Davey*, S. Field†*, R. Frank*, D.S.W. Tansley†

†BNR Europe Limited
London Road, Harlow, Essex, UK, CM17 9NA
paulb@bnr.co.uk, sf@bnr.co.uk, dswt@bnr.co.uk
** University of Hertfordshire*
College Lane, Hatfield, Herts, UK, AL10 9AB
comqrnd@herts.ac.uk, comqrjf@herts.ac.uk

Abstract

A novel neural network architecture, 'Dynamic Competitive Learning', has been applied to the software clone detection problem in telecommunications systems. A software clone is a copy (subsequently modified), of a piece of existing code; it is common practice in large systems to obtain similar code via reuse of fully tested code. This paper discusses a hierarchical network with dramatically reduced training time compared to a Self Organising Map, but with at least as good a classification performance.

1 Background

1.1 Clones

Many large software systems have evolved over a long period of time and have made use of relatively unsophisticated software engineering techniques, such as 'cloning'. Here a unit of existing code is adapted for a new purpose, usually by copying and minor modification. Whilst this represents reuse of resources it can also result in software which is difficult to enhance and maintain. One of the major problems with cloned software in a large system is simply that of locating clones of a unit of code and it is this issue which is addressed in the body of work reported here and in our earlier paper [2]. This task must be accomplished from the source code alone as historical documentation may not be sufficiently detailed in all cases to enable an historical audit to be accomplished.

1.2 Overview

A solution to this problem is, firstly, to represent software units by some numerical feature vector and then to use an unsupervised neural net to identify clusters in these feature vectors. The neural net should also organise the clusters so that it is possible to locate less likely clones than those found in the immediate cluster of the unit being examined.

In view of this latter requirement a natural network architecture to use is Kohonen's self organising map (SOM) [4], in which the clusters found are organised in a two-dimensional topographic map. This was the approach that was taken in our first system, as reported in [2]. Whilst the SOM approach performed reasonably well, in a search for improved performance, a novel neural network architecture, Dynamic Competitive Learning [5] has been investigated.

2 Representation

The requirement here is that a block of source code, written in a standard block structured high level language, should be represented by a numeric vector in such a way that similar blocks of code have similar representational vectors. It is important to note here that the notion of 'similar' is different in either case; for the source code similarity relates to the probability of the code being cloned, and for the corresponding vectors that they are close to in Euclidean space. More specifically we require that if software unit A is represented by vector $\mathbf{v}_A$ and B by $\mathbf{v}_B$ then if B is a clone of A $\|\mathbf{v}_A - \mathbf{v}_B\|_2$ should be relatively small.

In fact the set of source code units can be thought of as a set of indexed taxonomic trees, where the root of each tree corresponds to the first occurrence when a unit was cloned and branching points further down the tree correspond to further instances of cloning — the whole structure is analogous to a set of phylogenetic trees. The index of each tree denotes the degree of divergence of the clone from its parent. Due to the equivalence of indexed hierarchical trees and ultrametric [6] spaces it can be seen that the task of detecting clones becomes one of inducing an appropriate ultrametric in the space of representational vectors. Clustering the source code vectors with a SOM provides a similarity metric, whereas our newer model produces the desired ultrametric - the vectors are organised into a tree structure.

2.1 Representation Vector Definition

To represent all the information in a source code unit would not be feasible due to the resulting size of the vector and the need to represent a variable length structure in a fixed length format. Some degree of abstraction is therefore required. Our first simplification was that the user chosen tokens (e.g. identifiers or operators), should be largely ignored; this gives a big reduction to the size of the problem. We were therefore left with the problem of capturing the information in the parse tree of the software unit. We do this in three ways: firstly the frequency of keywords in the unit are accumulated, secondly the indentation pattern is represented and lastly the length of each line is recorded.

The method is applicable to any source code language (it has even been suggested that compiled code may be usefully examined).

As all the code is printed in a standard format (or can be easily filtered to a standard format), which is generated by the programming environment, the indentation pattern is isomorphic to the structure of the parse tree. The problem with representing it is that the number of lines in a unit of software, and therefore the number of indentations, is variable. To map this to a fixed length vector we first took the raw indentation values and viewed them as ordinates on a graph; we then sampled one hundred points from this graph, using linear extrapolation where necessary. This coding method is relatively stable against minor modifications to the source code, such as the addition or removal of a line.

The line length gives some indication of the user defined tokens; it was coded in a similar way to the indentation pattern, again being mapped to one hundred points across the software unit.

Finally each field in the vector was normalised, so that each field had roughly the same importance. Each keyword frequency was divided by the maximum frequency for that keyword averaged over a large set of source code units. Each indentation and line-length value was divided by a mean value calculated as before.

3 Results Using the SOM architecture

The results produced in [2] used 1775 input vectors taken from 5 million lines of source code. A SOM was trained using these vectors. At the end of training 100 test procedures were randomly chosen and the clone database was used to access the nearest procedure to it subject to a cloniness threshold. The two procedures were viewed side by side by a software engineer familiar with PROTEL[1] source code who then decided whether the neighbouring procedure was or was not a clone of the test procedure. A total numerical value of success was then calculated by adding the cloniness value if the procedure was a clone and subtracting it if it were not. The final result was then scaled to lie in the range 0 — 100. The best results being 68.7 under this measure.

[1] BNR's proprietary programming language.

4 Dynamic Competitive Learning

4.1 Background

The DCL model [5] differs in two respects from the standard competitive learning paradigm. Firstly the number of nodes in the network is assigned dynamically without any preimposed upper or lower bound and secondly the nodes in the net are organised in a tree structure. The dynamic creation of nodes has been proposed several times in the context of competitive learning and is best known from the ART type networks [3]; the arrangement of nodes in a tree like structure has also been suggested more than once, see [1] for a detailed discussion of this type of method. In this work we make use of the DCL model with a number of refinements identified in the next section.

4.2 Network Specification

The DCL network

The nodes in the DCL network are arranged in a tree structure with a dummy node as root. When an input is presented to the network it is passed to the first layer of the tree, where the standard competition takes place between nodes at this level. If the winner is not sufficiently close to the input, that is the distance is greater than a **quality** value, a new node is created at this level to classify the input. This process is analogous to the use of a vigilance parameter in ART. The quality value determines the radius of the hypersphere, that is the classification volume of a node; this value must therefore decrease as the tree grows deeper, so that lower levels provide a finer classification than higher levels.

If a new node is not required then the winner moves towards the input, mediated by a learning rate, as in standard competitive learning. The learning process is then recursively applied to the children of the winner, until a leaf node of the tree is reached.

At any stage of the learning process a winning node without children may procreate. This occurs if the relative frequency of wins of the node against its parent exceeds a **threshold** value. In order to prevent unbounded growth in the tree the threshold value is increased for lower levels in the tree.

More formally the algorithm is:

Initialise learning rate, r, threshold, t , quality q and Tree to be the root node
Repeat until convergence criterion met
    Set **x** to the next input vector
    Select the child of Tree that is closest to **x**, **w** say
    If | **w** - **x** | < q *then*
        move w towards x, according to $\mathbf{w}' = \mathbf{w} + r(\mathbf{x} - \mathbf{w})$

Increment the win count of **w**
If w has children *then* set Tree to **w**, increment t, decrement q
and recurse
else if the win count of **w** / win count of Tree > t
generate for **w** a child with weight vector = **x**

Else

Create a new child of Tree that has weight vector identical to **x**

End Repeat

The shape of the resulting tree is determined by the input vectors and the choice of learning parameters. The following factors must be specified:

- The quality value of the top level nodes
- The rate at which the quality factor decreases for successive layers as a proportion of the previous value
- The threshold value of the top level nodes
- The rate at which the threshold value increase for successive layers as a proportion of the previous value
- The schedule for the learning rate — which should decrease after each epoch to promote stability as a proportion of the previous value

The Quality Factor was arrived at with the aim of producing a reasonably large initial cluster. With this setting the initial cluster contains 20 nodes with 4 developing a sub classification. So the initial space is divided up into 16 outlier clusters and 4 larger clusters which are sub-divided.

The Quality Reduction rate was chosen so that the network developed the appropriate amount of decomposition and clustering. The very low initial value of the threshold allows the tree to rapidly deepen, but with an increment rate of 1.15, the threshold quickly becomes large: at ten ply the threshold is 0.12, so that a unit must account for 12% of its parents wins before it can gain a successor.

4.3 Training

With these parameters the network converged rapidly; with 10,257 296–ary vectors only two or three epochs were required for reasonable convergence. The training time was less than forty minutes on an HP 712/80.

4.4 The Resulting Network

The network produced consisted of between 7000 and 10,000 units, which even allowing for the redundancy in a tree representation, is obviously excessive. In fact the majority of these nodes were completely redundant; they fell into one of the following two classes:

* Units that did not win during the test phase. These units had been generated during training, but due to migration of nodes in the tree were now redundant
* Runaway branches. Here a branch of the tree would consist of a sequence of only children, so that passage down the branch did not provide any additional classification

When these redundant node were removed the tree consisted of around 1500 active nodes. The shape of the final tree can be characterised as follows:

No of nodes	1500
Maximum Ply	19
Average Branching Factor	2.84
Widest point	298 nodes, at depth 5

The leaf nodes of the tree classified, on average, about 20 input vectors.

5 Results and Evaluation

5.1 Comparison of Classification of SOM and Dynamic network

The dynamic net was tested alongside the equivalent SOM network on the 10,257 vectors.

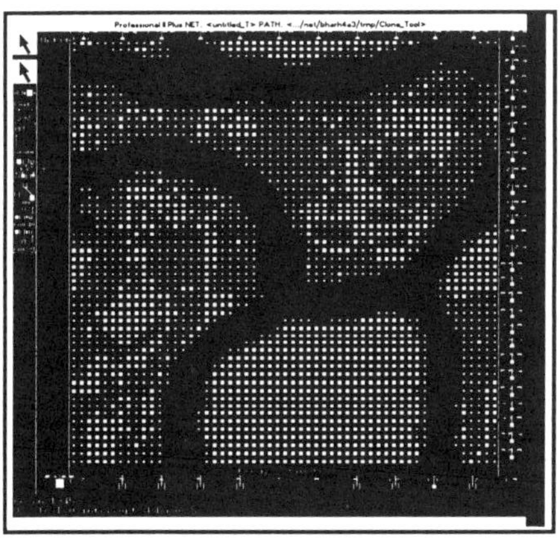

Figure 1. Frequency activation of the nodes contained in the network.

The SOM network used 3025 nodes (55 x 55) with wrap around, an initial neighbourhood of 24 reducing to 1. The SOM produced 4 areas in the output space (see Fig 1).

Similar results were obtained from both networks. Both networks allow the user to navigate their way through the clusters formed; in the SOM by looking at increasing neighbourhoods and in the tree by looking at relations at progressively greater distance, that is increasing the ultrametric neighbourhood.

5.2 Evaluation of the Dynamic Network as a detector of clones

In order to test how well the network found clones, procedures from the test set were modified in ways that a software engineer might do if s/he was cloning the code. The changes made were:

- Adding comment lines
- Changing the indentation by increasing the left hand margin
- Adding a operator, such as 'bind'
- Adding an additional single line statements
- Adding an additional block structures such as 'IF...THEN...ENDIF'

With all of these changes, except the addition of a blocked statement, the classification in the net remained unchanged. The addition of the blocked statement caused a major movement of the classification.

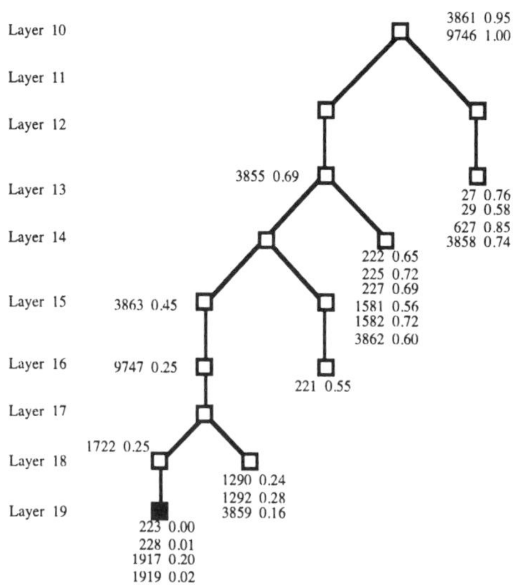

Figure 2 Classification performed by the DCL net.

Figure 2 shows the input (left) and the Euclidean distance (right) from a given test input. For the test input 223 procedure 228 is the closest.

5.3 Conclusions

Generally the DCL model performs as accurately as the SOM on this classification problem and has some significant advantages:

- Training/Recall time: the DCL takes roughly half the time to train and recall is ten times faster
- Size: The trained DCL is about 4Mb and the SOM 20Mb
- Quality of classification: where a tree like structure is anticipated from the classification, i.e. an ultrametric is expected, it is an obvious benefit for this type of structure to be produced, rather the flat topology of a SOM.
- Incremental training is more resilient; as additional areas of the tree are generated, rather than occupying previously assigned areas as in the SOM.

Using a DCL can be difficult however. The network is very sensitive to the selection of parameters, and obtaining the optimal setting can be problematic. It is important to remove redundant nodes as these constitute a significant part of a resulting net.

Acknowledgements

We gratefully acknowledge the financial assistance of the Department of Trade and Industry of the UK Government, partly funding the placement of Paul Barson on a Teaching Company Scheme at BNR Europe Limited, through the Teaching Company Directorate organisation, under grant no. TCS–1326.

References

[1] K. Butchart, "A Comparative Study of two Self Organising and Structurally Adaptive Neural Tree Networks," *University of Hertfordshire Technical Report*,1994.

[2] S. Carter, R.J. Frank, & D. S. W. Tansley, "Clone Detection in Telecommunication Software Systems: A Neural Network Approach," *Proceedings International Workshop on Applications of Neural Networks to Telecommunications* , pp273-280,1993.

[3] G. Carpenter, & S. Grossberg, "The Art of Adaptive Pattern Recognition by a Self-Organising Neural Network", *IEEE Computer* , Vol 21(3), pp77-88. 1988.

[4] T. Kohonen, "The Self-Organising Map", P*roceedings of the IEEE*, Vol. 78, No 9, 1990.

[5] J. Racz, & T. Klotz, "The Dynamic Competitive Learning Method", *Computers in Industry*, 17, pp155-158, 1991

[6] R. Rannal, G. Toulouse, & M. A. Virasoro, "Ultrametricity for Physicists", *rev. mod. phy.*, 58, pp765-788, 1986.

Neural Network Application for External Telephone Network Management

Dr. Eduardo Bayro-Corrochano
Instituto de Electrónica Aplicada. Departamento de Ingeniería Electrónica
Universidad Mayor de San Andrés. La Paz-Bolivia
edb@informatik.uni-kiel.d400.de

Ing. Raúl Espinoza-Soliz
Instituto de Ciencias y Tecnología Apropiada
Universidad Autónoma Juan Misael Saracho. Tarija-Bolivia
admin@ictams.bo

Abstract

This paper presents neural network application useful for external telephone network management. When a new telephone connection is needed the computer locates the required area from the digital map. A quick program search traces the perimeter of the blocks and finds the potential connections in the area. Here a hybrid program which combines a 3-level back-propagation, an orientation estimator, logical decision rules, and a set of orientated masks which trace the perimeter of the block. A second program search identifies the distribution box with a pair of free cables by using the perimeter tracer and after locating the customer's house, the computer quickly finds the best connection. This work has been tested in a Latin American city without rigid urban planning. The developed method allows one to quickly trace the perimeter to find the best telephone connection, even when there are many perimeter irregularities in the neighborhood.

Keywords: Intelligent Systems, Image Processing, Neural Network, Back-propagation Multi-Layer Perceptron, External Telephone Network Management

1. Introduction

In order to find the best connection in an external telephone network, the method presented in this article uses digitized map processing, recognition of patterns using neural networks[1], an orientation estimator, logical decision rules, and a set of orientated masks.

The method traces the perimeter of the blocks, using a set of reconigtion patterns to detect the perimeter image of the different blocks, using a neural network back-propagation.

The order of this paper is: in section 2, a summary is made about external telephone network management. In section 3, the structure of the neural back-propagation multi-layer perceptron network is described. In section 4, the algorithm for finding the best telephone connection is described. In section 5, contains the conclusions and recommendations of the work.

2. Computer Aided External Telephone Network Management

Nowadays the management aided by computers is indispensable for the efficient control of the telephone networks

For such purpose, the Telephone Administration System (TAS) which permits one to integrally run and control an external telephone network has been developed. In this way, the system will establish itself as a utility tool for the technicians in charge of the external plant. The external maintenance plant which is made up of engineers and technicians, will be able to consult the TAS about the tasks and/or problems which occur in the telephone network, and the answers provided by the system will help in carrying out the tasks and solving the problems with greater efficiency and speed, rendering, in such way, a better service to the users. Once the system is functioning, it should be continually fed with the information about the changes which are being carried out in the external network, in this way, the system is permanently up to date.

To carry out the management of an external network, one should know the location and the size of the network, also all the information should be stored and sorted out in the computer by means of maps. The maps should show us the layout of the telephone network, the number of primary cables and their influence areas, also the detailed secondary network plans in the areas. The secondary network maps per area detail the location of the blocks and the distribution boxes which contain the twin cable telephone connections for their assignment to the subscribers. All the maps are digitized and stored in the computer. The maps with a wider access frequency belong to the districts.

Among the tasks of greater importance in the telephone network management, we can include: installation of new subscribers, changes of subscribers from one district to another, elimination of subscribers, follow-up of network faults, etc.

The tasks related to the assignment of twin cable telephone connections, and subscriber moves make up the application part of the neural network for the external telephone network management.

The objective of this system is to render a high quality service to the users, and the cost reduction in the laying down of the subscriber's lines, considering the best connection according to the minimum distance between the distribution box and the subscriber.

3. Neural Network Perceptron

The perceptron was developed by Rosenblatt [2]. and is the first neural network that has been shown. This perceptron uses the artificial neuron concept McCulloch and Pitts. Each neuron processes the present input by means of a correlation function. The result passes on to be evaluated by a non-linear transference function. The perceptron presents a forward moving process and can extend itself by means of multiple levels.

Every second level neuron combines the input with its respective connection weight and subtracts the given fixed threshold value. This value is made up of a parameter for the transference function, in the case of the perceptron, the transference function is a binary type, see figure 1(a) and 1(b).

The function of the perceptron is very simple. It starts by presenting a pattern at the input level, this can have *n* neurons, therefore there is a *n* longitude vector in the input.

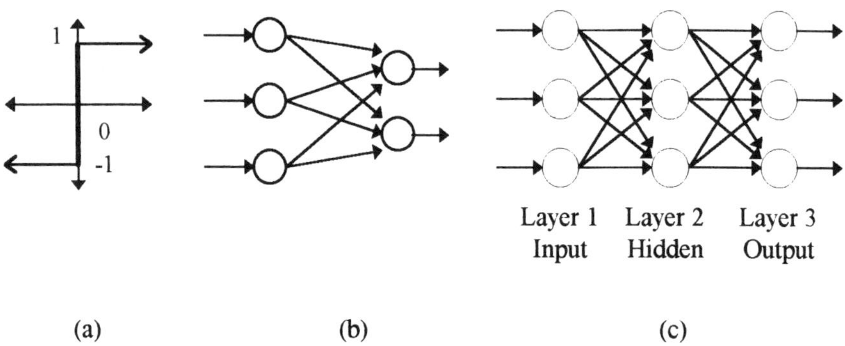

Fig. 1. a) Transference function, b) perceptron structure, c) BMLP.

The neuron activation at the second level can be expressed in the following way.

$$O_{activ} = f\left(\sum_{i=0}^{n} w_{ji}\, x_i - \theta_j \right)$$

Where: O_{activ} is the activation of the jth neuron at the second level.

- f is the transference function.
- w_{ji} is the weight of the connection between the ith neuron at the input level and the jth neuron at the second level.
- x_i is the input of the neuron at the second level.
- θ_j is the threshold value of the jth neuron at the second level.

The perceptron employs a supervised learning. There is a previous neural network training, using definite patterns. This data should be good examples so that the network learns.

The neural network Back-propagation Multi-Layer Perceptron (BMLP) is probably one of the most well-known and widely used networks, this network includes a hidden level, see figure 1(c). Rumelhart, Williams and Parker developed an extension of the delta rule which is called the rule of learning with backward propagation error. In fact, this neural network is a perceptron with a differing transference function. It is very strong and with great learning capacity. The rule propagation error has supervised learning which spreads the output error backwards from the output level up to the input level [3]. Among the applications which are taken into account for the BMLP neural network are: recognition of patterns, signal filtering, noise removal, signal/image segmentation, classification, mapping, adaptive control, data compression, etc.

4. Algorithm for finding the best telephone connection

The use of artificial intelligence techniques, like the neural network, allows us to process the image patterns and to extract the perimeter of the blocks for finding the best connections. Very satisfactory results have been obtained; demonstrating a practical and realistic application of the neural network in the management of telephone systems.

In order to trace the perimeter of the blocks, the program needs to know the location of the subscriber's house in the block. This location establishes the initial reference point to carry out the tracing. The first five points can be found by using the basic masks until the first binary model, which will feed the BMLP neural network for its processing, is obtained. With the active binary model pixels, the orientation estimator computes the contour orientation angle with the slopes which form the adjacent active pixels. With the response of the neural network and the orientation angle, the logical rules will look for the next pixel of the block's perimeter, if the logical rules don't find this pixel, one again resorts to the basic masks. The pixel which is found will form part of the new binary windows, discarding the window's previous first pixel. The program will continue searching until it returns to the initial reference point, indicating that the perimeter has been completely traced.

The BMLP neural network, enables the extraction of perimeter characteristics of the blocks, using a number of patterns and/or models which are appropriate for carrying out this task. Thus, when analyzing images of the block's perimeter regardless of their size and shape, we can find the shortest distance between the customer's house and the distribution box. In order to carry out the extraction of the block's perimeter, we have used the BMLP neural network with the following structure, see figure 2.

The input level has 25 neurons, one neuron for each pixel of the binary window. The hidden level contains 80 neurons, this number guarantees an acceptable learning in the process of the neural network training. The network output level contains 20 neurons, this number is determined by the amount of training models that have been used.

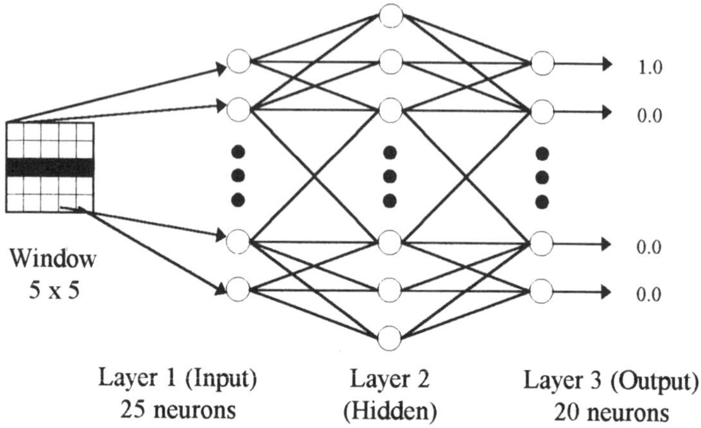

Fig. 2. BMLP neural network structure that was used

The BMLP neural network is used in two modes: training and recall.

In the training mode the neural network learns how to recognize the perimeter images by means of a set of models and/or patterns established by geometrical forms. The number of these patterns is 20, each one of them represented by a binary 5 x 5 window, each element of the window is one pixel wide and each pixel is represented by "1" when it is active and "0" when it is inactive, see figure 3 and 4.

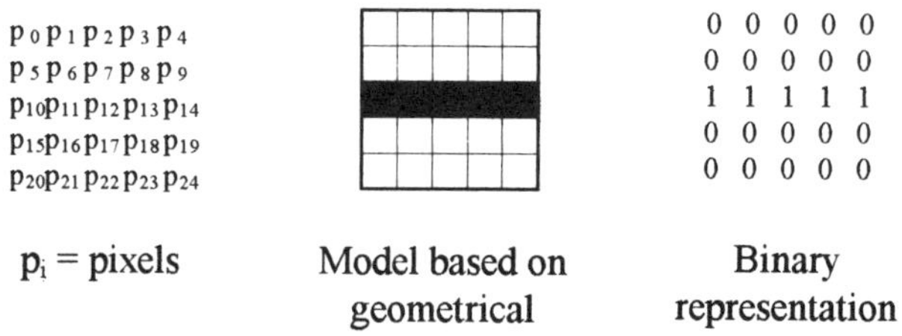

Fig. 3. Window of 5 x 5 pixels, binary model

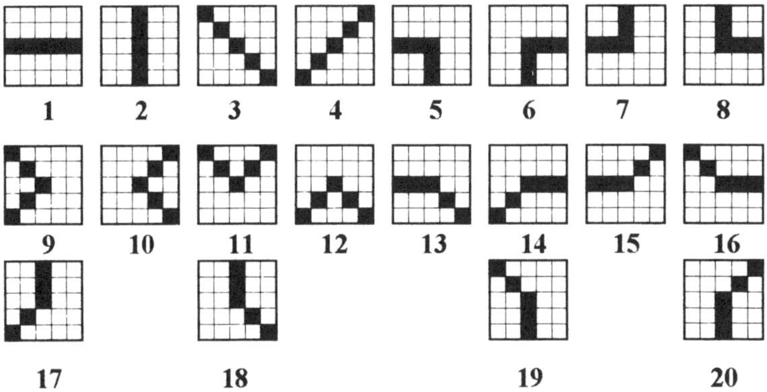

Fig. 4. The set of training models for the BMLP neural network

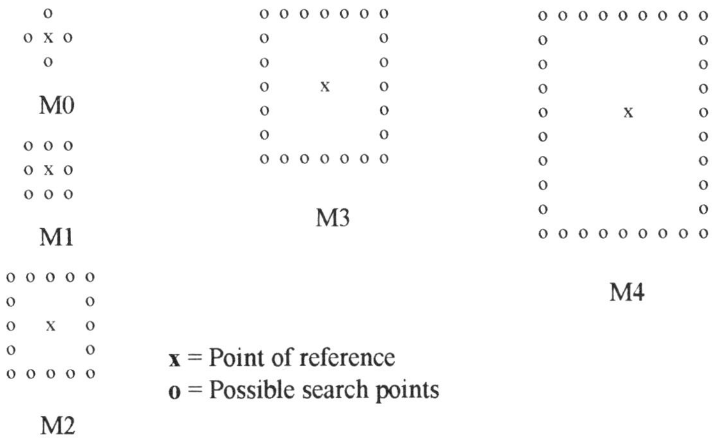

Fig. 5. Basic Bidimensional Masks for pixel searching

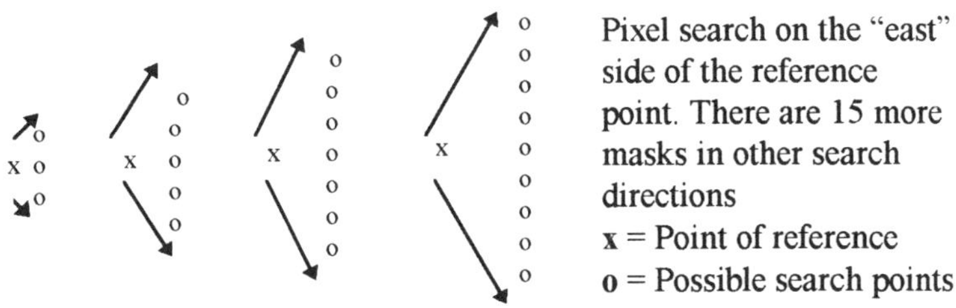

Fig. 6. Oriented Bidimensional Mask for pixel search

The angle of image orientation is obtained through a simple orientation estimator.

$$\alpha_o = \frac{\sum_{i=(n+1)/2}^{n} \alpha_i}{(n-1)/2}$$

where:

n represents the size of the window, n=5

α_i angle between adjacent points

α_o orientation angle of the image in the 5 x 5 window

In the recall mode, the neural network processes the image patterns which show up at the input level, for the classification of the pattern, the highest value of the output neurons is selected.

In a hybrid program which combines the 3-level BMLP neural network, an orientation estimator, logical decision rules and a set of oriented masks enables one to continue the tracing of any block's perimeter. The distribution boxes that are on the houses are located by this program, and the customer's house is identified in the block, then the computer will quickly find the best connection. The details of the algorithm are given belongs:

Stage 1.- The program is started with the initial point of reference. From this point on, the perimeter of the block will be followed. In order to find the following point of the block's perimeter the basic masks are used M0, M1, M2, M3, M4, see figure 5, the search starts in the 4 adjacent pixels using mask M0, if this pixel can't be found, one reverts to mask M1, and in this way, successively until M4; if there is a greater distance than 4 pixels between adjacent points of the block's perimeter, then it is considered to be a mistake in the tracing of the block. The search continues until the completion of the first 5 points which enables one to form the binary model in the 5 x 5 window, only when you have the first binary model, may you feed the BMLP neural network, with this image.

Stage 2.- The given image in a 5 x 5 pixel window will be presented in the input of the BMLP neural network, the network processes the present information in the neurons at the input level and, as a result of this process, a response is obtained in the neurons at the output level. This is interpreted as a response of the network to the maximum value in one of the neurons of the output level.

Stage 3.- The orientation estimator computes the contour orientation angle using the slopes of the pixels within the 5 x 5 window.

Stage 4.- With the information given by the neural network and the orientation estimator, we proceed to use a set of logical rules which will enable us to find the next pixel that belongs to the perimeter of the block, this point is found by making use of orientated masks, see figure 6; take note that a minimum number of bits are used here for the search. In the case that the next pixel is not found, which may happen because of the sudden change in the shape of the block, one resorts to using the basic masks until the next pixel is found.

Stage 5.- Once the pixel is found, it becomes part of a new 5 x 5 pixel window, discarding the first pixel, which means the next window was moved one pixel, with this new image you return to stage 1 and repeat the process.

5. Conclusion

This application of the neural network enables the best connection between a distribution box, which contains free twin telephone cables, and the customer's house which are in the same block in the area, to be accomplished.

The work is carried out by means of tracing the perimeter of the block, regardless of its shape and size.

This application was tested by tracing the more complex irregular perimeters which occur in real situations, obtaining the same results.

6. Acknowledgments

The authors are grateful to the Universidad Mayor de San Andrés, La Paz-Bolivia, Instituto de Ciencias y Tecnología Apropiada (ICTA) - Universidad Autónoma Juan Misael Saracho, Tarija-Bolivia, Cooperativa de Servicios Telefónicos de Tarija (COSETT) and SIEMENS for the cooperation given in carrying out this work.

References

1. Pham D.T. and Bayro-Corrochano E.J. Neural Computing for Noise Filtering, Edge Detection and Signature Extraction. Cardiff. UK. 1992
2. Maren Alianna J., Harston Craig T., Pap Robert M. Handbook of Neural Computing Applications. Academic Press, Inc. 1990
3. Karna Kamal N., Breen David M. An Artificial Neural Networks Tutorial Computer Communications and Graphics Associates, Inc. USA.
4. Simpson Patrick K. Artificial Neural Systems Foundations, Paradigms, Applications, and Implementations. Pergamon Press, Inc. 1990
5. Belleza Eduardo, Szymanczyk Oscar. Diseño de Planta Externa. FICUM S.A. 1987

An Intelligent Monitoring Hierarchy for Network Management

Cynthia S. Hood and Chuanyi Ji

ECSE Department
Rensselaer Polytechnic Institute
Troy, NY 12180
`hoodci@ecse.rpi.edu, chuanyi@ecse.rpi.edu`

Abstract

To improve network management in today's increasingly complex communication networks, we propose an intelligent monitoring hierarchy. As demonstrated on real network data, this hierarchy can detect and correlate abnormal behavior at various levels, thereby providing the network manager a complete picture of the nodes health. The hierarchy is comprised of Hidden Markov Models (HMMs) and neural networks.

1 Introduction

As communication networks increase in size and complexity, understanding and managing network behavior becomes increasingly difficult. To maintain network availability and reliability, the network manager (human or automated computer system) must recognize problematic or anomalous behavior at an early stage and choose the appropriate actions to prevent the situation from snowballing into a catastrophic failure.

The temporal relationship between events or alarms must be understood so the problem can be identified at the highest level and the appropriate action(s) can be taken. If the symptoms are identified as the problems and the higher-level issue is not addressed, the actions taken may actually degrade the network further. Currently, due to the way data is collected and reported [7], temporal information is lost making correlation extremely difficult.

The correlation problem is further complicated by the fact that it is not feasible to identify every abnormal situation (what and how to correlate for anomaly detection?). Previous research approaches include expert systems [1], Finite State Machines [4] and advanced database techniques [6]. These approaches consider temporal relationships, but require network and fault specifications to set up the required constructs. They do not address the fact that anomaly information is not available in most cases.

We propose an intelligent network monitoring hierarchy at each node to address these issues. There are two key ideas incorporated into this hierarchy. The first is that the state probabilities at various levels in the hierarchy are determined using only low-level information so fault models are not required at the higher levels. This is critical since complete fault information is only available at the lowest level (usually in the form of thresholds). The second is that the hierarchy correlates the information at the node, where the information is more accurate and complete. The hierarchy consists of Hidden Markov Models (HMMs) and neural networks. The neural networks are used to learn the normal behavior of the network, and the HMMs provide temporal context.

Our results demonstrate on real network data that this type of hierarchy can provide a full picture of a node's health to the network manager, thereby providing the alarm correlation locally. The network manager can then more easily determine if the alarm is a problem itself (no higher-level problems exist) or a symptom (higher-level problems also exist) of a bigger problem.

2 Intelligent Monitoring Hierarchy

We have taken advantage of the breakdown of network architectures into a set of hierarchical layers to construct a network monitoring hierarchy. The three-level hierarchy detects anomalies in a Managment Information Base (MIB) variable, within a protocol layer, and spanning both protocol layers (defined as network anomalies in this paper)(Fig. 1(a)). This allows both low- and high-level information to be passed on to the network manager.

The detection is accomplished by defining two states (normal and abnormal), corresponding to the health of the element in the hierarchy (i.e MIB variable, protocol layer, or network). The state of each element is estimated using the state probabilities from the previous or lower level as input. Elements identified as being in the abnormal state are considered to be anomalies.

At each level, the state estimation involves an HMM. We have assumed that the underlying process is Markov in nature, but we cannot observe the states (i.e. they are hidden), therefore an HMM is chosen. This model provides a means for considering temporal context which is very important in the network monitoring application.

2.1 Lowest Level of the Hierarchy

At the lowest level (MIB variable), each MIB variable is the input into a hybrid neural network-HMM as proposed by Smyth [5](Fig. 1(b)). A feed-forward network is trained using the standard backpropagation algorithm to provide the a posteriori state probabilities $p(s_i|\theta(t))$ [3], where s_i is the state and $\theta(t)$ is the time-series data of the MIB variables at time t. Given these probabilities, the HMM then provides temporal context by estimating the state probabilities as

$$p(s_i|\theta(t)) = \frac{\alpha_i(t)}{\sum_{j=1}^{m} \alpha_j(t)},$$

where

$$\alpha_i(t) = p(\theta(t)|s_i) \sum_{j=1}^{m} a_{ij}\alpha_j(t-1).$$

We assume a_{ij}, the state transition probabilities to be known ahead of time. By Bayes' rule

$$p(\theta(t)|s_i) = \frac{p(s_i|\theta(t))p(\theta(t))}{p(s_i)},$$

where $p(s_i)$ are the known prior state probabilities. The outputs of the HMMs (state probabilities) are used to estimate the state of the MIB variable and also passed up to the second level. The state is estimated by selecting the state with the highest posterior state probability. If the state is determined to be abnormal, an alarm will be sent to the network manager (along with the states of the next two levels).

2.2 Higher Levels of the Hierarchy

At the each of the next two levels (Fig. 1(b)), the outputs from the previous level are combined linearly to estimate the higher-level state probabilities $p(S_i|\Phi(t))$, where S_i is defined as a higher-level state probability and $\Phi(t)$ is $\theta(t)$ and the outputs from previous level HMM(s).

$$p(S_1|\Phi(t)) = p(s_{11}, s_{12}|\Phi(t)) + 0.5[p(s_{11}, s_{22}|\Phi(t)) + p(s_{21}, s_{12}|\Phi(t))]$$

$$p(S_2|\Phi(t)) = p(s_{21}, s_{22}|\Phi(t)) + 0.5[p(s_{11}, s_{22}|\Phi(t)) + p(s_{21}, s_{12}|\Phi(t))]$$

where s_{ij} represents the ith state of the jth output from the previous level. In the cases where one of the inputs is normal and the other is abnormal, the probability mass is split evenly. The variables we monitored were assumed to be independent (a reasonable assumption based on the data), so the joint probabilities can be easily calculated. These state probabilities are then used as input into the HMM as described in Section 2.1. Once again the state is estimated and the state estimate is included in the information sent to the network manager if an alarm is warranted (i.e. an abnormal state has been detected within the hierarchy).

3 TCP/IP Network Monitoring

The proposed hierarchy is applied to the TCP and IP layers of a single node on the Internet. Time-series data is collected by polling the MIB of the node. The MIB was chosen as a data source due to the clear separation of data by protocol layer and the availability. Each Internet node serving as a Simple Network Management Protocol (SNMP) agent collects and maintains a standard set of information in its MIB [2]. The Management Information Tree (MIT) is analogous to the MIB in OSI networks [7], making this approach viable on OSI networks as well.

4 Experimental Results

4.1 Data

The concepts of the intelligent monitoring hierarchy detailed in Section 2 were tested using data collected from a single node on the Internet. The MIB of the node was polled every 15 seconds to collect data. The data was gathered over a period of fourteen days. During this time, we created a network congestion situation by flooding the network with broadcast packets.

A total of four MIB variables were monitored. In the IP layer, the number of input datagrams received from interfaces (ipInReceives) and the number of IP datagrams supplied to IP by local users for transmission (ipOutRequests) were monitored. In the TCP layer, the number of segments received (tcpInSegs) and the number of bytes retransmitted (tcpRetransBytes) were monitored.

4.2 Design of Monitoring Systems

The neural networks used had one input, a hidden layer consisting of three nodes and two outputs. They were trained by backpropagation with 6000 training samples. Artificially generated abnormal data (approx. 10%) was added to the training set to compensate for the lack of abnormal samples in the network data (approx. 0.5%). The abnormal samples were drawn from a uniform distribution.

The HMM transition probabilities were chosen based on discussions with network managers and the collected data. HMMs at the same level in the hierarchy had the same transition matrices. We assumed that the prior probabilities of both states were equally likely.

4.3 Results

Figure 2(a)-(g) illustrates the performance of the intelligent monitoring hierarchy on the network congestion situations we created. At each level the state estimation (state 1 = normal, state 2 = abnormal) and the probability of being in an abnormal state is shown. Alarms would be generated in places where state 2 has been reached. At any given time, an alarm message sent to the network manager would include the state of the variable, the protocol layer and the network.

Starting from the lowest level, Figs. 2(d)-(g) show the alarms generated after each MIB variable is temporally correlated through the hybrid neural network-HMM. When you go to the next level, the amount of time in the abnormal state decreases as expected. There are only four intervals in which the IP layer is identified as abnormal (Fig. 2(b)) and only one where the TCP layer is identified as abnormal (Fig. 2(c)). At the highest level of the hierarchy, there are two intervals where the entire network is in an abnormal state (Fig. 2(a)).

These results demonstrate that the system is capable of generating temporally correlated alarms at each level, thus providing the network manager with a full picture of the node's health. The power of this system can be illustrated by examining the second abnormal state interval for the network (Fig. 2(a)). During this interval, the IP layer is also in an abnormal state (Fig. 2(b)), but the TCP layer is not (Fig. 2(c)) although the probability of TCP being in an abnormal state is non-zero indicating that things may not be totally normal in that layer. Given the structure of the protocol, we know that when IP is abnormal, problems may "snowball" up to the TCP layer. Therefore, in this case, the monitoring hierarchy has detected the network problem before the TCP layer has totally become abnormal. We have thus demonstrated the system's ability to notify the network manager of a "snowballing" problem.

5 Conclusion

The intelligent network monitoring hierarchy proposed in this paper has demonstrated the ability to detect unknown anomalies at various levels, thereby creating a complete picture of the health of a node. This information can be of great assistance to the network manager by identifying the problem at the highest level so the appropriate actions can be taken. It is important to note that this can be achieved without specific information about anomalies at the protocol layer and network layers.

Future work will address incorporating more of the MIB variables into the monitoring system. As these variables are not all independent, this will involve investigation of other methods of combining the outputs of the HMM.

Acknowledgements

We would like to thank David Durham for his work in generating the network congestion situations and Isabel Nirenberg for her helpful discussions on Internet network management.

References

[1] G. Jakobson and M.D. Weissman, "Alarm Correlation," *IEEE Network,* November 1993, p. 52-59.

[2] K. McCloghrie and M. Rose, "Management Information Base for Network Management of TCP/IP-based internets: MIB-II," RFC 1213, March 1991.

[3] M.D. Richard and R.P. Lippmann, "Neural network classifiers estimate Bayesian a posteriori probabilities," *Neural Computation,* 3(4), pp.461-483, 1992.

[4] I. Rouvellou, "Graph Identification Techniques Applied to Network Management Problems," Ph.D dissertation, Columbia University, 1993.

[5] P. Smyth, "Markov Monitoring with Unknown States," *IEEE Journal on Selected Areas in Communications* vol.12, pp.1600-1612, 1994.

[6] O. Wolfson, S. Sengupta, and Y. Yemini, "Managing Communication Networks by Monitoring Databases," *IEEE Transactions on Software Engineering,* vol.17, no.9, Sept. 1991.

[7] Y. Yemini, "A Critical Survey of Network Management Protocol Standards," Telecommunications Network Management Into the 21st Century, S. Aidarous and T. Plevyak (eds.), New York, NY, IEEE Press, 1994.

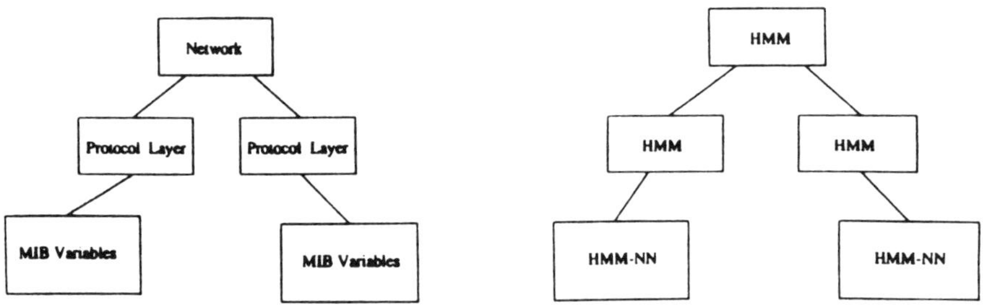

(a) (b)

Figure 1: (a) Logical structure of hierarchy and (b) Implementation of hierarchy.

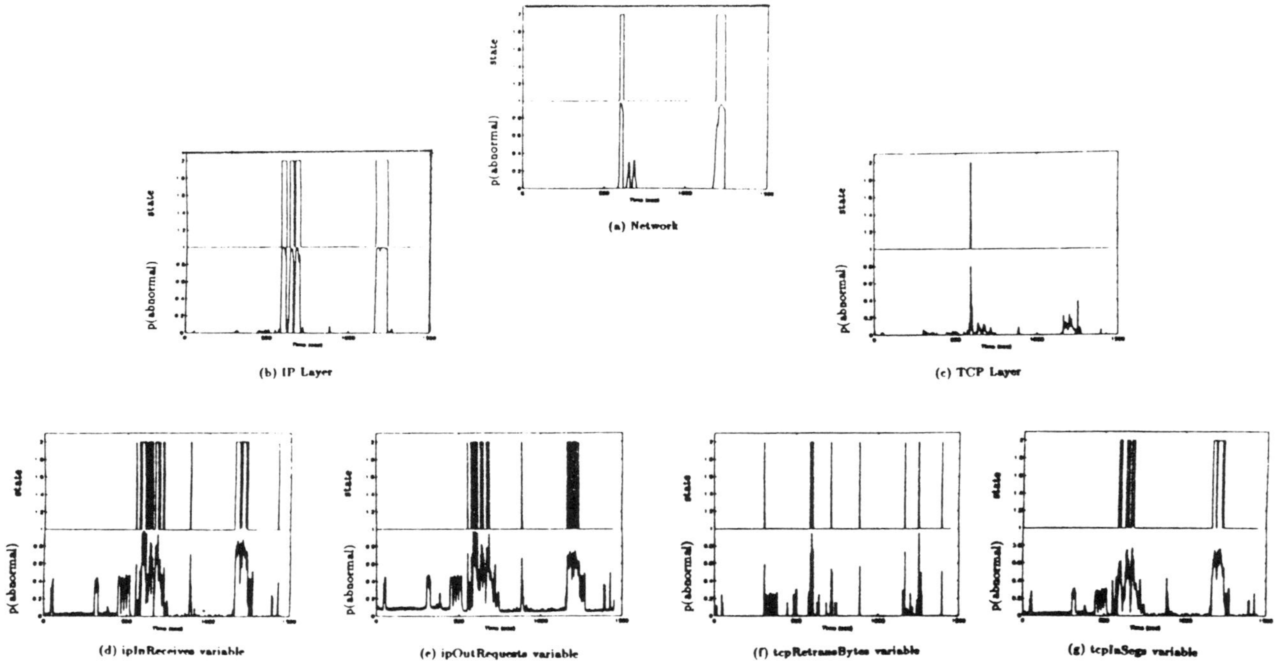

Figure 2: Hierarchy output at each level for a network anomaly

Learning Telephone Network Trunk Reservation Congestion Control using Neural Networks

R. M. Goodman and B. E. Ambrose

California Institute of Technology
Pasadena, CA91125, USA
`rogo@micro.caltech.edu`

Abstract

The paper reports on the use of a neural network to learn the trunk reservation problem, which is a form of congestion control. The telephone network is symmetric. The error for backpropagation purposes is the network blocking, after simulating light, medium and heavy traffic and appropriately weighing the results. The results indicate that we can be very confident that the neural network can learn the task and that it performs better than schemes with fixed trunk reservations. The results indicate a 13% decrease in blocking for this assumed traffic pattern.

1 Introduction

Congestion control is an important topic in the design of telephone networks and will be even more so in design of ATM networks. In this paper, neural networks are used to decide the level of trunk reservation to apply in congestion situations. The communication networks investigated are symmetric fully connected telephone networks.

2 Trunk Reservation

A simple example of a small network is given in Figure 1. This shows a simple five node network, with nodes labelled A to E. Each node has a set

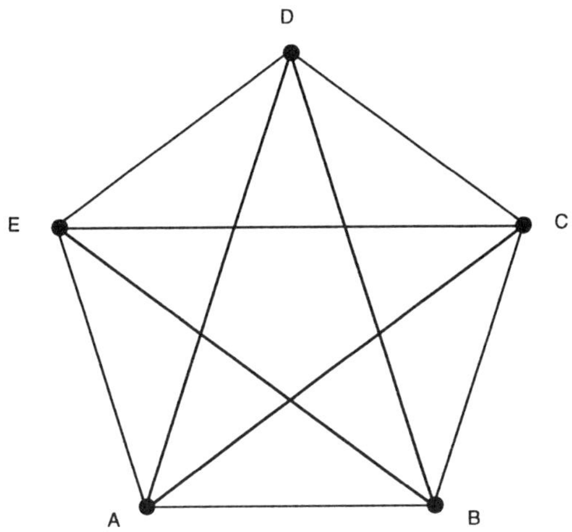

Figure 1: Example of a Symmetric Five Node Network

of links between them. This small network can be used to demonstrate the concepts underlying trunk reservation.

Calls are usually setup using either one or two hops. For example, a call from A to E could be setup using either the trunks from A to E, or else A to B and then B to E. In lightly loaded traffic conditions, using two hops allows more opportunities to complete telephone calls. However in heavily loaded traffic conditions, the two trunk groups used by a two hop call could instead be used to set up two one hop calls and reduce the network blocking.

One hop traffic is called *direct routed* traffic. Two hop traffic is called *alternate routed* traffic.

Trunk reservation is a policy whereby in each trunk group, alternate routed traffic is blocked if there is less than R circuits free on the trunk group. R is known as the trunk reservation parameter. This has the effect of favoring the direct routed traffic at the expense of alternate routed traffic, increasing the network efficiency in time of high load. The parameter R can be very small and still have a substantial effect as is illustrated in Figure 2. A fixed point model [1, 2] was used to generate this plot.

Akinpelu in [2] shows that trunk reservation prevents networks having two stable states at high loads. Kelly in [3] references a result by Hunt and Laws that a policy that chooses the least busy alternative for routing

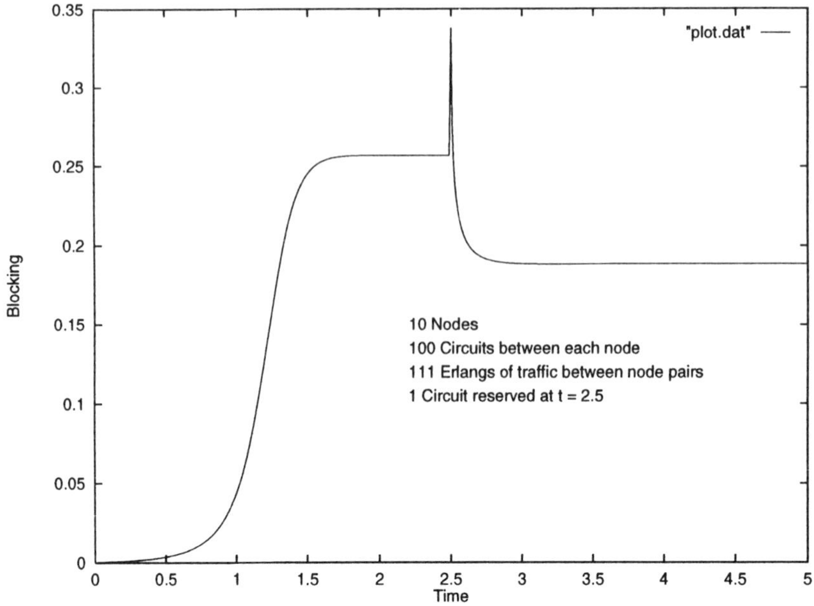

Figure 2: Effect of turning on trunk reservation at $t = 2.5$

and implements trunk reservation is an asymptotically optimal policy in minimizing blocked traffic, as the number of network nodes increases.

Closed form solutions for the correct value of the trunk reservation parameter, R, to use as a function of network load, traffic mix on the trunk group and number of circuits on the trunk group are not known [1]. In this paper, a neural network is used to decide on the value of the trunk reservation parameter as a function of switch load and trunk group traffic mix.

The remainder of the paper talks about the constraints for this problem, training the neural network, the architecture of the neural network, the traffic mix, results and conclusions.

3 Constraints

In a large network, it is unrealistic to expect each node to have a global view of the network. Therefore it was decided to require the inputs to the neural network to be local information, in other words, statistics about the route or statistics about the switch to which the route is attached.

The main benefit of using a neural network on this problem is the ability of network to adjust its weights to reflect changes in the nature of the traffic in the communications network. An unsupervised learning approach was taken. In this case, the neural network is not provided with training examples taken from analytic studies of trunk reservation. Instead simulation is used to provide the training set. The neural network is required to adjust its weights to minimize the overall network blocking in the simulated network.

4 Neural Network

The neural network had two inputs and two hidden units. A linear output unit was used to aid in function fitting. Since good results were obtained with two hidden units, the number of hidden units was not varied.

The inputs were the switch loading and the traffic mix. The *switch loading* was defined to be the number of occupied trunks attached to the switch divided by the total trunks attached to that switch. The *traffic mix* was defined to be the number of direct routed calls on the trunk group divided by the total number of calls on the trunk group. This second parameter was used experimentally. The results indicate it does not have a strong influence on the value of the trunk reservation parameter chosen and could be left out. See Figure 3.

The neural network will output fractional values for trunk reservation. These were interpreted as follows. If the trunk reservation parameter from the neural network was 0.3, then 0.3 of the time a trunk reservation parameter of 1 was used, and the remainder of the time a trunk reservation parameter of 0 was used.

Each simulation cycle represented a 3000 minute or 50 hour traffic simulation. This generated about 18000 test cases and 18000 training cases. Each training case or test case fell into one of the following classes:

- a direct routed call was blocked because the trunk reservation parameter was not high enough. The training data contained a new higher value of trunk reservation.

- an alternate routed call was blocked indicating that the trunk reservation parameter was too high. The training data contained a new lower value of trunk reservation.

- to keep the neural network output steady in regions of the input space that rarely experienced blocking, 1% of the simulated call arrivals were used as training data, keeping the value of the trunk reservation parameter unaltered.

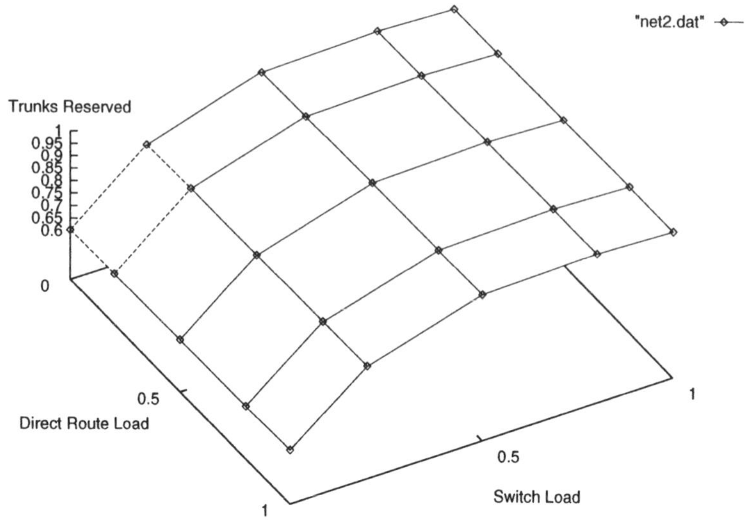

Figure 3: Neural Network Output

Quickprop [4] was then run for 200 iterations to tune the neural network weights. About 10 such simulation cycles were necessary to get good results. See Figure 4.

5 Traffic Mix

For the traffic simulation, a mixture of light (4.4 Erlangs), medium (6.7 Erlangs) and heavy (8.9 Erlangs) traffic was used. The simulated network had 10 nodes and 10 circuits between each node. Because light traffic occurs far more often in real life than medium or heavy, a weight of 0.8 was given to the light, 0.15 to the medium and 0.05 to the heavy traffic for the purpose of computing blocking.

The light, medium or heavy traffic level was indirectly input to the neural network through the switch load input variable. As the traffic level increases, the expected switch load will increase also.

It is important to note that online training in a "live" network could be substituted for the traffic simulations. Once enough new training cases were available, the neural network weights could be tuned. In this way the neural network could continually adjust itself to the traffic mixes found in

	Blocking	±
Neural Network (train)	0.0129164	±0.0000348
Fixed Res of 0 (train)	0.0148592	±0.0000568
Fixed Res of 3 (train)	0.0158059	±0.0000341
Neural Network (test)	0.0129246	±0.0000368
Fixed Res of 0 (test)	0.0148622	±0.0000477
Fixed Res of 3 (test)	0.0157703	±0.0000417

Table 1: Blocking for Neural Network and Fixed Reservation Parameters

the real network.

6 Results

Figure 4 shows how the training and test error decrease with the number of simulation iterations. For comparison, fixed reservation schemes with 3 trunks and 0 trunks reserved are also shown.

The results are given in tabular form in Table 1. The results indicate a 13% decrease in blocking compared to not using any trunk reservation. An even greater decrease is found compared to using a value of 3 for the trunk reservation parameter.

7 Conclusions

In this paper, neural networks have been applied to telephone network congestion relief and done better than the conventional technique of a fixed reservation scheme. In addition, it has been shown that:

- The neural network can be applied to a difficult unsupervised learning problem.
- In the test network, a large decrease in average blocking was seen.
- By substituting live traffic for the simulations in this paper, the neural network can learn in real-time.

It should be possible to extend the results in this paper to networks containing trunk groups of different sizes. This would require an extra input to the neural network containing this information, and longer simulation runs to make sure all parts of the input space were well represented.

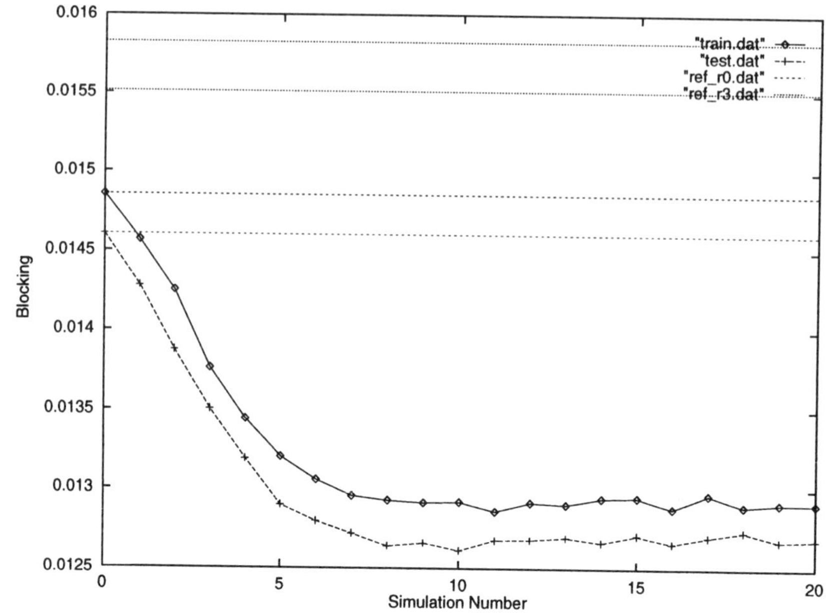

Figure 4: Learning the Trunk Reservation Problem

References

[1] D. Mitra and J. B. Seery, "Comparative evaluations of randomized and dynamic routing strategies for circuit-switched networks," *IEEE Transactions on Communications*, vol. 39, pp. 102–116, Jan. 1991.

[2] J. M. Akinpelu, "The overload performance of engineered networks with nonhierarchical and hierarchical routing," *Bell System Technical Journal*, pp. 1261–1280, Sept. 1984.

[3] F. P. Kelly, "Bounds on the performance of dynamic routing schemes for highly connected networks," *Mathematics of Operations Research*, vol. 19, pp. 1–21, Feb. 1994.

[4] S. E. Fahlman, "Faster-learning variations on back-propagation: An empirical study," in *1988 Connectionist Models Summer School*, Morgan Kaufmann, 1988.

Virtual Lookahead - A New Approach to Train Neural Nets for Solving On-Line Decision Problems

András Faragó[1], Miklós Boda[2], Harald Brandt[2], Tamás Henk[1], Tibor Trón[1], József Bíró[1]

[1] *Department of Telecommunications and Telematics*
Technical University of Budapest
XI. Stoczek u. 2, Budapest, Hungary H-1111
Phone: +36 1 463 1861 Fax: +36 1 463 3107
E-mail: `farago@ttt-atm.ttt.bme.hu`, `henk@bme-tel.ttt.bme.hu`

[2] *Ellemtel Telecommunications Systems Laboratories*
S-12525 Älvsjö, Sweden
E-mail: `{Miklos.Boda,Harald.Brandt}@eua.ericsson.se`

Abstract

A training sample selection strategy, called Virtual Lookahead, is proposed to select the best samples for training the neural network to make on-line decisions. The essential idea is to mimic the off-line decision process and using it for generating training samples. The feasibility of the approach is demonstrated on the example of Call Admission Control.

1 Introduction

In modern Broadband Integrated Services Digital Networks (B-ISDN) the underlying Asynchronous Transfer Mode (ATM) raises a good number of new decision tasks that should be handled in an on-line manner. As an example, let us mention the decision problem generated by Call Admission

Control (CAC): a fast decision have to be made on the acceptance or rejection of a call, taking into account the required Quality of Service (QoS) parameters of the new call, as well as of all ongoing calls.

Another important example of a task when fast on-line decision is to be made is call routing. In this case the routing algorithm has to select a feasible route quickly from the set of all available routes.

The *on-line nature* of these decision tasks means that the decision has to be made immediately upon a request has arrived, without waiting for future requests.

Now let us imagine for a moment an ideal situation, when some "demon" tells us the future in advance. That is, remaining at the CAC example, we would know not just the present call request, but also future requests, together with their arrival time and duration. It is clear that in this ideal case we could achieve a better utilization of network resources. For example, if there is 100 Mbit/s unused transmission capacity on a communication link and a request arrives that needs 1 Mbit/s with a holding time of one minute, we would not accept this request if we knew that in the next second another request will come that needs 100 Mbit/s for an hour. On the other hand, in reality typically the first request is accepted and then, consequently, the second must be rejected, leading to a lower utilization of resources or lower network revenue.

Of course, we cannot use such a demon that tells us the future in advance, therefore, it is clear that the on-line decision situation involves some inevitable loss in resource utilization, compared to the off-line situation when everything is known in advance.

Now we raise the idea: let us use a neural network to replace the "demon", such that the net is trained by the off-line decisions. That is, in the training phase we do not have to make on-line decisions (e.g. in simulation). In the training phase we can run the simulation to see the "future" and then an off-line decision can be made for the present (a "demonic" decision in the above sense). Then the sequence of these "demonic" decisions are used as training samples for the neural net, which then will be trained to mimic the demon.

Thus, the main point here is not the problem of *how* to train the neural net. There are known algorithms for that, such as back propagation etc., if the samples are already available. The main point is *what* to "teach" to the net, that is, how to select the training samples, in order to achieve good performance when using the net for on-line decision.

We call the proposed training principle *Virtual Lookahead*, because the neural net is trained, in the above sense, to "look ahead" in the future. Of course, this is virtual in the sense that the exact future cannot be foreseen, we just use the information gained in the training phase on the behavior of the system.

An additional advantage of the Virtual Lookahead strategy is that the underlying statistical information can be utilized without any explicit or formal knowledge on it. The only thing we need is a good *off-line* algorithm for the problem, to generate the training decisions.

We introduce this strategy on the example of Call Admission Control. Although there are available methods for CAC, nevertheless, they concentrate only to the present, as well as to the ongoing requests and neglect the additional gain that could be obtained by using the statistical information on the overall behavior of the arrival process. We use this information in the form of the above Virtual Lookahead strategy.

2 A Simple Model for Link Call Admission Control

Let us consider a single communication link. Assume each call that wants to use the link is characterized by only two quantities: the *equivalent bandwidth* and the *holding time* (that is, the duration of the call which is not known in advance). The equivalent bandwidth or equivalent capacity [1] is a quantity that describes the bandwidth demand by a single number even for variable rate calls. This concept significantly simplifies CAC considerations, since the complexity is shifted to the problem of how to compute the equivalent bandwidth as a function of more detailed call descriptors. The way of this computation, however, can be assumed given as a predefined subroutine using known methods.

In this model the CAC task has a natural solving algorithm. If the link has capacity C, the ongoing calls have equivalent bandwidths $B_1, B_2, \ldots$ and the new request is characterized by equivalent bandwidth B, then the call is accepted if and only if

$$B + \sum_i B_i \leq C$$

holds. In other words, the new call can be fit in the link if there is at least as much unused capacity (in the equivalent bandwidth sense) as needed for the new call. For easy reference, let us call this algorithm *Simple Fit*.

Note that the holding time of any given call is not known in advance. For this reason the Simple Fit algorithm uses only the equivalent bandwidth.

The above natural Simple Fit algorithm is used as a comparison reference for the Virtual Lookahead strategy. The idea in this comparison is that in this way we can separate the effect of Virtual Lookahead in the following sense. Naturally, the quality of Simple Fit depends essentially on how the equivalent bandwidth is computed from the detailed call descriptors. If we do not specify this, by taking the equivalent bandwidth *given* as part of the call request, then we can concentrate the investigation on the effect of the new strategy. We only have to assume that the equivalent bandwidth is computed in the same way both for Simple Fit and for Virtual Lookahead, but the specific way is irrelevant from the viewpoint of our present investigation.

In what follows first we demonstrate the feasibility of finding the training set for a Virtual Lookahead based neural CAC algorithm by showing that an efficient and optimal *off-line* algorithm exists. Then we exhibit an example to show the advantage that can be gained by the new strategy.

3 Finding the Optimal Off-Line CAC Decisions

As mentioned above, in the off-line situation we can assume that a sequence of call requests is given in advance (obtained e.g. by simulation or from earlier recording). Let us describe each call in this sequence by a triplet (B, s, t), where B is the required equivalent bandwidth, s is the starting time and t is the termination time of the call (that is, $t - s$ corresponds to the holding time). Thus, the sequence of n calls is given by the sequence

$$(B_1, s_1, t_1), (B_2, s_2, t_2), \ldots, (B_n, s_n, t_n)$$

of triplets.

For easier computation and reduced complexity let us make the following assumptions:

- All the numbers are positive integers, that is, the equivalent bandwidth is always an integer multiple of some elementary unit and the time is also discretized.
- The time is measured in sufficiently fine (discrete) steps to make it assumable that at any instant at most one event (call start or termination) occurs. That is, $s_1, t_1, s_2, t_2, \ldots, s_n, t_n$ are all different numbers.

Let T be the time span of the whole arrival sequence and let C be the link capacity, both expressed by integers. Denote by S the state of the link, meaning the actual occupancy level in terms of occupied equivalent bandwidth. At any given discrete time instant t the following state transitions are possible:

- A new call request arrives and it is accepted. That is, $t = s_i$ for some i and the new state S' will be $S' = S + B_i$. Of course, this is possible only if $S + B_i \leq C$ holds.
- A new call request arrives and it is rejected. In this case no state change occurs, i.e. $S' = S$.
- An ongoing call ends, that is, $t = t_i$ holds for some i. Then the new state is $S' = S - B_i$.
- None of the above occurs, the link state remains unchanged.

Now the link states and possible transitions can be represented by a trellis-like graph, as shown in Fig. 1. Each column corresponds to a discrete time instant and each row represents an occupancy level. Thus, there are altogether TC points. For a given arrival sequence the possible state transitions can be represented in the figure according to the rules described above. For example, at a "general" point characterized by (S, t) the possible successors are:

- $(S, t + 1)$ if no state transition occurs.
- $(S + B_i, t + 1)$ if a call request arrives with bandwidth demand B_i and the call is accepted (assumed $S + B_i \leq C$).
- $(S - B_i, t + 1)$ if a call ends (assumed $S - B_i \geq 0$).

Each path from the first to the last column of he trellis along the allowed transitions represents a possible sequence of decisions. Let us now label each state transition by the *network revenue* generated by the transition. (Here we mean the revenue in an abstract sense, it can represent any objective to be maximized). For accepted calls this is the revenue generated by the call. For rejected calls and unchanged states the revenue can be taken 0.

Now if we find a path of maximum weight in the trellis, maximum in the sense of summed transition labels, from the initial state in the first column to the last column, then this path defines a *decision sequence of maximum network revenue.* Each decision (that is, acceptance/rejection for a request coming in a given state) in this sequence serves as a training sample, representing a "demonic" decision in the sense of the Introduction.

Finding a path in the trellis with the above requirements is a standard dynamic programming task that can be efficiently solved by the following recursion:

$$u(S, t+1) = \max_{(S', t)} \{u(S', t) + c(S', S)\}$$

where $u(S, t)$ is an auxiliary variable collecting the accumulated revenue, $c(S', S)$ is the revenue generated by the state transition and the maximum is taken over all trellis points that can be a predecessor of the considered point.

4 An Example

Assume there are two traffic classes with independent Poissonian arrival processes which offer the same average bit rate traffic of 300 Kbps each to a transmission link of $C = 150$ Mbps. The calls of the two classes are described by equivalent capacities $B_1 = 2$ Mbps and $B_2 = 20$ Mbps, respectively. Further, assume that each call generates one unit of revenue.

Let us apply first the Simple Fit algorithm for this case. Then the probability of call blocking in each class can be computed by known methods, see e.g. [2]. What we obtain for this case is a blocking probability of 0.6% for the first class and 9% for the second class. This can be interpreted in revenue terms as follows. Since second class calls need 10 times more bandwidth, but the two classes offer the same average traffic, therefore, the first class generates 10 times more calls. That is, on the average, from each 1100 calls 1000 belongs to the first class and 100 to the second. Out of the 1000 first-class calls 6 is blocked, on the average, and out of the 100 second-class calls 9 is blocked. Thus, altogether $9 + 6 = 15$ calls are blocked from 1100. This means that $15/1100 = 1.363\%$ of the revenue is lost, compared to the ideal case when no blocking occurs.

Now we show that in the optimal off-line decision sequences, computed according to the previous section, we can achieve higher average revenue.

In the Simple Fit algorithm each first-class call is accepted if the link occupancy is at most 148 Mbps, while second-class calls are accepted only if the occupancy is not higher than 130 Mbps, since they require 20 Mbps capacity. In this way second-class calls experience higher blocking probability. Now if we equalize the classwise blocking probabilities by introducing the rule that *any* call is rejected if the occupancy level exceeds 130 Mbps, then it can be computed by known methods [2] that each class will experience a blocking probability of 7%. If we count the number of lost calls, similarly to the previous paragraph, then we obtain that on the average 14 calls are lost from 1100. This yields a lost revenue of $14/1100 = 1.272\%$, smaller than what we had for Simple Fit.

¿From the above reasoning we can draw the conclusion that *there exists* a strategy that overperforms Simple Fit in terms of revenue. Since the *optimal* off-line algorithm cannot be worse than the above one, therefore, the optimal off-line must consequently be better than Simple Fit.

Thus, we know that the training set, generated by the off-line algorithm, represents a decision that is definitely better than Simple Fit. The question that remains is whether we can train a neural net such that its decision will be sufficiently similar to the one represented by the training sample?

The answer to the above question is nontrivial, since e.g. back propagation does *not* guarantee this behavior. In some cases it can perform quite poorly, as shown in [3]. Nevertheless, the answer is positive if take into account that a neural net, already with one hidden layer, can realize a *universally consistent* decision rule. That means, if the training set and the number of neurons grow, then the decision made by the net asymptotically approaches the best possible decision rule, for *any* probability distribution imposed on the model. This is a highly nontrivial fact, rigorously proven in [4]. It requires, however, another training algorithm, based on choosing the parameter setting that results in the smallest number of errors on the training sample, rather than minimizing a mean squared error. For details see [4].

References

[1] R. Guérin, H. Ahmadi and M. Naghshineh, "Equivalent Capacity and Its Application to Bandwidth Allocation in High-Speed Networks", *IEEE J. Selected Areas of Communications,* vol. 9., pp. 968-981, 1991.

[2] M. Ritter, P. Tran-Gia, "Multi-Rate Models for Dimensioning and Performance Evaluation of ATM Networks", *COST 242 Interim Report,* June 1994.

[3] M.L. Brady, R. Raghavan and J. Slawny, "Back Propagation Fails to Separate Where Perceptrons Succeed", *IEEE Trans. Circuits and Syst.,* vol. 36, pp. 665-674, 1989.

[4] A. Faragó and G. Lugosi, "Strong Universal Consistency of Neural Network Classifiers", *IEEE Trans. Information Theory,* vol. 39, pp. 1146-1151, 1993.

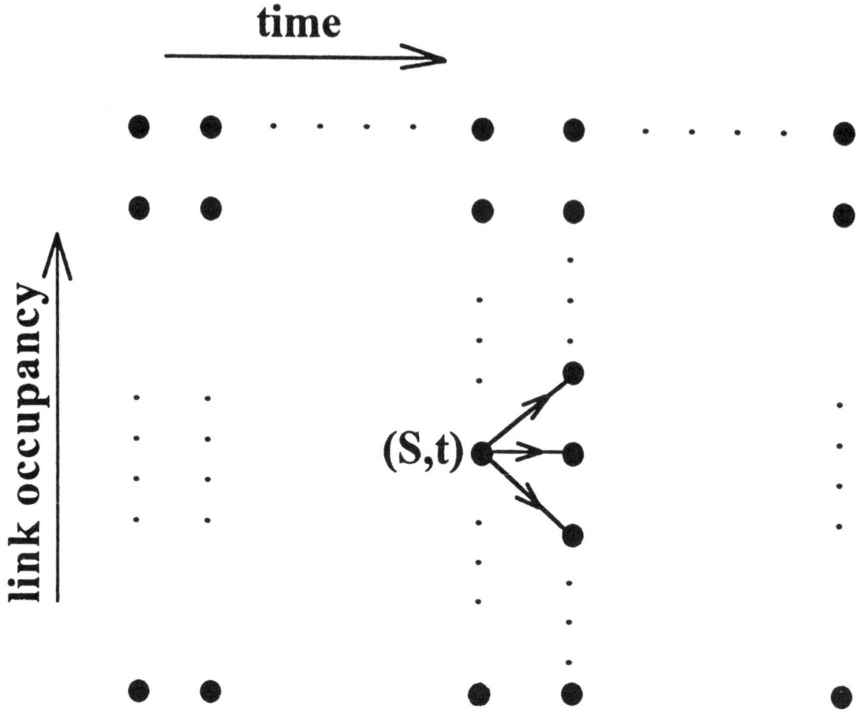

Fig. 1: Trellis for finding the optimal off-line decision sequence

Neural Networks in Real Time Decision Making

P. Campbell, H. L. Ferrá, A. Kowalczyk, C. Leckie and P. Sember

Telecom Research Laboratories
770 Blackburn Road Clayton, Vic. 3168, Australia
p.campbell@trl.oz.au,h.ferra@trl.oz.au,a.kowalczyk@trl.oz.au
c.leckie@trl.oz.au, p.sember@trl.oz.au

Abstract

This paper outlines two applications of neural networks to telecommunications utilising primarily their ability to deliver output in real time. In both applications, one for evaluation of telecommunications structures supporting radio antennas and the other for re-configuration of an SDH network, supervised training was used with a corpus of data labelled by computer implemented complex algorithms.

1 Introduction

Among many virtues of neural networks are their efficiency, in terms of both execution time and required memory for storing the structure, and their practical ability to approximate complex functions. A typical drawback is the usually "data hungry" training algorithm. However, this is not a problem if data can be computer generated. If the generating algorithm turns out to be too complex to implement under practical restrictions (e.g. execution time or memory limitation constraints) a neural network substitute can become crucial for the feasibility of the project. In this paper we present two such telecommunication applications proving that neural networks can also be practical for novel implementation of

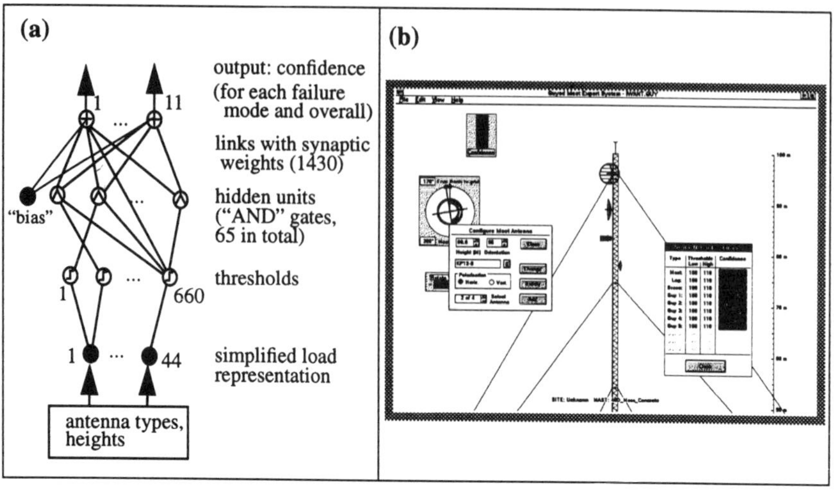

Figure 1: (a) Example of a neural network architecture (mask perceptron) used for guyed mast evaluation. (b) Working screen of the Guyed Mast Expert System.

mature software techniques.

2 Evaluation of Guyed Masts

This particular task was to develop a PC based, interactive tool for configuration of antennas on guyed masts. The project intended to provide end users in the field (radio-designers, planners) with the capability to roughly evaluate the feasibility of proposed modifications and to eliminate in this way one of the serious bottlenecks in the expansion and maintenance of Telecom Australia's extensive microwave network. Adaptation of the typical software used by structural engineers which evaluate stresses in hundreds of elements and joints composing the structure as well as in several supporting guys was ruled out, mainly due to their complexity but also due to some other (e.g. legal) reasons. Usage of the neural network classifier instead was proposed. It is worth mentioning that the structural evaluation of guyed masts, especially tall ones, is a non-trivial exercise due to the complex interaction between wind, guys and the structure itself which is highly tuned, with little redundancy. For instance, in some cases the re-allocation of an antenna by less than a meter may have a dramatic impact on structural stresses, whilst in others masts are so highly tuned that they have to be loaded with some antennas (even dummy ones) in order to avoid overstressing.

Twenty four thousand load cases randomly generated according to statistics from existing sites were evaluated for each of 30 mast types (between 30 and 150 m high) and split evenly into training and test sets. For each of them two neural networks (mask perceptron and fuzzy mask perceptron, c.f. Fig. 1) were developed as follows. First we generated a mask perceptron [7, 8]. The input of such a network consists of the wind load configuration transformed into 44 separate numbers. These are then passed through a few hundred binary threshold units. Outputs of some of these units are connected to the network outputs via weighted links, some become inputs to the hidden units ("AND" gates) whose outputs are again connected via weighted links to the network outputs. Network outputs are designed to evaluate different modes of failure of the structure (e.g due to the excessive stresses in the guys, or in the 'legs' or in the 'bracing'). The normalised maximum of these network outputs was used as a final single figure of merit, but the particular modes of failure are also used as an indication of the possible ways of upgrading the structure. The mask perceptron training routine task [7, 8] is to select a set of useful thresholds and "AND" gates out of thousands of possibilities, then to set weights to minimize the mean-square-error on the training set. In a typical case 20-70 thresholds and 20-70 "AND" gates were selected. Such networks are very compact in terms of disk space required and easy to simulate since the only operations required are comparison (thresholds) and additions. Indeed, the final product covering the whole family of 30 masts fits on a single distribution diskette together with the graphical interface (requiring under 0.5 Mb). The evaluation results are obtained in real time, without visible delay when antennas are being dragged with the mouse on the computer screen (implementation on PC486DX2/66 with the graphical user interface and mask perceptron simulator written in Visual Basic). In order to provide smoother operation a fuzzified version of the mask perceptrons was also developed. This was achieved by replacing hard thresholds with ramp functions and the "AND" gates with minimum (i.e. "fuzzy AND") gates and recalculation of the weights. The average test error of the overall system is below 1% in either case (in the critical range). The typical training time (on 12000 cases) was below typically 10 minutes of user time on a SPARCStation 2.

The system is subject to the Provisional Patent application: PL6503/92, "A Structure Design Method and System".

3 Rapid Service Restoration in SDH Network

The Synchronous Digital Hierarchy (SDH) is a new standard for digital transmission over optical fibres [3] adopted for Australia & Europe (c.f. Figure 2.a). Nodes at the periphery of the network are switches that handle individual calls. Each switch concentrates traffic for another switch into a number

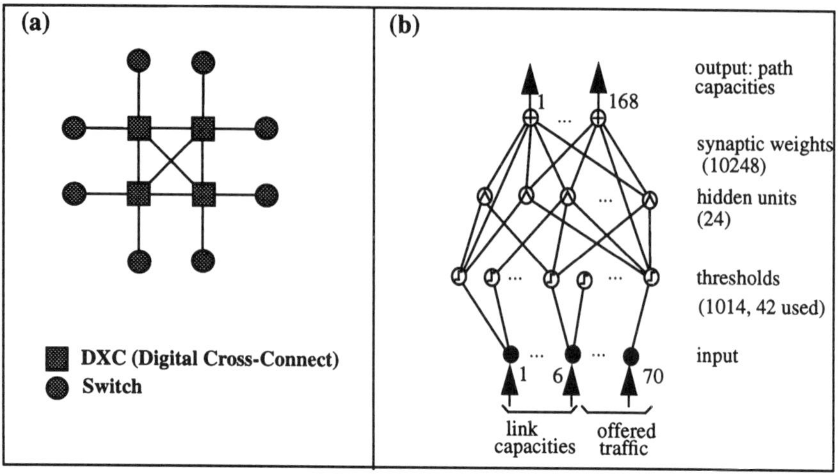

Figure 2: (a) Example of a simple metropolitan SONET/SDH Network used in experiments. (b) Example of an architecture of the mask perceptron generated in experiments.

of streams. Each stream is then transferred to a Digital Cross-Connect (DXC) for switching and transmitted to its destination by allocating to it one of several alternative virtual paths. A particular problem we are concerned with here is the dynamic allocation of capacities (sizes) to these virtual paths in order to maximize network throughput. This is a non-linear optimization task since the virtual path capacities and the constraints, i.e. the capacities of physical link between cross-connects, are quantised (between 0 and 64 in the case of Fig. 2), and the objective function (erlang blocking) depends in a highly non-linear fashion on the allocated capacities and demands.

One of the major features of an SDH network is that it can be remotely reconfigured using software controls. Reconfiguration of the network can become necessary when traffic demands vary, or when failures occur in the DXCs or the links connecting them. Reconfiguration in the case of failure must be extremely fast, with estimated restoration time < 60ms [1]. The problem of optimal network reconfiguration has been researched already. In particular Gopal et. al. have proposed a heuristic greedy search algorithm [4] to solve this nonlinear integer programming problem, Herzberg in [5] reformulated this nonlinear integer optimization problem as a linear programming (LP) task, Herzberg and Bye in [6] investigated application of a simplex algorithm to solve the LP problem, whilst Bye [2] considered an application of a Hopfield neural network for this task, and finally Leckie [9] used another set of AI inspired heuristics to solve the optimization task. All these approaches have practical deficiencies: the linear programming is slow while the heuristic approaches are relatively inaccurate and the Hopfield neural network method (practically simulated on a serial computer) suffers from both deficiencies.

In this paper we apply neural networks to the task but differently than in Bye's approach. We do not use the neural network to solve the optimization task, e.g. as a substitute for the simplex algorithm, but teach it to replicate the "optimal" LP solution provided by it. We decided to use a two stage approach in our experiments. For the first stage we developed a feedforward network able to produce an approximate solution. More precisely, we used a collection (2000) of random examples for which the linear programming solution of capacity allocations had been pre-computed to develop a feedforward neural network able to approximate these solutions. Then, for a new example, such an "approximate" neural network solution was rounded to the nearest integers and used to seed the second stage providing refinement and enforcement of constraints. For the second stage experiments we have used a heuristic module based on the Gopal approach [4] (alternatively, a recurrent neural network can be used here). The heuristic firstly reduces the capacities assigned to paths which cause a physical link capacity violation, then secondly increases the capacities assigned to paths across the physical links which are being under-utilized. The reductions are completed before any path has its capacity increased, as one path can traverse a number of links.

In the initial experiments to date several SDH architectures have been investigated. In this paper we concentrate on results for the simple, fully meshed, SDH network shown in Figure 2(a) (c.f. [5] for further details); the results for other networks were very similar and will be reported elsewhere. The first stage feedforward neural network architecture used, shown in Figure 2(b), is again a mask perceptron. The network, whose inputs are link capacities (6 inputs) and offered traffic (64 inputs), has the appropriate path capacities as outputs (168 in total). This neural network was trained in the supervised mode, using a classical optimization algorithm (linear programming) to label random examples of input patterns (2000 for training and another 2000 for test).

For comparison, it was decided to use this heuristic algorithm by itself to see what sort of results we could obtain. As well as starting from scratch (i.e. having no traffic streams assigned to begin with, and use the incrementing part of the algorithm), it was decided to also see what sort of values could be obtained by starting with the steady-state solution - i.e. the solution proposed by linear programming for the network's original link capacities and traffic - and then decrementing and incrementing traffic stream sizes, just as was done with the neural network solution. For conciseness, these two alternatives will be referred to as the Gopal-0 and the Gopal-S algorithms, for a zero (0) and Steady state starting point respectively.

The results in terms of percentages of the optimal throughput of the SDH network, as determined by the linear programming solution for a given capacity/demand configuration, are as follows. The form of the results is: *mean throughput*

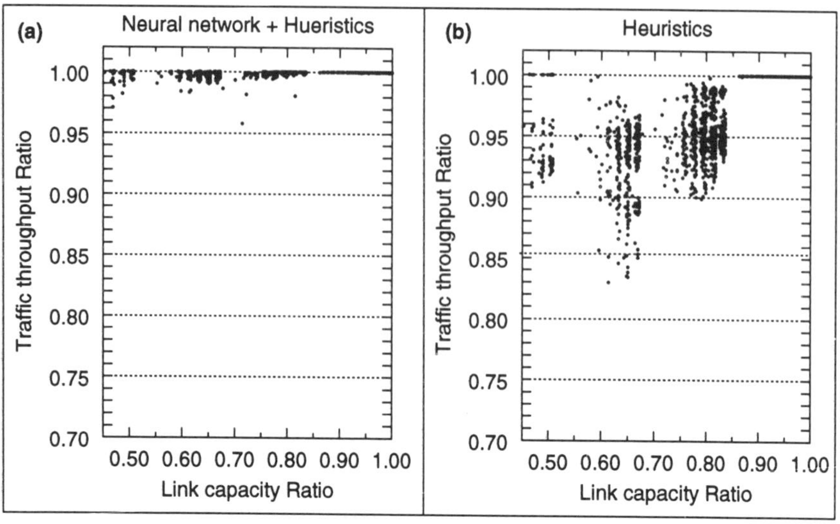

Figure 3: Experimental results for simple SDH network (Fig. 2(a)) on the independent test set of 2000 random cases. The ordinate shows the ratio of the carried traffic to the optimal carried traffic (given by the linear programming solution) as the function of the degradation of the network, i.e. the ratio of the sum of the capacities of all links to such a sum in the steady state. Fig. (a) shows results for the neural network combined which the heuristic second stage and Fig. (b) for the heuristic only initialised by the steady state (Gopal-S).

± *standard deviation%*. For the training set, the neural network achieved a throughput of 99.94 ± 0.17%, the Gopal-S algorithm 96.68 ± 4.75%, and Gopal-0 78.58 ± 21.72%. The result for the neural network on the test set was very similar; 99.93 ± 0.24%, with 96.85 ± 4.66% for Gopal-S and 78.66 ± 21.67% for Gopal-0. Some results are shown graphically in Figure 3, with each dot representing one of the 2000 test cases. Note that the neural network achieves better results than the Gopal-S heuristic with increased network degradation (a decreased link capacity ratio); this trend is even more pronounced in the case of the Gopal-0 heuristic.

3.1 Timings

The neural network generation took about 8 minutes on a SPARCStation 20. Results of the neural network simulation are as follows.

With the help of a profiling tool, average times for the various algorithms were calculated. The majority of experiments were run on a SPARCStation 2 using SunOS 4.1 when few people were around to ensure nearly full usage of the processor, and averaged over three runs to make sure the results were consistent.

On this machine, it took an average of only 35.7 ms for the neural network to calculate the network capacity assignments, but a further 145.8 ms for the second stage heuristic algorithm to ensure that physical link capacity was not exceeded or wasted, for a total of 181.5 ms.

As a comparison, the Gopal-S solution took an average of 221.0 ms, and Gopal-0 174.8 ms; slightly faster than the overall dual-stage neural solution, but with ~20% worse results, on average.

On a recently acquired SPARCStation 20 using Solaris (SunOS 5.3) much better times were achieved. The neural network solution took an average of 9 ms per test case with a further 36.5 ms for the heuristic, for a total of 45.5 ms, with the Gopal-S algorithm requiring 55.3 ms and the Gopal-0 solution 43.7 ms.

The optimal solutions calculated using the linear programming algorithm took between 2 and 60 seconds per case. These were calculated on a SPARCStation 10 running SunOS 4.1 (different machines were needed for different parts of the experiment due to licensing conditions on software and/or available disk space). For those further interested in the technical details, all programs were written in C++ and compiled with the GNU projects C/C++ compiler gcc/g++, version 2.6, on optimization level 2, to ensure a quick implementation of the algorithms.

4 Discussion and Conclusions

The neural network used in these experiments, the mask perceptron, is well suited for digital implementation: the operations required are only comparisons, calculation of logical "AND" and the addition of synaptic weights (no multiplications or any other non-linear transformation are required). Thus its simulation time can be relatively easily improved, if necessary, with a more efficient coding.

In the SDH re-configuration experiment, the fact that the second stage heuristic is so much slower than calculating the initial solution using the neural network led to the idea of using a recurrent neural network instead in future experiments. Furthermore, note that the neural net solution can be speeded up dramatically by implementing it in parallel, e.g. using dedicated hardware, something that cannot be done easily with the serialized heuristic.

It is planned to continue the SDH network experiment in the future; especially with the recurrent network for the second stage and also more complex SDH architectures.

Finally, the results of this paper demonstrate that artificial neural networks can be practically used in applications requiring real-time response.

Acknowledgments

All the research and development reported has the active support of many sections and individuals within and outside Telecom Research Laboratories (TRL). The Guyed Mast Expert System has been developed under contract for Network Products, Telecom Australia, in active co-operation with Structel Pty. Ltd. In particular we acknowledge the contribution and enthusiasm of Mr. T. Tan, now in Mobile Networks, Telecom Australia and Mr. R. Coxhill, TRL, in this project.

The SDH application was possible due to co-operation of a number of our colleagues in TRL, in particular Dr. L. Campbell (who suggested this particular application), Dr. M. Herzberg, Mr. A. Herschtal, and finally Mr. S. Bye and Mr. J. Ginger, both currently with OPTUS.

The permission of the Managing Director, Research and Information Technology, Telecom Australia, to publish this paper is acknowledged.

References

[1] E. Booker, Cross-connect at a Crossroads, *Telephony*, Vol. 215, 1988, pp. 63-65.

[2] S. Bye, A Connectionist Approach to SDH Bandwidth Management. Proc. of the 191 International Conference on Artificial Neural Networks (ICANN-93), Brighton Conference Centre, UK, 1993). pp.286-290.

[3] R. Gillan, Advanced Network Architectures Exploiting the Synchronous Digital Hierarchy, *Telecommunications Journal of Australia* 39, 1989, pp. 39-42.

[4] G. Gopal, C. Kim and A. Weinrib, Algorithms for Reconfigurable Networks. In: Proceedings of Thirteenth International Teletraffic Congress (ITC-13), Copenhagen, Denmark, 1991, pp. 341-347.

[5] M. Herzberg, Network Bandwidth Management - A New Direction in Network Management, 6-ATRS, Wollongong, Australia, 218-225.

[6] M. Herzberg and S. Bye, Bandwidth Management in Reconfigurable Networks, *Australian Telecommunications Research* 27, 1993, pp 57-70.

[7] A.Kowalczyk, G.Aumann, H.L. Ferra, and J.Cybulski. *Associative mappings with positive bounded coefficients.*In T. Kohonen at.al, eds, Artificial Neural Networks, North-Holland, Amsterdam, 1991.

[8] A. Kowalczyk and H. Ferra. Developing Higher Order Networks with Empirically Selected Units, IEEE Trans. on Neural Net., 698-711, 1994.

[9] C. Leckie, A Connectionist Approach to Telecommunication Network Optimisation, in Complex Systems: Mechanism of Adaptation, R.J. Stonier and X.H. Yu, eds., IOS Press, Amsterdam, 1994.

Analyser Neural Networks: An empirical study in revealing regularities of complex systems

A.F. Nejad and T.D. Gedeon

School of Computer Science and Engineering
The University of New South Wales
Sydney 2052 AUSTRALIA
{tom | akbar}@cse.unsw.edu.au

Abstract

The causal index method is one of the most applied techniques in extracting meaning from neural networks. We have derived a measure of the significance coefficients for input variables using the weight distribution in the first layer of back-propagation trained multi-layer perceptrons, called *hidden indexes* which we compare to causal indexes. We have also studied finding a degree of importance for each input value depending on its coefficient in a system equation represented as in a multivariate linear regression model, by passing the weight matrix of a trained network to another network. This is an *analyser* neural network.

1 Introduction

One of the most successful attribute based learning methods are artificial neural networks (ANNs). In spite of their advantages, if we accept data transparency as a comparison component, they do not provide any rule or explanation for the decision for an input pattern. That is one of the main reasons for which ANNs are criticised by researchers in other fields. Thus, extracting rules and explanations is an important area of research in neural networks.

Previous work on extracting meaning from neural networks can be group categorised in three groups: (i) generating a set of rules; (ii) interpreting a trained neural network by visualisation techniques; and (iii) calculating a significance factor for input variables.

The first group of researchers have tried to generate or couple ANNs with a set of consistent and generalised rules [1-6]. Their aim has been to produce a sequence of rules as produced by algorithms like CART [7], or M5 [8] which can be consulted by users to explain why a decision has been made. The second group have worked on analysing weight matrices and activation of hidden units by visualising their changes during training [9-11], while the third group attempt to assign some level of significance to each input pattern variable by calculating the derivative of an output neuron with respect to an input variable (causal index) [12-13], or measuring the change in an output unit by a small change in an input variable (sensitivity analysis) [14]. They assign some coefficients to each independent input variable to the ANN which may be compared with statistical methods like standardised canonical discriminant function coefficients.

The causal index has some special importance in that these results may also be used in other methods [eg. 6, 12]. However, these indexes are sensitive to the hidden layer size and to the error criterion used to terminate training.

Figures 1 and 2 show the causal index (CI) dependency on the structure and parameters in an ANN trained to predict final students marks by the error back propagation algorithm.

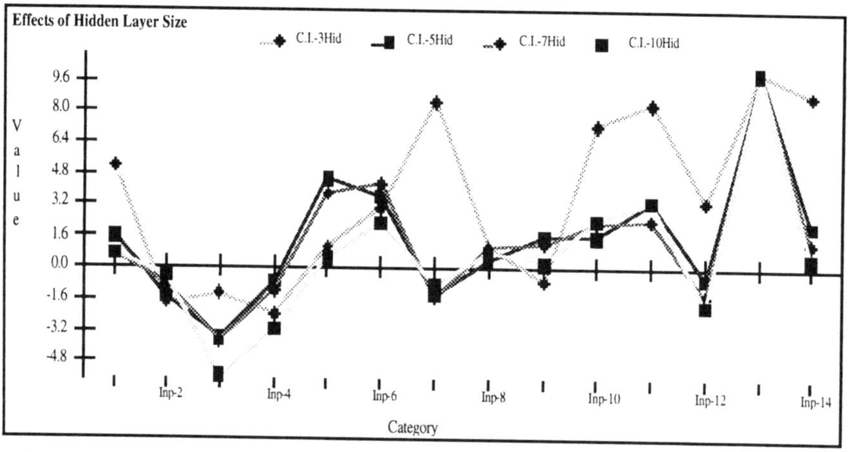

Figure 1. Difference among Causal Indexes of trained neural networks to predict student final marks with 3, 5, 7, and 10 hidden units.

This paper uses a new method to find significant coefficients. Comparing this method with causal index and sensitivity analysis (SA) not only measures its correctness but also may reveal some regularities of the behaviour of complex systems. We also investigate using an ANN in assigning significance indexes to each input of any other ANN by using its raw weight matrix as training input.

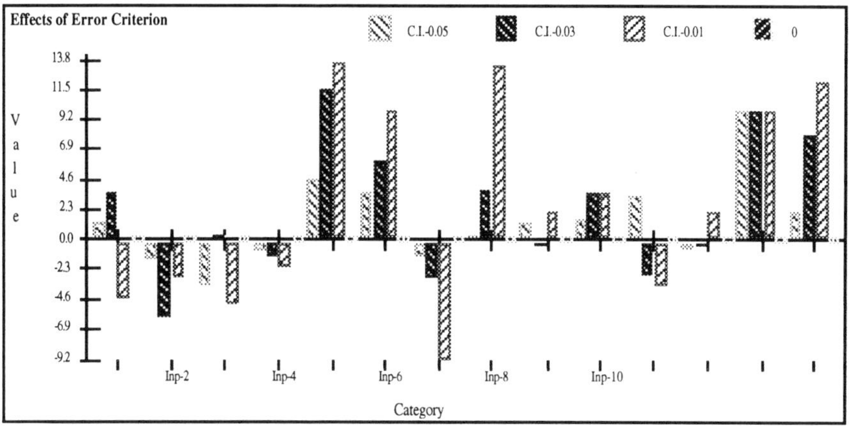

Figure 2. Relative magnitude of causal indexes may indicate the equivalent point in trade-off between overfitting and generalisation.

2 Training Analyser Neural Networks

For extracting meaning from artificial neural networks, we try to find some way to analyse the network by assigning a multi-variable equation to input patterns. Much of the previous work in this area was done on extracting logical rules from neural networks. We will find a degree of importance for each input value depending on the coefficient which will be shown in the equation.

Suppose we have found a multivariate regression equation like $\underline{y = a1\ x1 + a2\ x2 + a3\ x3}$ for our network. Using this equation we will be able to describe the behaviour of the output value and the degree of positive or negative dependency on each of the inputs.

Limiting the coefficients $\underline{ai}$ to the range of real numbers in [-3, +3], we trained a 3-3-1 multi-layer perceptron neural network to find the value of y for every possible combination of cardinal $\underline{ai}$ coefficients. Then, we extracted a vector of size 16, being the number of weights and thresholds of each of these networks, normalised with respect to the length of each pattern vector. These normalised vectors form the 343 input patterns for training our analyser network.

The size of analyser network for this example was 16-5-3. Once such a network is trained the extracted weight vector of a network to be analysed can be input. The three output nodes of the analyser network will find the $\underline{ai}$ coefficients for the particular equation of your network.

We did a number of experiments, for producing the 343 trained ANNs from which to extract weight vectors to use in training the analyser network. Each of the 343 networks was trained using back-propagation to learn a particular equation, by inputting a training set of size 1331 (11*11*11) generated to cover the possible range of input patterns by steps of 0.1 for each variable. The program to train ANNs of size 3-3-1 and 3-4-1 took 37 and 29 hours run time on DEC machines to prepare weight matrices for training sets for our experiments.

We had no difficulty in training our analyser neural network and could analyse some simple networks immediately. We noted the complexity of the task in handling systems with a lot of variables, due to the many possibilities to adjust weights in very different forms (depending on weight initialisation, learning rate, momentum, overall error, number of epochs, and so on). This complexity was especially apparent in more extended domains. Thus we found we needed to do some pre-processing on the weight matrices before inputting into the analyser neural network.

This need lead to the following experiments in analysing the weight matrix of trained neural network s, which resulted in our hidden index approach.

3 Hidden Indexes

One possible way to prepare the raw weight matrix of a trained neural network as input to the analyser neural network is to multiply the weight matrix (M) by its transpose (MT) which results in a symmetric matrix (C), see Figure 3. At the first stage the experiments were done on a one hidden layer neural network with three hidden units. The experiments were repeated on neural networks with different numbers of hidden units, and the method worked consistently.

$$M = \begin{matrix} w_{11} & w_{11} & \cdots & w_{11} \\ \vdots & \vdots & \vdots & \vdots \\ w_{n1} & w_{n2} & \cdots & w_{nm} \\ th_1 & th_2 & \cdots & th_m \end{matrix} = \begin{vmatrix} v_1 \\ \vdots \\ v_n \\ v_{n+1} \end{vmatrix} \qquad C = \begin{matrix} v_1 v_1 & v_1 v_2 & \cdots & v_1 v_{n+1} \\ v_2 v_1 & v_2 v_2 & & v_1 v_{n+1} \\ \vdots & \vdots & \vdots & \vdots \\ v_{n+1} v_1 & v_{n+1} v_2 & \cdots & v_1 v_{n+1} \end{matrix}$$

Figure 3. Weight matrices

It appeared from considering the result matrix that each column of the matrix shows the magnitude of coefficients of input variables and measures the amount of effect each input unit has on the output variables. We call these coefficients *hidden indexes*. A comparison of correctness of hidden and causal indexes for the 343 different ANNs is shown in Table 1, each column measures the correctness of the found coefficients by the Euclidean estimation criterion.

Table 1. Correctness of found coefficients by the Euclidean estimation criterion

Err. Cr.	HI@1	HI@2	HI@3	HI@T	AVG	HI@avg	differ.	CI
0.02	74	36	40	56	25	25	26	23
	5.91	10.33	16.23	1.49	7.44	8.22	2.71	2.74
0.002	18.14	11.08	24.58	2.19	2.19	9.6	1.04	1.26
0.00075	6.99	4.23	4.52	0.81	5.88	3.36	1.20	1.43

HI@1 to HI@3 show Hidden Indexes found by input weights, HI@T shows the indexes by thresholds and HI@avg contains average of indexes which are calculated by HI@1 to HI@3. The *differ.* column contains HI@avg - HI@T.

For a better understanding of these phenomenae, we can compare the convergence curves in ANN, CI, and HI. These are shown in Figures 4 and 5.

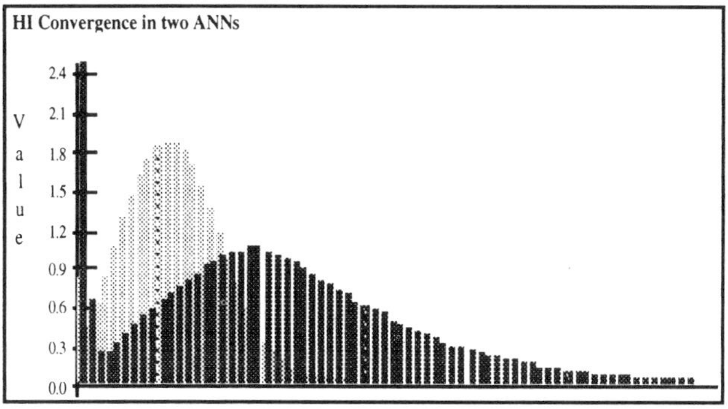

Figure 4. Comparing HI convergence of different ANNs.

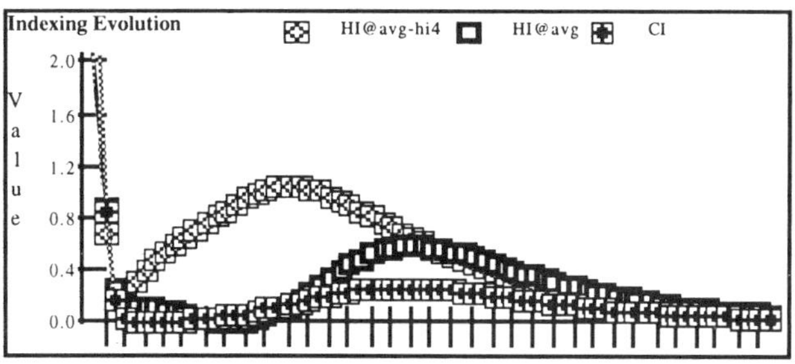

Figure 5. Comparing HI, CI convergence curves.

4 A Real World Example

In figure 6 we compare the obtained significance coefficients of input variables in a network trained to estimate the final mark of students by four methods: causal index, hidden index, sensitivity analysis, and multivariate linear regression. For hidden indexes we have calculated indexes by HI@avg - HI@4.

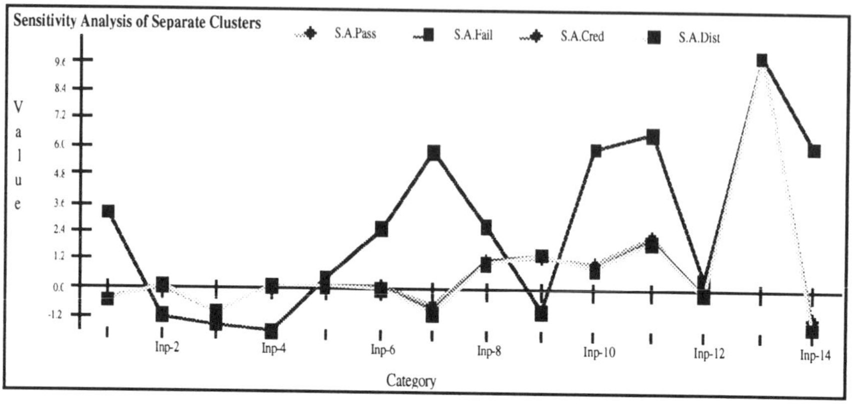

Figure 6. First order Sensitivity Analysis on each input of a trained ANN, with all other inputs held constant at canonical of 4 separate clusters.

We argue that in spite of the differences in obtained coefficients, all of them are acceptable, the benefits of our method are described below.

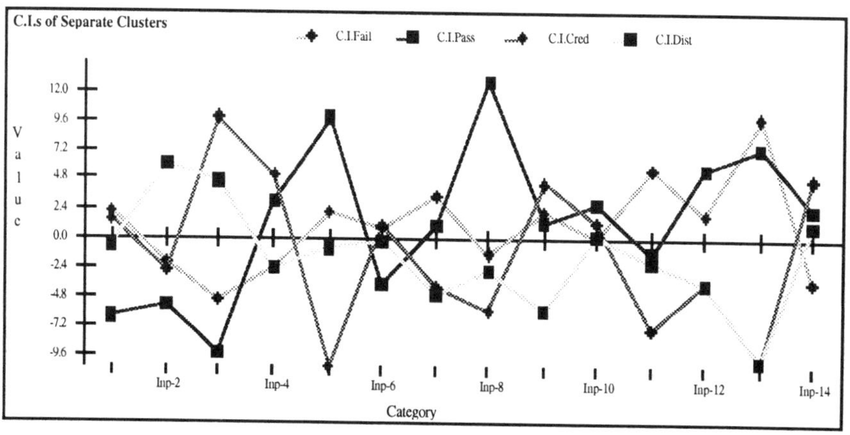

Figure 7. Causal index method on inputs of trained ANN for 4 clusters.

Back-propagation trained ANNs can be considered as non-parametric models which can adapt network weights to estimate any function [15]. They can be compared with multivariate multiple nonlinear regression [16]. In a multivariate linear regression, entering a new variable into a system can have significant effects on the coefficients of previously entered independent variables or may fail the T-significance test hypothesis of a desired significance level. Adding insignificant variables will not increase the explanatory power of the equation, and will lead to the loss of one degree of freedom.

One way of computing sensitivity analysis (SA) is to perturb one input slightly with other inputs held constant at 0, 0.5, or 1. This approach is inefficient because SA usually leads to completely different results in separate regions. Figures 7 and 8 show the results of CI and SA for four separate classes.

Another way of computing sensitivity analysis is to compute SA for one input over the range of values for one or a few other inputs. But this method is also inefficient because having a separate perturbation for each input over its range is time consuming and makes the analysis of the result difficult.

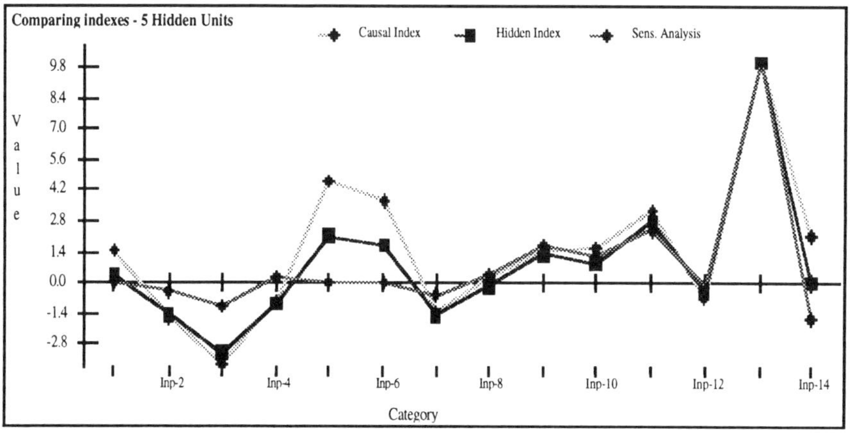

Figure 8. Hidden Indexes vs CIs and SA approaches applied on a trained neural network with 5 hidden units to predict student final marks.

As one appropriate solution to this problem we have used the canonical of each separate cluster which we obtained by training a Bidirectional Neural Network [17]. We also averaged the effects of two different positive and negative dithers. We show in Figures 9 & 10 that results of HI and SA are less dependent on structure of ANNs than CI, and HI results are usually closer to SA than CI.

In another experiment we trained networks to learn the relation between three dependent variables. We saw that if input variables are not mutually exclusive then no pattern was seen on the coefficient of independent variables and they will change in each training regime (eg. for different learning rate

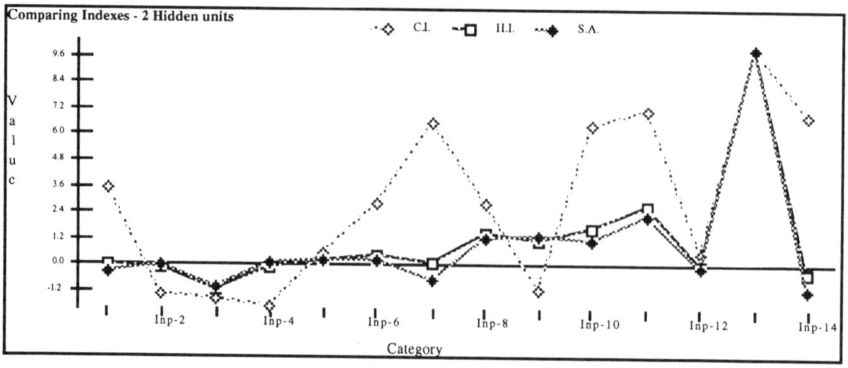

Figure 9. Comparing Hidden Indexes vs CIs and SA approaches applied on a trained neural network with 2 hidden units.

parameters). This may be compared with some problems of multi-collinearity in statistics with intercorrelations between supposedly totally independent variables.

5 Conclusion

We trained supervised error back propagation neural networks as analyser neural networks. Using this technique it is possible to develop a toolkit of analyser networks trained for recognising linear and non-linear regressions. The process of discovering the rules obeyed by a specified network is very quick. Since the analyser networks are already trained, we only need to input the normalised weight and bias vector of the specified network, and then propagate through to produce a result which can be read off directly.

There is little regularity in the raw weight matrix of ANNs and this would make it very difficult to analyse, whether by our analyser neural nets or other techniques. However, we have shown that there are underlying regularities which govern weight distributions in ANNs, and extracting these regularities using our technique of hidden indexes is much more effective than using raw weight matrixes in analysing ANNs. In other work not reported here, comparing the results of different approaches showed that they may also be used in determining the optimal number of epochs as well as optimal number of hidden units in order to have more robust networks. There is still a lot of work to be done on using these regularities in optimising the structure and training methods.

References

[1] Gallant, SI "Connectionist expert systems," *Communications of the ACM*, vol. 31, no. 2, pp. 152-169, February 1988.

[2] Gallant, SI *Neural Network Learning and Expert Systems*, MIT press, 1993.

[3] Bochereau, L and Bourgine, P "Expert systems made with neural networks," *Int. Joint Conf. on Neural Nets*, vol. 2, pp. 579-582, 1990.

[4] Saito, K and Nakano, R " Automatic Extraction of Classification Rules," *International Neural Network Conference*, Paris, pp 9-13, July 1990.

[5] Towell, G, and Shavlik, JW "Extracting refined rules from knowledge-based neural networks," *Machine Learning*, 13(1), pp. 71-101, 1993.

[6] Gedeon, TD and Turner, H "Explaining student grades predicted by a neural network," *Proc. Int. Joint Conf. Neural Networks*, 609-612, Nagoya, 1993.

[7] Breiman, L, Friedman, J, Olshen, R and Stone, C *Classification and Regression Trees*, Monterrey, Caiwadsworth, 1984.

[8] Quinlan, JR "Learning With Continuous Classes," *Proceedings of 5th Australian Joint Conference on Artificial Intelligence*, 1992.

[9] Hinton, GE "How Neural Networks Learn from Experience," *Scientific American*, pp. 105, Sep, 1992.

[10] Good, RP and Gedeon, TD "Network Analysis Techniques as visualisation tools," *Proc. Int. Conf. on Non-Linear Theory*, pp. 945-952, Hawaii, 1993.

[11] Pratt, L, Nicodemus, S "Case studies in the use of a hyperplane animator for neural network research," *Proc. of the IEEE Int. Conf. on Neural Networks*, vol. 1, pp.78-83, 1994.

[12] Hora, N, Enbutsu, I and Baba,K "Fuzzy rule extraction from a multilayer neural net," *Proc. IEEE*, vol. 2, pp. 461-465, 1991.

[13] Yoda, M, Baba, K and Enbutsu, I "Explicit representation of knowledge acquired from plant historical data using neural networks," *Internat. Joint Conference on Neural Networks*, San Diego, vol. 3, pp. 155-160, 1991.

[14] Hashem, S "Sensitivity Analysis for Feedforward Artificial Neural networks with Differentiable Activation Functions," *IEEE*, 1, pp 419-424, 1992.

[15] White, H *Artificial Neural Networks: Approximation and Learning Theory*, Oxford, UK: Blackwell, 1992.

[16] Sarle, WS "Neural Networks and Statistical Models," *Proceedings of the Nineteenth ANNual SAS Users Group Internat. Conference*, April, 1994.

[17] Nejad, AF and Gedeon, TD "Bidirectional MLP Neural Networks and Their Applications," *Proc. Int. Symp. Artificial Neural Networks*, Taiwan, 1994.

Applications of Genetic Algorithms in Telecommunications

P. Chardaire, A. Kapsalis, J. W. Mann*, V. J. Rayward-Smith and G. D. Smith

School of Information Systems
University of East Anglia, Norwich, UK
`gds@sys.uea.ac.uk`
* *BNR Europe Ltd, London Rd., Harlow, UK*
`J.Mann@bnr.co.uk`

Abstract

This paper describes some of the genetic algorithm, or *GA*, work undertaken by the *combinatorial optimisation* group at the University of East Anglia on a range of telecommunications applications. The group have developed a *GA* toolkit, **GAmeter**, which they have applied to the solution of problems including frequency assignment, facility location, static and dynamic traffic routing and other general network applications. In general, the results have been very encouraging and promise fruitful ground for current and future studies in telecommunications and in industry in general.

1 Introduction

Telecommunications is one of the most important and fastest growing industries in the modern world. The continuing ability to provide services in a market of ever-increasing demand relies heavily on, among other things, the solution of a wide range of optimisation problems in network planning, network management and resource allocation in general.

It is therefore essential that the industry is ready and able to use the whole range of optimisation techniques from exact methods such as linear and integer programming through to the modern heuristic techniques of artificial neural networks, genetic algorithms, simulated annealing and tabu search, as well as hybrids thereof [15, 14].

In this paper we describe briefly the application of genetic algorithms to some optimisation problems in telecommunications, namely the Steiner Tree Problem in Graphs, a Radio Link Frequency Assignment problem and a Concentrator Location problem. Members of the group are also developing intelligent routing applications using genetic algorithms; this is reported in [17].

In the following section, we describe the general *GA* and particular features of the *GA* toolkit, **GAmeter**, which has been developed by the group at UEA and provides an easy to use environment for the development and experimentation of optimisation problems using genetic algorithms.

2 The Genetic Algorithm

The *genetic algorithm*, or *GA*, is a stochastic search algorithm developed by John Holland at the University of Michigan [6]. The algorithm is based on the mechanisms of evolution, wherein a species is under continual development by the combined processes of natural selection, sexual recombination and mutation. Although *GAs* were first used by Holland and his associates in developing machine learning systems, it was not long before their potential in function optimisation was realised [3].

```
GA
    {objective is to maximise value(p) such
    that p ∈ U }
    P, Q, R : multi-set of solutions ⊂ U;
    initialise(P);
    while not finish(P) do
        begin
        Q := select(P);
        R := create(Q);
        P := merge(P, Q, R)
        end
end GA
```

Figure 1: The Genetic Algorithm Paradigm

In function and combinatorial optimisation, the *GA* starts with a (typically random) population, P, of chromosomes, each chromosome representing a solution to the problem and each evaluated according to some objective function. This is done in the *initialise* function - see Figure 1. In **GAmeter**, there is an option to seed this initial population using a solution derived using another heuristic. Some of these chromosomes are then *selected* to join a gene pool, Q. A mechanism for the *select* function which simulates *survival of the fittest* is often used. A number of standard *select* options are available in **GAmeter**.

These selected solutions/chromosomes are then recombined, using genetic operators such as crossover and mutation, to produce a new set, R, of chromosomes, the *offspring* pool. Typically, crossover exchanges substrings between two parents to create two new children, while mutation randomly alters individual values within each chromosome to generate a solution with new 'genetic' material. Various options for selecting and dynamically controlling these operators are available in **GAmeter**.

The new population pool, P, is finally reconstructed from the original pool, P, the gene pool, Q, and the offspring pool, R. This process continues until some specified *finishing* criterion is satisfied. This general genetic paradigm is shown in Figure 1, in which U represents the set of all chromosomes, equivalently the search space, and p a member of U; see [8, 13] for details of the basic *GA*, and a description of the toolkit **GAmeter**.

In general, the *GA* is a global search mechanism, capable of converging very rapidly to good regions of the search space, and in many test problems it converges very quickly to the known optimal, see for example [2, 10].

3 The Steiner Tree Problem in Graphs

The Steiner Tree Problem in Graphs, or *SPG*, is that of determining a minimal spanning tree solution for a subgraph consisting of a set of special nodes which must be connected, together with a set of (unknown) Steiner nodes to minimise connectivity costs. Formally, let $G = (V, E)$ be an undirected graph with a finite set of vertices, V, an edge set, E and a cost function assigning a positive, real cost, to each edge in E. Given a subset, S, of the vertices in G, the *SPG* is that of finding a subgraph, $G' = (V', E')$, of G such that V' contains all the vertices in S, G' is connected and the sum of the cost of all edges in E' is minimal. It is the fact that the set of Steiner nodes is unknown which makes the decision problem associated with the *SPG* NP-complete.

A comprehensive overview of the *SPG* is given by Winter in [19]. Applications of the problem to areas such as topological network design, printed circuit design, multiprocessor scheduling and phylogeny are described by Foulds and Rayward-Smith in [4] and by Winter in [19]. This problem also has direct relevance to multicast routing, see [12, 7, 18].

We can represent a chromosomal solution to the *SPG* by a bit string of size equal to $|V - S|$ in which each bit position i corresponds to a node $v_i \in |V - S|$ and a 1 means that vertex $v_i \in |V - S|$, a 0 means that $v_i \notin |V - S|$.

Given a chromosome representing a set, U, of vertices, the evaluation proceeds as follows. Firstly, we construct the subgraph of G induced by $U \cup S$. Say this subgraph comprises $k(\geq 1)$ components. We compute the MST of each component using Prim's algorithm and sum these values. If $k > 1$, we add a large penalty linearly dependent on k. A more detailed description of the use of *GAs* to solve the SPG can be found in [10]. This also contains details of experimental results for a popular set of problems used by many researchers in the field. This set consists of 18 problems and is known as Beasley's [1] B-problem set (it can be obtained via anonymous ftp to *mscmga.ms.ic.ac.uk*).

We point out here that the *GA* was able to find the optimal solution for each of the 18 problems, in times varying from 5 seconds to 90 seconds on an HP 700 series workstation, and typically converging to within 5% of the optimal solution within 5 seconds. These results show that GAs provide a remarkably successful method for finding good quality solutions to the Steiner Problem in Graphs. The technique was found to be extremely robust in that the solutions to the set of test problems were invariably found even with relatively large variations in the values of the parameters used. Work on the Steiner Tree Problem in Graphs is ongoing and includes parallel implementation of **GAmeter** to solve Beasley's C-problem set (1,000 vertices and more), see [11].

4 The Frequency Assignment Problem

The UEA group is responsible for the *GA* work programme in a EUCLID-funded multinational project (Combinatorial Algorithms for Military Applications - *CALMA*) investigating the relative performance of a wide range of exact and heuristic search techniques for the resolution of highly combinatorial problems occurring in military applications. One particular problem chosen as a representative difficult problem is that of assigning a set of frequencies to a large number of transmitters in such a way as to minimise the connectivity costs and the interference, i.e. the Radio Link Frequency Assignment Problem, or *RLFAP*.

In the *RLFAP*, we are given a set of links that have to be assigned frequencies. Typically, there may be of the order of 200 to 1000 links and approximately 50 discrete frequency values, so the latter are assigned to many different links. As a result, interference considerations generate many thousands of constraints. We have to find a **feasible** assignment if possible, i.e. one which satisfies all given constraints. In some cases, we have to find an **optimal**, feasible assignment, i.e. a feasible assignment that minimises some objective function such as the distinct number of frequency values used.

Formally, each link $i \in L = \{1, 2, 3, ..., n\}$ is assigned a frequency value f_i from a given domain D_i, which is a subset of the set of all available frequency values. Typically, the domains have non-null intersections. In the data sets with which we are working, there are six domains, each containing between 7 and 50 frequency values. For the particular problem type we are studying, there exist a large number of interference constraints of the type $|f_i - f_j| > \varepsilon_{ij}$ or $|f_i - f_j| = \varepsilon_{ij}$ for a given pair of links i and j and some frequency separation ε_{ij}. In our examples, there are between 200 to 500 equality constraints and 1000 to 5000 inequality constraints.

The *RLFAP* is easily shown to be NP-complete and generates an enormous solution space. For example, a particular problem with 80 frequency values and 1000 links has a search space of 80^{1000}, containing mostly infeasible solutions. Finding an algorithm which can explore this size of space effectively is therefore very desirable.

If we let N be the number of variables in the problem, a possible assignment is represented by a string of frequencies, $f_1 f_2 f_3 \cdots\cdots f_N$, where f_i is a frequency in Domain D_i, i.e. the domain associated with link/variable i.

For each given assignment, it is then possible to determine, among other things, the number of distinct frequency values assigned, D, the number of equality constraints violated, EC and the number of inequality constraints violated, IC. The resulting fitness function used to determine the suitability of a particular assignment is then made up of weighted terms drawn from these various measures:

$$fitness = w_1 * D + w_2 * EC + w_3 * IC + Penalties. \tag{1}$$

In equation (1), the value of the term *Penalties* is determined from, among other things, the number and type of constraints violated.

The resulting *GA* was tested on a range of test problems, the CELAR data sets. In addition to using standard genetic operators, it was found necessary to introduce a number of specialised genetic operators in order simply to find feasible solutions. This was due to the high degree of interaction between the variables through the constraints. Typically, each link is involved in 10 or more constraints, each with another link. This high level of *epistasis* created enormous problems for the standard operators to the extent that no feasible solutions were found using the standard *GA*. Specialised operators which attempted to simulate the desired effects of crossover and mutation, but without their destructive properties, were tried and, as a result, feasible and occasionally, optimal solutions were found.

A full report of the data sets used and the results obtained, as well as detailed descriptions of the many new genetic operators implemented in order to improve the overall performance of the standard *GA*, is given in [9].

5 The Concentrator Location Problem

Members of the group have also applied a *GA* to the solution of the *Concentrator Location Problem*, or *CLP*. Here, the objective is to connect a set of remote terminals to a host computer site in such a way as to achieve an effective network with minimum connectivity costs. There are many topologies possible to achieve connectivity [5], but one of the most common topologies in practice is the *centralised concentrator-based network with star subnetworks*, and is achieved by inserting concentrators between each terminal and the host computer in such a way that one concentrator is linked to many terminals and uses a single, (higher-capacity) link to the host.

Consider a finite set I of terminals with a prescribed demand for a single service; consider also a finite set J of possible concentrator locations to supply this service. These concentrators may, in the first instance, be assumed to have unlimited capacity in that any one has the potential to satisfy all the client demands. The uncapacitated *CLP*, or *UCLP*, is that of determining the number and location of the concentrators in order to minimise the overall costs. Note that neither the number of concentrators established nor the actual connections are given beforehand; these are determined as part of the solution.

For the problems under study, I is a set of coordinates in Euclidean space. We restrict attention to cases in which a concentrator $j \in J$ is located at a site of a terminal, and not at a separate location. Thus the *UCLP* can be stated succinctly as follows:

$$Minimise \sum_{i \in I} \sum_{j \in J} c_{ij}\, x_{ij} + \sum_{j \in J} f_j\, y_j, \tag{2}$$

subject to

$$\sum_{j \in J} x_{ij} = 1, \quad \forall\, i \in I, \tag{3}$$

$$x_{ij} \leq y_j, \quad \forall\, i \in I \; and \, \forall\, j \in J, \tag{4}$$

$$x_{ij} \in \{0,1\} \text{ and } y_j \in \{0,1\}, \quad \forall\, i \in I \; and \, \forall\, j \in J, \tag{5}$$

where

$$x_{ij} = \begin{cases} 1 & \text{if terminal i is served by concentrator j} \\ 0 & \text{otherwise} \end{cases}$$

$$y_j = \begin{cases} 1 & \text{if concentrator j is established} \\ 0 & \text{otherwise} \end{cases}$$

and where f_j is the set-up cost of establishing concentrator j, and c_{ij} is the total cost of serving terminal i from concentrator j.

Constraints (3) and (5) together ensure that each terminal is served by one and only one concentrator. Constraints (4) and (5) ensure that each terminal can only be assigned to a concentrator if that concentrator is open (available).

In using a *GA* to solve the *UCLP*, the chromosome represented the bit string y_j, i.e. the vector of bits determining whether a terminal site was also a concentrator site. Once y_j is specified, each terminal is assigned to the cheapest concentrator (shortest Euclidean distance), thus determining the x_{ij}, and the fitness of that particular solution.

The *UCLP* can be extended to include capacity constraints where each concentrator is restricted in the number of terminals to which it can be linked, the capacitated *CLP* or *CCLP*. We have investigated the use of a genetic algorithm to solve both the *UCLP* and the *CCLP*. To test the *GA*, we generated 100 random data sets, each having either 300 or 400 terminals randomly located in the unit square, and ran the *GA* until it converged. We compared the results obtained, in terms of quality of solution, to those obtained using a dual-based approach in which feasible solutions are obtained by assigning terminals to the set of concentrators y_j obtained using Lagrangian relaxation.

The results are reported in full in [16], but it is worth noting here that, although the dual-based approach finds better quality solutions for the *UCLP*, the *GA* compares more favourably for the *CCLP*, particularly when a local optimiser was switched in every 500 generations. We note from Figure 2, which shows the frequency distribution of the solutions of the 100 data sets used in the *CCLP* that, although the *LR* approach finds higher quality solutions more often, the *GA* performance is more consistent.

Current work includes comparative studies for the multi-level, capacitated concentrator location problem, and early results indicate that the *GA* approach is an effective technique for this more difficult problem.

6 Conclusions

One of the main advantages of using genetic algorithms in any application area is the ease with which the evaluation function can be altered to solve variations of the problem. Since the major part of the *GA* application is problem independent, and hence already coded and tested, users need only code and test the *evaluate* function necessary to determine the fitness of

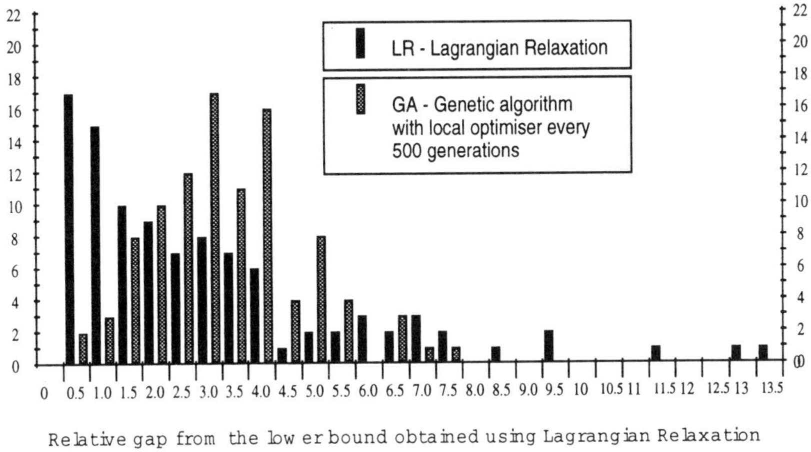

Figure 2: The Capacitated Concentrator Location Problem - comparative results.

a typical solution. Thus development costs are also reduced. This, along with the often surprisingly good results that emerge, may explain why many companies are beginning to investigate the genetic algorithm as a means to solving large, troublesome optimisation problems.

In general, the authors conclude that the Genetic Algorithm is potentially a very powerful global search tool that, in problems with little interaction between the variables, can converge to a neighbourhood of a very good near-optimal, if not optimal, solution very quickly. However, in problems with a high degree of interaction between the variables, such as in the *Radio Link Frequency Assignment Problem*, the *GA* needs to be adapted by the inclusion of some other heuristic, often one which relies on problem domain knowledge. These hybrids have been shown to be very effective search engines, and it is in this area that further development is recommended.

References

[1] J. E. Beasley, "Or - library: distributing test problems by electronic mail," *J. Opl Res.*, vol. 41, pp. 1069–1072, 1990.

[2] R. H. Berry and G. D. Smith, "Using a genetic algorithm to investigate taxation-induced interactions in capital budgeting," in *Proceedings of the International Conference on Neural Networks and Genetic Algorithms*, (Innsbruck), pp. 567–574, Springer-Verlag, 1993.

[3] A. D. Bethke, "Genetic algorithms as function optimisers," Logic of Computers Group Technical Report 197, University of Michigan, 1978.

[4] L. Foulds and V. Rayward-Smith, "Steiner problems in graphs: Algorithms and applications," *Engineering Optimization*, vol. 7, pp. 7–16, 1983.

[5] B. Gavish, "Topological design of computer communication networks - the overall design problem," *European J. Operational Res.*, vol. 58, pp. 149–172, 1992.

[6] J. H. Holland, *Adaptation in Natural and Artificial Systems*. Ann Arbor, MI: University of Michigan Press, 1975.

[7] X. F. Jiang, "Distributed path finding algorithm for stream multicast," *Computer Communications*, vol. 16, no. 12, pp. 767–775, 1993.

[8] A. Kapsalis, V. J. Rayward-Smith, and G. D. Smith, "Fast sequential and parallel implementation of genetic algorithms using the GAmeter toolkit," in *Proceedings of the International Conference on Neural Networks and Genetic Algorithms*, (Innsbruck), Springer-Verlag, 1993.

[9] A. Kapsalis, V. J. Rayward-Smith, and G. D. Smith, "Using genetic algorithms to solve the radio link frequency assignment problem," in *Proceedings of the Second International Conference on Artificial Neural Networks and Genetic Algorithms*, Springer-Veralg, 1995. (Accepted).

[10] A. Kapsalis, V. Rayward-Smith, and G. Smith, "Solving the graphical Steiner tree problem using genetic algorithms," *J. Oper. Res. Soc.*, vol. 44, no. 4, pp. 439–448, 1993.

[11] A. Kapsalis, G. Smith, and V. Rayward-Smith, "Unified paradigm for parallel genetic algorithms," in *Lecture Notes in Computer Science 865*, Springer Verlang, 1994.

[12] X. L. Lin and L. M. Ni, "Multicast communication in multicomputer networks," *IEEE Transactions on Parallel and Distributed Systems*, vol. 4, pp. 1105–1117, 1993.

[13] J. Mann, A. Kapsalis, and G. D. Smith, "The GAmeter toolkit," in *Applications of Modern Heuristic Techniques* (V. J. Rayward-Smith, ed.), pp. 195–209, Alfred Waller, 1995.

[14] V. J. Rayward-Smith, ed., *Applications of Modern Heuristic Techniques*. Alfred Waller, 1995.

[15] C. R. Reeves, ed., *Modern Heuristic Techniques for Combinatorial Problems.* Blackwell Scientific, 1993.

[16] G. D. Smith, P. Chardaire, and J. W. Mann, "The location of concentrators using genetic algorithms," submitted to *Studies in Location Analysis.*

[17] G. D. Smith, V. J. Rayward-Smith, and J. W. Mann, "Traffic routing - a case study in the use of genetic algorithms," Tech. Rep. *S&SE*/GA/tr/20652/1.0, BNR Europe Ltd, 1994.

[18] Y. Tanaka and P. C. Huang, "Multiple destination routing algorithms," *IEICE Transactions on Communications*, vol. E76B, no. 5, pp. 544–552, 1993.

[19] P. Winter, "The Steiner problem in networks: A survey," *Networks*, vol. 17, pp. 129–167, 1987.

Reinforcement Learning by Construction of Hypothetical Targets

Olle Gällmo and Lars Asplund

Department of Computer Systems, Uppsala University,
Box 325, S-751 05 Uppsala, Sweden
Fax: +46 18 55 02 25,
http://www.docs.uu.se/docs/ann/
{Olle.Gallmo, Lars.Asplund}@DoCS.UU.SE

Abstract

A general approach to delayed reinforcement learning by the use of supervised training algorithms is proposed. The approach is monolithic, direct and involves minor modifications to any supervised learning algorithm. Each connection has two weight change registers, one for eventual success and one for eventual failure, and the network is trained on self-generated *hypothetical* target vectors. The method is a close, but more general, relative to selective bootstrap adaption and is tested on an abstract model of the Link Allocation problem in Asynchronous Transfer Mode (ATM) telecommunication networks.

1 Introduction

In real world control applications, there is often a need for an agent which can be trained on evaluative feedback in a closed loop with the environment, i.e. from its success and/or mistakes rather than from examples of input-output pairs. This is called *reinforcement learning* (RL), and its main property is that the feedback is an evaluation of the environmental state and, therefore, only an indirect evaluation of the agent's actions. Supervised learning is more specific and instructive.

In a *delayed* RL task the feedback is considered to be an (indirect) evaluation of a sequence of actions, not only of the most recent one (as in *immediate* RL). Here, the feedback is assumed to be scalar. (See [5] for a learning task with non-scalar feedback which could still be considered a RL task in this respect.)

The most popular RL approaches today are so called temporal difference methods [8], related to dynamic programming, where neural networks are sometimes, but not always, used as building blocks. Other methods have been proposed [4, 5, 6], perhaps more obviously related to neural information processing, where conventional supervised techniques such as the back propagation algorithm [7] are used to train neural networks on evaluative feedback from the environment. However, the latter approaches often require going back in time through all input-output pairs in history for every network update. Furthermore, often two separately trained networks are needed, one which learns the actual task and one which models the environment. The control network depends on the model network being a good model of the environment.

This paper presents a general method to train single feed forward neural networks on scalar evaluative feedback using supervised training algorithms. The method is based on the automatic generation of hypothetical target vectors and has its closest relatives in Michie's work on MENACE [3], and in the *selective bootstrap adaption* algorithm [9].

No storing of passed input-output pairs is needed, and the only modification to the network is that two weight change registers are needed for every weight. The method is implemented here using the back propagation algorithm and tested on an abstract model of a difficult telecom problem: ATM link allocation.

2 The Application: ATM Link Allocation

In Asynchronous Transfer Mode (ATM) broadband networks [2], an efficient link allocation algorithm is an essential part. Link allocation is a stochastic bin packing problem, where connection requests (e.g. phone calls) of unknown arrival rates and life lengths are to be allocated to one of a number of links in such a way as to maximize the long term utilization of all the links. The link allocator should adapt to changes in traffic flow and learn to predict what connection requests to expect in the near future.

This is a typical and very general reinforcement learning problem and its solution should be interesting for a wide audience outside the ATM field. A simplified model of link allocation has therefore been developed in order to bridge the gap between different application areas. In this model, called *Bubbles*, the connection requests are modelled as soap bubbles and the links as buckets (Figure 1).

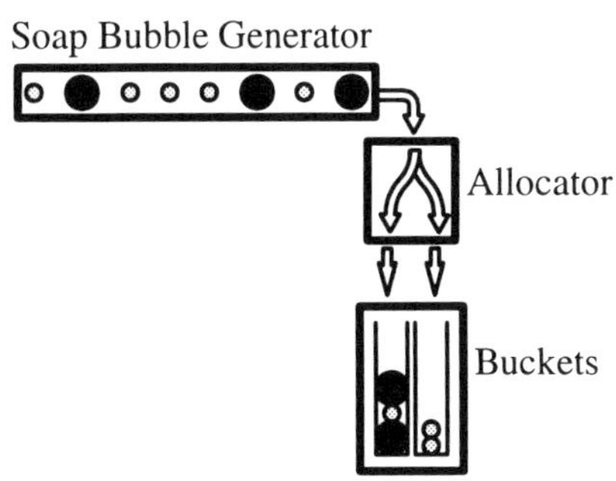

Figure 1: The bubble allocation problem

The model consists of two parts, a bubble generator and an allocator. The bubble generator creates a stream of soap bubbles of different sizes. The life length of a bubble is drawn from a given distribution, the parameters of which depend on the bubble size. Thus, all bubbles of equal size have a life length drawn from the same distribution and can be said to be members of the same class. For clarity, the classes are denoted by colours, e.g. black for big bubbles and white for small ones.

The task of the allocator is to allocate the bubbles to the buckets in such a way as to maximize the long term utilization of the buckets. The allocator has no prior knowledge of what bubbles to expect in the future, nor of their life lengths. The allocator only knows the current state of the buckets (typically the amount of free capacity in each of them) and the size of the current bubble to be allocated.

There are several conventional allocation algorithms for this type of allocation tasks. Three common algorithms are:

- First Fit: Search the buckets in a predefined order, e.g. from left to right, and put the bubble in the first one found with enough capacity.
- Random Search: Same as First Fit, but start the search at a randomly chosen bucket.
- Best Fit: Chose the bucket with least, but sufficient, capacity.

The three methods have one drawback in common in that they are not adaptive. They only consider the size of the current bubble and the remaining capacity in the buckets, and do not try to predict future bubbles, nor to estimate the expected life length of the current one. Nor do they treat the failure to allocate a bubble as an indication of bad allocation decisions in the past.

3 Neural allocation

In our experiments, a Multi Layer Perceptron (MLP) with one hidden layer is used as an allocator (Figure 2). State information from the buckets (e.g. a vector of free capacity values, $C_1 \dots C_N$) is given to the network together with information on the new arriving bubble (e.g. its size or colour). The MLP generates a real valued output vector $y_1 \dots y_N$, where N is the number of buckets. The output vector is then transformed to a binary action vector $a_1 \dots a_N$, where all values are 0 except the one corresponding to the chosen bucket, which has a value of 1.

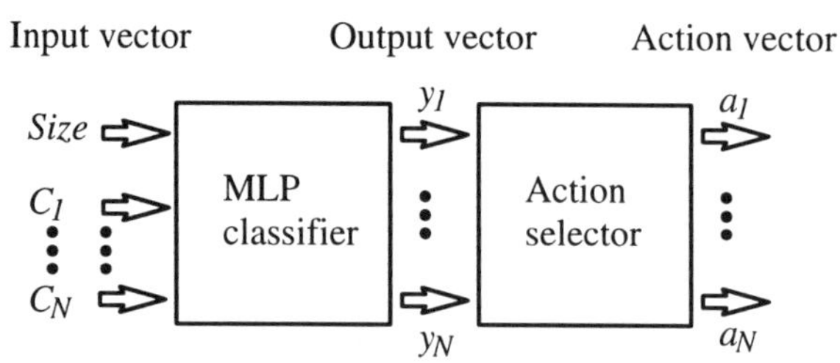

Figure 2: A neural allocator

A simple action selector is a deterministic one where the chosen bucket is the one with the largest corresponding output value from the network, but also stochastic action selectors are possible (and necessary in most applications).

The question is how to train the network. Allocation decisions not only affect the current state of the buckets, but also future states, which in addition depends on the bubbles to arrive after this one. Therefore, no target vectors for the network are known at the time of allocation. The only feedback is an evaluative signal from the environment, which may or may not be available for every action.

3.1 Training on Hypothetical Targets

The lack of target vectors does not disqualify the use of a supervised training algorithm, however [1, 9]. If it is assumed that the current action will eventually turn out to be a good one, the action vector in itself should be a good hypothetical target vector since training on it would enforce the action just taken (pushing the corresponding network output towards 1) and inhibit all others (pushing them towards zero). This is called an *optimistic* target vector and is denoted T^{+}.

Similarly, a *pessimistic* target vector, T^{-}, is a target vector constructed under the hypothesis that the current action will turn out to be bad. One way to construct T^{-} is to invert the optimistic one, i.e. pushing the output corresponding to the current action towards 0 and all others towards 1. However, better strategies can be found.

The network cannot update its weights before the feedback signal arrives, but it can use the hypothetical target vectors to accumulate weight change values, Δw, for each of its weights, using any supervised algorithm. Here, the back propagation algorithm [7] is used and the method is therefore called $\mathrm{BP_{HT}}$ (Back Propagation on Hypothetical Targets), though this is a somewhat misleading name since any supervised algorithm will do.

Each weight w_{ij} from node (or input) i to node j has two accumulating weight change registers, Δw_{ij}^{+} (optimistic) and Δw_{ij}^{-} (pessimistic), updated after every decision, k, by:

$$\Delta w_{ij}^{+}(k) = \gamma \Delta w_{ij}^{+}(k-1) + \eta \delta_j^{+} x_i \quad (1)$$

$$\Delta w_{ij}^{-}(k) = \gamma \Delta w_{ij}^{-}(k-1) + \eta \delta_j^{-} x_i \quad (2)$$

where η is a gain term, δ_j is the error for node j (as calculated by conventional back propagation on the corresponding hypothetical target) and x_i is the current value of node (or input) i. The parameter $\gamma \in [0,1]$ is a discount factor deciding how far back in time decisions are to be credited for the outcome (called the *credit horizon*). In effect, each of the two accumulated weight changes form an exponentially decreasing function:

$$\Delta w(k) = \gamma^{k} c(0) + \gamma^{k-1} c(1) + \cdots + \gamma^{1} c(k-1) + \gamma^{0} c(k) \quad (3)$$

where $c(i)$, i=1...k, is the weight change contribution $\eta\delta x$ caused by decision i.

When the feedback signal $r \in [-1,1]$ arrives, after decision k', its absolute value acts as an extra gain parameter and its sign decides which weight change register to apply when updating the weight:

$$w_{ij}(k') = w_{ij}(k' - 1) + |r| * \begin{cases} \Delta w_{ij}^{+}(k'), & \text{if } r > 0 \\ \Delta w_{ij}^{-}(k'), & \text{if } r < 0 \end{cases} \tag{4}$$

Note that the gain parameter, η, can be eliminated from (1) and (2) by constraining the feedback signal to the interval $[-\eta, \eta]$ instead of $[-1,1]$.

Some applications require finite credit horizons. For example, in learning how to play chess, each game is an isolated session and the Δw registers should therefore be cleared when the game has ended. However, in bubble allocation the 'game' has no end and the feedback is (at least in theory) an evaluation of all previous decisions since the beginning of time. Therefore, the weight change registers are never cleared, but continuously updated after every action.

BP_{HT} makes the implicit assumption that if the feedback is positive, all actions leading to it were good, and vice versa. This is, of course, not a very good assumption in each single case, but in the long run the bad effects cancel out. A notable, and very similar, example of reinforcement learning under the same assumption is a Tic-Tac-Toe machine called MENACE (Matchbox Educable Noughts And Crosses Engine), built in the early 60's by D. Michie [3] out of matchboxes and coloured beads.

Another example can be found in [9], where an algorithm called *selective bootstrap adaption* is used to train a single binary neuron to play Blackjack. The latter is in effect a special case of BP_{HT} where $\gamma=1$, the weight change registers are cleared after use (since the application is a game) and the patterns are stored and recycled (instead of maintaining two weight change registers).

That BP_{HT} calculates weight change values 'on the fly', instead of recycling stored input- and output vectors, makes an infinite credit horizon feasible. Furthermore, in BP_{HT} the weight change contributions $c(k)$ are calculated with respect to the weights at the time of the corresponding decision (k). A recycling scheme, on the other hand, calculates the contribution w.r.t. the set of weights at the time of feedback (k'). If the weights have already been updated sometime between k and k', these two sets of weights are not the same. Also, BP_{HT} only has to calculate the contribution once, while a recycling scheme will recalculate it for every update.

The $A_{R\text{-}P}$ algorithm [1] is also similar to BP_{HT}, where $\gamma=0$ (i.e. no history), only two actions are possible, evaluative feedback is available for every action and a stochastic action selector is used.

4 Experiments

Preliminary experiments have been made, applying BP_{HT} to the bubble allocation problem and to compare it with the conventional algorithms First Fit, Best Fit and Random Search defined in section 2. The goal of this experiment was to verify that the neural allocator converges to a reasonable level of performance, i.e. showing results comparable to the other approaches but not necessarily better.

The setup was chosen to be a small instance of the bubble allocation problem, with only two buckets. There were two bubble types (colours), black and white. Black bubbles had size 5 and lived for 5-9 time steps, while white bubbles had size 2 and lived for 7-12 time steps. Both life length distributions were rectangular. White bubbles were twice as common as black ones and a new bubble was produced at every time step.

All the allocators to be tested were given the same bubble stream of 300,000 bubbles and a moving average of the utilization (defined as the average load over all buckets) over the latest 50,000 time steps was measured continuously and plotted for every thousand time steps (Figure 3).

The conventional allocators were given the amount of free capacity in each bucket and the size of the incoming bubble. Since the bubble stream was stationary, and since the conventional methods are not adaptive, their level of performance was almost constant. The average utilization over the last 50,000 time steps was 74.4 % for both First Fit and Best Fit, and 73.3 % for Random Search.

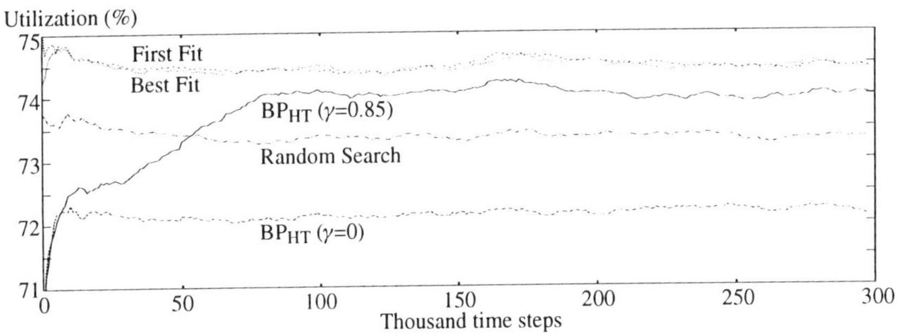

Figure 3: Moving average over a 50,000 time step window of the utilization.

The neural allocator was a 4-10-2 MLP. Two of the inputs were fed with the load in each of the two buckets. The other two inputs were given binary values (1,0) or (0,1) corresponding to the incoming bubble being black or white. The gain and discount parameters were η=0.2 and γ=0.85 and the action selector was deterministic.

The reinforcement strategy was the simplest possible: Feedback was available for every action, equal to +1 if the bubble fitted where it was placed by the allocator, and –1 otherwise. In other words, the network was only told of the rules of the game, not how to play it good.

After 55,000 time steps, the performance of the BP_{HT} allocator surpassed that of Random Search. It then continued to improve and leveled out after about 80,000 time steps. The average utilization over the last 50,000 bubbles was 0.740, which is not as good as First Fit and Best Fit, but the difference is small and much can be done to help the network finding a better solution (discussed in the next section).

Though feedback is available every time step, this is a *delayed* reinforcement learning task in that history is important and not only the latest decision should be credited for the outcome. To show this, an experiment was made under the same initial conditions but with γ=0. The result can be seen in Figure 3. The allocator converged very quickly (after 5,000 decisions) to a level of 72.1 % utilization.

5 Conclusions and Future Work

A general approach to *delayed reinforcement learning* by the use of supervised training algorithms has been introduced. The approach is *monolithic* (only one network is used), *direct* (a direct mapping from states to actions, without explicit evaluation functions) and involves only minor modifications to any supervised learning algorithm.

Here, the back propagation algorithm [7] was used, and has in this incarnation been named BP_{HT} (Back Propagation on Hypothetical Targets). The idea is to let the network create its own target vectors under two assumptions, one optimistic and one pessimistic, akin to the selective bootstrap adaption algorithm [9]. BP_{HT} is more general, however, in that it can cope with infinite credit horizons.

To verify the feasibility of this approach, a simple model of the link allocation problem in asynchronous transfer mode (ATM) telecommunication networks has been proposed. This is a stochastic bin packing model where connection requests are modelled as soap bubbles, to be allocated to one of a number of buckets. Preliminary experiments on a small allocation problem show that the proposed method is able to learn this task, reaching a level of performance comparable to conventional (non-adaptive) methods.

Much remains to be done, however. In these experiments, the bubble stream was stationary and the network's ability to adapt to changes in the generating distributions was not tested. Furthermore, a minimum of information was given to the network. Having a neural network as an allocator has the advantage of making other (larger) state descriptions feasible. For example, the allocator could benefit from knowing not only how full the buckets are, but also for how long the bubbles already in the buckets have been there.

Another important property which should be exploited is that the network is trained with (a slight modification of) a supervised algorithm. This implies that target information, if available at some time during training, can be used directly. This ability to switch between the two forms of learning is an advantage in many applications. For example, the neural bubble allocator could be given an initial guess of a good solution by first training it (supervised) to imitate the First Fit algorithm and then setting it free to explore, by reinforcement learning.

Furthermore, the pessimistic target vectors, T^{-}, can be used to inject rules of the application. For example, instead of setting only the target value corresponding to the latest action to 0, as described in section 3.1, the target values corresponding to all illegal actions (e.g. allocations to full buckets) could be set to 0 as well.

Acknowledgements

The authors would like to thank Jakob Carlström and Ernst Nordström, who have both contributed to this work in many ways. The work is financially supported by ELLEMTEL Telecommunication Systems Laboratories and by NUTEK, the Swedish National Board for Industrial and Technical Development, which is also greatly acknowledged.

References

[1] A.G. Barto and P. Anandan, "Pattern-Recognizing Stochastic Learning Automata", *IEEE Transactions on Systems, Man, and Cybernetics*, vol. 15, no. 3, pp. 360-375, 1985.

[2] R. Händel and M. N. Huber, *Integrated Broadband Networks: An Introduction to ATM-based Networks*, Addison-Wesley, 1991.

[3] D. Michie, "Trial and error", in S.A. Barnett and A. McLaren (Eds.), *Science Survey*, part 2, pp. 129-145, Penguin Books Ltd., 1961. Republished in D. Michie, *On Machine Intelligence*, 2nd edition, Ellis Horwood, 1986.

[4] P.W. Munro, "A Dual Back-Propagation Scheme for Scalar Reinforcement Learning", in *Proceedings of the Ninth Annual Conference of the Cognitive Science Society*, pp. 165-176, Lawrence Erlbaum, 1987.

[5] D. Nguyen and B. Widrow, "The truck backer-upper: An example of self-learning neural networks", in *Proceedings of the International Joint Conference on Neural Networks (IJCNN-89)*, vol. 2, pp. 357-363, Washington, 1989.

[6] A.J. Robinson and F. Fallside, "Dynamic Reinforcement Driven Error Propagation Networks with Applications to Game Playing", in *Proceedings of the Eleventh Annual Conference of the Cognitive Science Society*, pp. 836-843, Lawrence Erlbaum, 1989.

[7] D. Rumelhart, G. Hinton and R. Williams, "Learning internal representations by error propagation", in D. Rumelhart, J.L. McClelland and the PDP Research Group, *Parallel Distributed Processing*, vol. 1, pp. 318-362, MIT Press, 1986.

[8] R.S. Sutton, "Learning to Predict by the Methods of Temporal Difference", *Machine Learning*, vol. 3, pp. 9-44, Kluwer Academic Publishers, 1988.

[9] B. Widrow, N.K. Gupta and S. Maitra, "Punish/Reward: Learning with a Critic in Adaptive Threshold Systems", *IEEE Transactions on Systems, Man, and Cybernetics*, vol. 3, no. 5, pp. 455-465, 1973.

A Connectionist System for Medium-Term Horizon Time Series Prediction

Samy Bengio, Françoise Fessant, and Daniel Collobert

France Télécom,
Centre National d'Études des Télécommunications,
LAB/RIO/TNT,
2, avenue Pierre Marzin,
22307 Lannion, FRANCE.
e-mail: {bengio,fessant,collober}@lannion.cnet.fr

Abstract

In this paper, we propose some improvements for the problem of time series prediction with neural networks where a medium-term prediction horizon is needed. In particular, the ionospheric prediction service of the french Centre National d'Études des Télécommunications needs a six-month ahead prediction of a sunspots related time series which has a strong influence on wave propagation in ionosphere. The proposed improvements consist in two different modular architectures and a way to increase the size of the training set. Experimental results are compared to those of a simple multi-layer perceptron.

1 Introduction

Suppose we have a given one-variable time series represented by the N values $\{x_1, x_2, \cdots, x_N\}$, where x_t is the series value sampled at time t. Prediction then consists to find the future values $\{x_{N+1}, x_{N+2}, \cdots\}$. It has been shown in [6] that if the series is deterministic, there exists an integer d (which is called the *embedding dimension*), an integer τ (which is an arbitrary delay) and a function $f(\cdot)$ such that for every $t > (d \cdot \tau)$:

$$x_t = f(x_{t-\tau}, x_{t-2\tau}, \cdots, x_{t-d\tau}) \tag{1}$$

Unfortunately, there exists no exact method to find neither d, τ or $f(\cdot)$ when the series is too small (less than 10^d samples for d and τ).

The ionospheric prediction service of the Centre National d'Études des Télécommunications (CNET) publishes monthly prediction reports about ionospheric propagation of radio-electric waves addressed to radio-diffusion services.

Ionospheric state depends directly on solar activity, so in order to make a prediction of its future state we have to predict the solar activity, which is here represented as the number of sunspots R [2]. The sunspots time series is known to be difficult to predict and has served as a benchmark in the statistics literature [9].

2 Problem Description

The CNET ionospheric prediction service has given us the $IR5$ time series which is a non-centered five-month mean of the monthly sunspots number mean MR:

$$IR5_t = \frac{1}{5}(MR_{t-3} + MR_{t-2} + MR_{t-1} + MR_t + MR_{t+1}) \tag{2}$$

where MR_t is the mean sunspots number of month t. The CNET needs a six-month ahead prediction of the $IR5$ index in order to publish and distribute a report to its users. The time series used in the present work starts in 1849 and ends in 1991, which gives us 1712 monthly data, over which the last 238 are kept for testing and comparison with their own heuristic. Figure 1 shows the $IR5$ time series.

Sunspots related time series such as the $IR5$ index are suspected to be non-stationary but there is not enough data to remove tendency and/or seasonality. This is very annoying because, when using learning techniques such as neural networks, to expect a good generalization (prediction) error while minimizing a training set error, one has to hypothesize the test set is drawn from the same probability distribution as the training set. As the series may have a tendency, this may not be the case. This is the reason why our test set is kept small in comparison to the training set size.

3 A Simple Connectionist Solution

The simplest way to use neural networks for prediction, which has already been used in many applications [4, 8], is to use a multi-layer perceptron

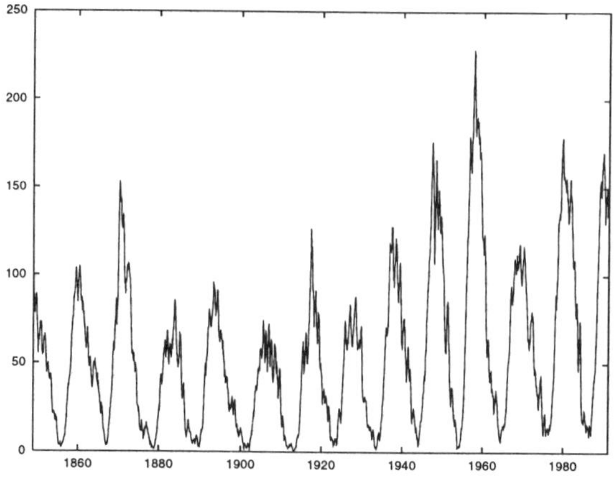

Figure 1: The *IR*5 Time Series

with one hidden layer, trained with backpropagation [5]. All input units are connected to all hidden units and all hidden units are connected to all output units. Given a series with N values, an embedding dimension d, a delay τ and an horizon h, training data consists of $N - (d \cdot \tau) - h$ input/output pairs such that:

$$\{x_{t-\tau}, x_{t-2\tau}, \cdots, x_{t-d\tau}\} \longrightarrow \{x_t, \cdots, x_{t+h}\} \tag{3}$$

where x_i is the ith normalized value of the series[1]. The embedding dimension d, as well as the hidden layer size are usually set by cross-validation techniques. In most applications, h is set to 0 and τ is set to 1.

For the *IR*5 series, the best network we found had 40 input units, 23 hidden units, and 6 output units. τ was set arbitrary to 1. Hidden and output units used a nonlinear squashing function with output in $[-1, 1]$. Figure 2 summarizes this architecture. Results are reported later in the paper.

4 Improvements Over a Simple Use of Neural Networks for Prediction

In this section, we present three improvements over this simple use of a neural network for time series prediction. All three are based on the fact that generalization (prediction) error is related to training set error, network capacity (which itself is related to the number of free parameters), training

[1]The series is usually normalized to get a zero mean with values in $[-1, 1]$.

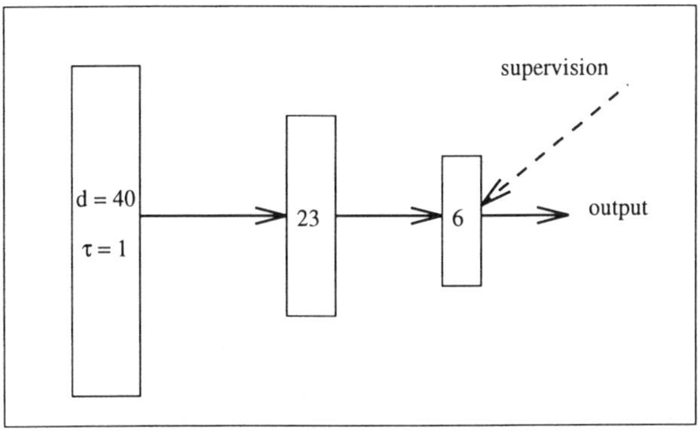

Figure 2: The Simple Multi-Layer Perceptron Solution

set size and the use of prior knowledge in system design [7]. The first two improvements are architectural: a better system design and less parameters. The last one shows how to use more training data.

4.1 Mix of Different Output Schemes

When one wants to predict x_{t+h} given $\{x_{t-\tau}, x_{t-2\tau}, \cdots, x_{t-d\tau}\}$ with $h > 0$, there are at least three simple ways to do it:

- a multi-layer perceptron (MLP) with one output unit x_{t+h},
- an MLP with $h + 1$ output units $x_t, x_{t+1}, \cdots, x_{t+h}$,
- an MLP with one output unit x_t. In that case x_{t+h} is predicted using as input previously predicted values $\hat{x}_t, \cdots, \hat{x}_{t+h-1}$, which are estimates of $x_t, \cdots, x_{t+h-1}$.

Experimental results tend to show that the latter method is better when h is big ($h > 10$) whereas the first two are better when h is small.

We now propose a modular architecture which is composed of three small interconnected networks: the first two networks try to find independently a solution and the third one combines their results.

The first network tries to find x_{t+h} while the second tries to find also intermediate values $x_t, \cdots, x_{t+h-1}$, using the same input data. The outputs of the first and the second network are then used as input to a third network, which try to find x_{t+h}. The three networks are trained simultaneously and supervision is provided to all output units. This means that weights of the first two networks are influenced both by supervision provided by their respective network but also by the third network.

Since all three networks are small (the hidden layer sizes for experiments reported here were respectively 5, 5 and 11 for the three networks), the total number of free parameters in this model is still half the number of free parameters in the simple model. Once again , τ was arbitrary set to 1. Figure 3 shows this new architecture.

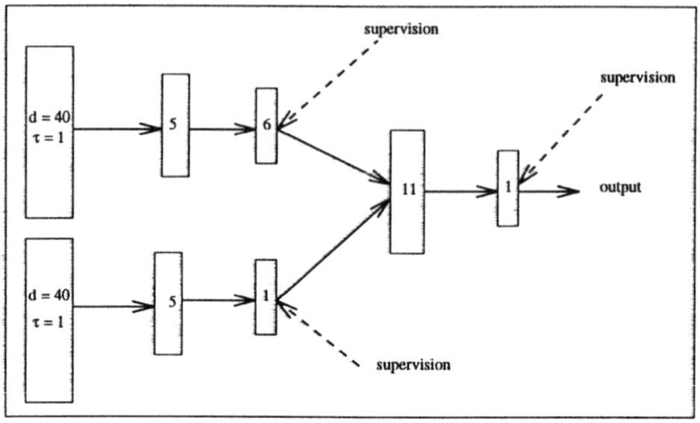

Figure 3: A First Architectural Improvement: Mix of Different Small Networks which have Different Outputs

4.2 Use of Different d/τ Combinations

The time series length being too small to determine adequately good values for d and τ, an alternate solution is to use a combination of many small networks each using different d and τ values. One way of doing this is to search for a new function which has the form:

$$\begin{aligned} x_{t+h} = g(& f_1(x_{t-\tau_1}, x_{t-2\tau_1}, \cdots, x_{t-d_1\tau_1}), \\ & f_2(x_{t-\tau_2}, x_{t-2\tau_2}, \cdots, x_{t-d_2\tau_2}), \\ & \vdots \\ & f_n(x_{t-\tau_n}, x_{t-2\tau_n}, \cdots, x_{t-d_n\tau_n}) \,) \end{aligned} \quad (4)$$

where $f_1(\cdot), \cdots, f_n(\cdot)$ are simple neural networks trained to predict x_{t+h} (and maybe intermediate values), each using its own d_i and τ_i, and $g(\cdot)$ is another network which is in top of the others and simultaneously trained. In the experiments we report here with the *IR*5 time series and a prediction horizon $h = 5$, we had three functions $f_i(\cdot)$:

- $f_1(\cdot)$ used $d = 40$, $\tau = 1$ and tried to predict x_{t+5}.

- $f_2(\cdot)$ used $d = 34$, $\tau = 2$ and tried to predict x_{t+1}, x_{t+3} and x_{t+5}.
- $f_3(\cdot)$ used $d = 30$, $\tau = 3$ and tried to predict x_{t+2} and x_{t+5}.

Figure 4 gives the actual architecture used in this paper. Again, the resulting network still has around half the number of free parameters compared to the simple network.

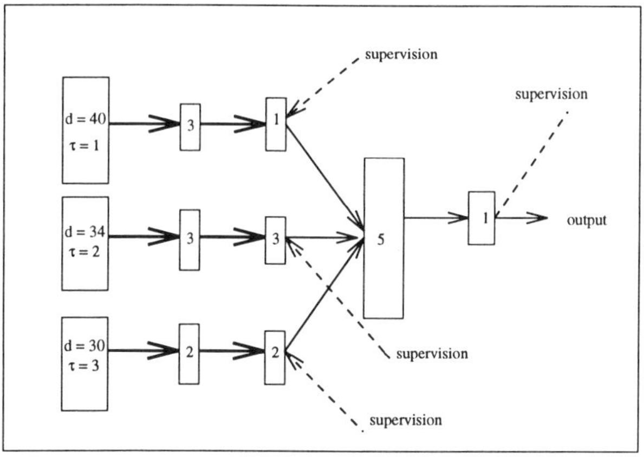

Figure 4: A Second Architectural Improvement: Use of Different d/τ Combinations

4.3 Increase of the Training Set Size

For the particular series of sunspots, data are in fact gathered every day, but only monthly means are usually used for prediction. We propose here a way to use more than just monthly means. The *IR*5 index, computed in equation (2) can be rewritten as follows:

$$NIR5_t = \frac{1}{152} \sum_{i=t-105}^{t+46} R_i \tag{5}$$

where $NIR5_t$ for a particular day t represents the *IR*5 index. The only difference with the *IR*5 series is that we now have a longer but daily series of the very same index. In fact, the *IR*5 series is a subset of the *NIR*5 one.

Setting τ to 30 implies we are searching for exactly the same kind of function as with the *IR*5 series with τ set to 1, except that the database size is around 30 times the *IR*5 one[2]. This is important, given that generalization

[2]And of course the fact that a month is not exactly equal to 30 days but this does not have any real impact on the goal.

error is related to training set size: the more data one have, the more free parameters one can handle without over-fitting, the more precise could be the resulting function.

In the experiments we report here, we used the *NIR*5 time series sampled every 7 days, which makes the database size around 4 times the original one.

5 Experimental Results

We compared the different connectionist models proposed in this paper for the *IR*5 time series to the actual heuristic used by the CNET prediction service.

For all networks, input dimension d, as well as hidden layers sizes, have been estimated using cross-validation. All experiments used the stochastic version of backpropagation algorithm, with a small weight decay to keep network capacity as small as possible. Results in Table 1 give the *Average Relative Cost* (ARV) obtained for the last 238 months of the series, which were not used to train the network. ARV is computed as follows:

$$ARV = \frac{1}{\hat{\sigma}^2}\frac{1}{N}\sum_{i \in P}(x_i - \hat{x}_i)^2 \qquad (6)$$

where P is the test set, N is the test set size, $\hat{\sigma}^2$ is the estimated variance of the series, $\hat{x}_i$ is the ith predicted value and x_i is the corresponding desired value. We also give the *Strong Error Percentage* (SEP) which represents the percentage of test set examples for which the distance between expected value and obtained value is more than 30 (for the *IR*5 time series, values range from 0 to 230). This statistic is important for the CNET because its users prefer to make many small errors rather than few important ones.

Table 1: Comparison of the Different Predictors

	ARV	SEP
actual CNET heuristic	0.113	5.04
Standard MLP	0.088	5.04
Modular Architecture (multiple output schemes)	0.075	1.68
Modular Architecture (multiple d/τ schemes)	0.073	1.68
Larger Training Set	0.071	1.68

6 Conclusion

In this paper we proposed three improvements over a simple use of multi-layer perceptron for time series prediction with a medium-term horizon. All gave better results for a particular and known difficult time series. We shall now try to mix some of these ideas, as well as use other models such as Elman recurrent networks [1] or hierarchical mixtures of experts [3].

References

[1] J. L. Elman, "Finding structure in time," *Cognitive Science*, vol. 14, pp. 179–211, 1990.

[2] A. Izenman, "J. R. Wolf and the Zürich sunspot relative numbers," *The Mathematical Intelligencer*, vol. 7, no. 1, pp. 27–33, 1985.

[3] M. I. Jordan and R. A. Jacobs, "Hierarchical mixtures of experts and the EM algorithm," *Neural Computation*, vol. 6, no. 2, pp. 181–214, 1994.

[4] D. C. Park, M. A. El-Sharkawi, and R. J. Marks II, "Electric load forecasting using an artificial neural network," *IEEE Transaction on Power Systems*, vol. 6, no. 2, pp. 442–449, 1991.

[5] D. E. Rumelhart, G. E. Hinton, and R. J. Williams, "Learning internal representations by error propagation," in *Parallel Distributed Processing* (D. E. Rumelhart and J. L. McClelland, eds.), vol. 1, MIT Press, 1986.

[6] F. Takens, "Detecting strange attractors in turbulence," in *Dynamical Systems and Turbulence* (D. A. Rand and L.-S. Young, eds.), vol. 898 of *Lecture Notes in Mathematics*, (Warwick 1980), pp. 366–381, Springer-Verlag, Berlin, 1981.

[7] V. N. Vapnik, *Estimation of Dependencies Based on Empirical Data*. New-York, NY, USA: Springer-Verlag, 1982.

[8] A. Varfis and C. Versino, "Univariate economic time series forecasting by connectionist methods," in *Proceedings of the International Neural Network Conference (INNC)*, (Paris, France), pp. 342–345, 1990.

[9] A. S. Weigend, B. A. Huberman, and D. E. Rumelhart, "Predicting sunspots and exchange rates with connectionist networks," in *Nonlinear modeling and forecasting* (M. Casdagli and S. Eubank, eds.), pp. 395–431, Addison Wesley, 1992.

Selective Presentation Learning for Forecasting by Neural Networks

Kazuhiro Kohara
NTT Information and Communication Systems Laboratories
3-9-11 Midori-cho, Musashino-shi, Tokyo 180, Japan
e-mail: kohara@nttaip.min.ntt.jp

Abstract

This paper proposes a selective presentation learning technique for improving the learnability and predictability of large changes by back-propagation neural networks. Daily stock prices are predicted as a complicated real-world problem. Training data corresponding to large changes of prediction-target time series are presented more often, and network learning is stopped at the point having the maximal profit. When this technique is applied to daily stock market prediction, the network's ability to make profits through experimental stock-trading, where the cost of trading is ignored, is about 70% better than that of conventional techniques.

1 Introduction

Predicting the future behavior of real-world time series using neural networks has been extensively investigated [8] [7], because neural networks can learn nonlinear relationships between inputs and desired outputs. Various attempts have been made to apply neural networks to financial market prediction (e.g., [1] [5]), electricity load forecasting (e.g., [4]), and other areas.

In the conventional approach, all training data are equally presented to neural networks (the number of presentations of all training data is the same) independently of the size of the changes in the prediction-target time series. Also, network learning is usually stopped at the point of minimal mean squared error between the network's outputs and the actual outputs. The following is an outline of the conventional learning (CVL) algorithm.

CVL Algorithm:

CVL1: Train backpropagation networks with equal presentation of all training data.

CVL2: Stop network learning at the point having the minimal mean squared error.

Let us consider two problems that occurr in CVL.

Problem 1: Generally, the predictability of large changes is more important than that of small changes. When all training data are presented equally as in the conventional approach, we assume that neural networks learn small changes as correctly as large changes and cannot learn large changes more effectively.

Problem 2: The target function for forecasting problems is not always to minimize the average of prediction errors. In financial market predictions, maximizing profits through financial trading by using predicted values is more important than minimizing errors. Minimizing the average of prediction errors does not always correspond to maximizing profits.

This paper proposes a learning technique that can learn about large changes more effectively and predict them more accurately. We present training data that correspond to large changes in the prediction-target time series more often than those corresponding to small changes.

This paper also investigates another stopping criterion for financial predictions. Daily stock prices are predicted as a noisy real-world problem. Network learning is stopped at the point having the maximal profit through experimental stock-trading.

2 Proposal of Selective Presentation Learning for Forecasting

To allow neural networks to learn about large changes of prediction-target time series more effectively, we separate the training data into *large-change data* and *small-change data*. *Large-change data* (*small-change data*) have next-day changes that are larger (smaller) than a preset value. *Large-change data* are presented to neural networks more often than *small-change data*. For example, all training data are presented in the first learning cycle, only *large-change data* are presented in the second cycle, and so forth.

The outline of the selective presentation learning (SEL) algorithm is as follows.

SEL Algorithm:

SEL1: Separate the training data into *large-change data* and *small-change data*.

SEL2: Train backpropagation networks with more presentations of *large-change data* than of *small-change data*.

SEL3: Stop network learning at the point satisfying a certain stopping criterion (e.g., stop at the point having the maximal profit).

3 Applying SEL to Stock Market Prediction

3.1 Multivariate Time Series

The following five indicators were used as inputs to the neural network in the same way as in our previous work [3]: Tokyo stock market index (TOPIX), dollar-to-yen exchange rate (EXCHANGE), interest rate (INTEREST), crude oil prices (OIL), and NY stock market prices (NY). These time series are plotted in Fig. 1. TOPIX was the prediction target.

The Pearson correlation coefficients between TOPIX and other time series are shown in Table I. EXCHANGE was positively correlated with TOPIX. INTEREST and OIL were negatively correlated with TOPIX. NY was slightly positively correlated with TOPIX.

Table I: Correlation Coefficient with TOPIX

	Correlation coefficient
Dollar-to-yen exchange rate (EXCHANGE)	0.510
Interest rate (INTEREST)	-0.830
Crude oil price (OIL)	-0.613
NY stock market prices (NY)	0.251

3.2 Neural Prediction Model

We input the daily changes in these five indicators (e.g., TOPIX(t) - TOPIX(t-1)) and present the next-day's change in TOPIX to the neural network as the desired output (Fig. 2). All the data of the daily changes have been scaled to the interval [0.1, 0.9]. A 6-6-1 multi-layered neural network was used (6 neurons in the input layer, 6 in the hidden layer, and 1 in the output layer).

The back-propagation algorithm [6] was used to train the network. The algorithm is described here briefly. The output from neuron i, O_i, is connected to the input of neuron j through the interconnection weight W_{ij}. Unless neuron k is one of the input neurons, the output of the neuron k is:

$$O_k = f(\Sigma_i W_{ik} O_i + \theta_k), \tag{1}$$

where $f(x) = 1/(1 + e^{-x})$, θ_k is a threshold of the kth unit in the hidden layer, and the sum is of all neurons in the adjacent layer. If neuron k is one of the input neurons, its output is the same as the input.

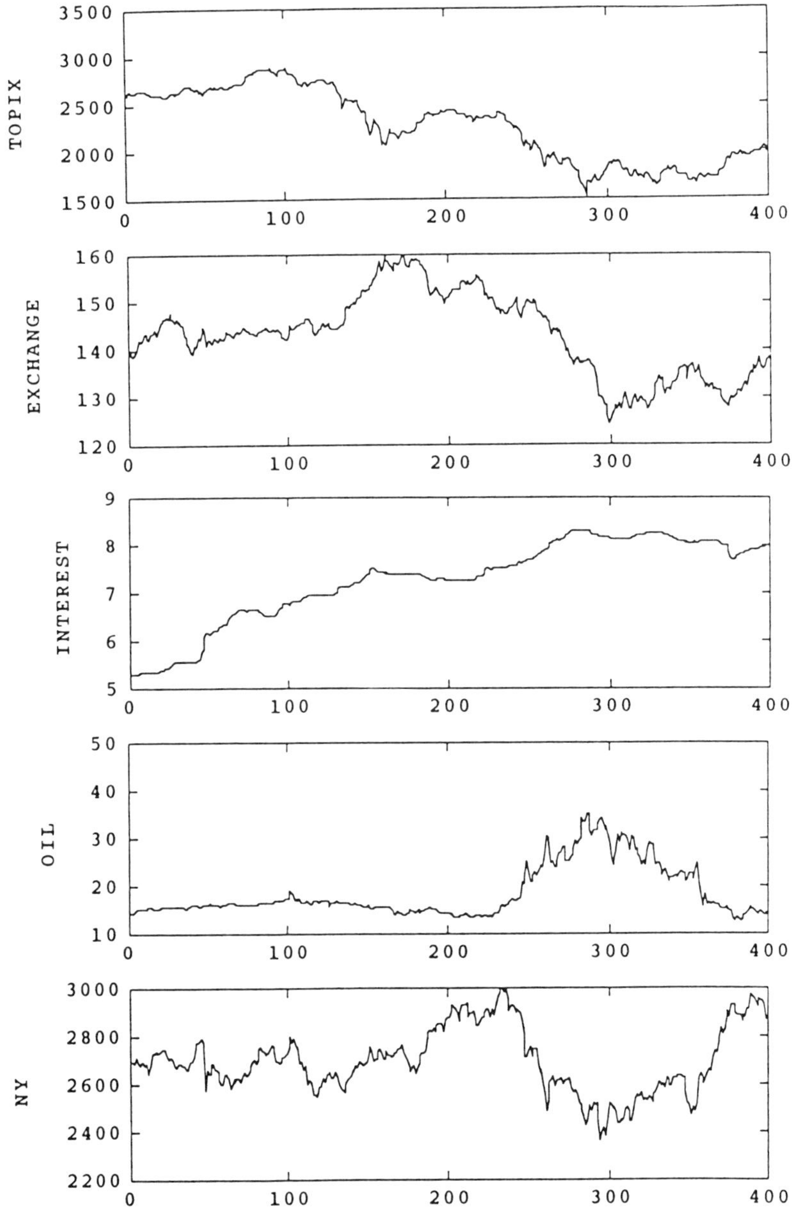

Figure 1: Multivariate time series

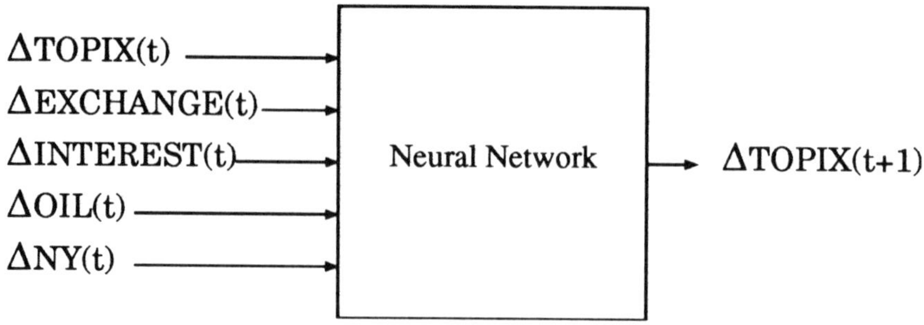

Figure 2: Neural Prediction Model

The error at the output neuron can be defined as

$$E = \frac{1}{2}(D_k - O_k)^2, \tag{2}$$

where D_k is the desired output value of the output neuron and O_k is the actual value. The connection weight between the kth unit in the output layer and the j th unit in the hidden layer is modified according to

$$\Delta W_{kj} = \epsilon \delta_k O_j, \tag{3}$$

$$\delta_k = (D_k - O_k)O_k(1 - O_k), \tag{4}$$

where ϵ is a scalar quantity which determines the learning rate. The connection weight between the j th unit in the hidden layer and the i th unit in the input layer is modified according to

$$\Delta W_{ji} = \epsilon \delta_j O_i, \tag{5}$$

$$\delta_j = O_j(1 - O_j)\Sigma_k \delta_k W_{kj}. \tag{6}$$

To improve the convergence characteristics, a momentum term with momentum gain α can be introduced as follows:

$$\Delta W_{ij}(n+1) = \epsilon \delta_j O_i + \alpha \Delta W_{ij}(n), \tag{7}$$

where n represents the iteration index.

3.3 Experiments

In the experiments, the learning parameter ϵ was 0.7 and the momentum parameter α was also 0.7. All the weights and biases in the neural network were initialized randomly between -0.3 and 0.3. We ran the neural network four times with the same training data in each experiment and took the average.

We used data from 409 days (from August 1, 1989 to March 31, 1991): data from 300 days for training, data from 30 days for validation (making decisions on stopping the network learning), and data from 79 days for making predictions.

In the CVL experiments (Ex. 1), network learning was stopped at the point having the minimal mean prediction error. The mean prediction errors for both the training and validation data were monitored. As long as both errors decreased, training continued. When either error increased, we stopped training to avoid overfitting. We also examined the case where equal presentation learning was stopped at a point having the maximal profit based on the validation data during 8000 learning cycles (Ex. 2). The prediction error and profit were monitored after every hundred learning cycles.

In the SEL experiments (Ex. 3), learning was stopped at the point having the maximal profit based on the validation data. Here, a *large-change* was a change larger than 14.78, which was the median of TOPIX daily changes in the training data. In the SEL experiments, *large-change data* were presented five times as often as *small-change data*.

When a *large-change* was predicted, we estimated "Profit" as follows: when the predicted direction was the same as the actual direction, the daily change was earned, and when the predicted direction was opposite the actual direction, the daily change was lost. This calculation of profit corresponds to the following experimental stock-trading system. A buy (sell) signal is given when the next-day predicted TOPIX is more than 14.78 points (about US$1.40) higher (lower) than today's TOPIX. If a buy (sell) signal is given, the system will buy (sell) shares at the current price and sell (buy) the shares back at the next-day price. "Profit" is the profit and loss balance at the end of the period. The cost of trading was ignored in calculating the profit.

The experimental results are shown in Table II, where "Prediction error" is the mean absolute value of the prediction error on test data. Using maximal profit as the stopping criterion improved profit in every case: for training, validation, and test data. Applying our selective learning technique further improved every profit. Using SEL, the profit on the test data was 70% higher than when using CVL. Although the average of prediction errors by SEL was larger than that by CVL, a small average of prediction errors did not always correspond to large profits.

3.4 Other Predictors

We examined multiple regression analysis (MR) as follows:

$$\Delta\mathrm{TOPIX(t+1)} = c_0 + c_1\Delta\mathrm{TOPIX(t)} + c_2\Delta\mathrm{EXCHANGE(t)} + c_3\Delta\mathrm{INTEREST(t)} + c_4\Delta\mathrm{OIL(t)} + c_5\Delta\mathrm{NY(t)}, \quad (8)$$

Table II: Experimental Results on Neural Prediction Models

	Ex. 1 (CVL)	Ex. 2	Ex. 3 (SEL)
Presentation method	equal	equal	selective
Stopping criterion	minimal error	maximal profit	maximal profit
Profit on training data	1275	1823	2317
Profit on validation data	154	245	315
Profit on test data	329	422	550
Prediction error	16.2	16.3	17.0

where c_0, c_1, c_2, c_3, c_4, and c_5 are regression coefficients. Data from 330 days were used to estimate these coefficients (Ex. 4). We also examined the case where we made the number of *large-change* data five times as large as before and estimated the coefficients (Ex. 5).

The following two trivial predictors (TR1 and TR2) were also evaluated: TR1 always predicts a change of 0 (Ex. 6) and TR2 always predicts that tomorrow's change is the same as today's (Ex. 7). We chose these predictors as fair comparisons according to the efficient market hypothesis [2].

The experimental results are shown in Table III. The profit in Ex. 5 was 43% larger than that in Ex. 4. The basic idea of SEL was also useful in MR. The profit in Ex. 3 (SEL) was 45% larger than that in Ex. 5 (MR). The profit of TR1 was zero, because TR1 always predicted no change and did not trade any stocks. The prediction errors by TR1 correspond to the actual average of TOPIX daily changes. Although the prediction errors by TR2 were much larger than those by TR1, TR2 earned a small profit. The profit of TR2 was about 30% that in Ex. 1 (CVL).

Table III: Experimental Results on Other Predictors

	Ex. 4 (MR)	Ex. 5 (MR)	Ex. 6 (TR1)	Ex. 7 (TR2)
Profit on test data	265	379	0	109
Prediction error	16.4	16.7	18.9	25.5

4 Conclusions

We proposed a selective presentation learning technique for forecasting. Training data corresponding to large changes in the prediction-target time series are presented more often, and network learning is stopped at the point having the maximal profit on the validation data. The results of several experiments on stock market prediction showed that our approach outperforms the conventional approaches.

We will examine experimental stock-trading, where the cost of trading is included. We also plan to study a forecasting method by integrating statistical analysis and neural networks.

Ackowledgements

The author would like to thank Messrs. Yukihiro Nakamura and Yoshimi Fukuhara for their encouragement in this work.

References

[1] E. Azoff, Neural Network Time Series Forecasting of Financial Markets, John Wiley and Sons, 1994.

[2] E. Fama, "Efficient Market Hypothesis: A Review of Theory and Empirical Work," in *Journal of Finance*, Vol. 25, pages 383-417, 1970.

[3] K. Kohara and T. Ishikawa, "Multivariate Prediction Using Prior Knowledge and Neural Heuristics," in *Proceedings of International Symposium on Integrating Knowledge and Neural Heuristics*, pp. 179-188, Pensacola, USA, 1994.

[4] D. Park, M. El-Sharkawi, R. Marks II, L. Atlas, and M. Damborg, "Electric Load Forecasting Using An Artificial Neural Network," in *IEEE Transactions on Power Systems*, Vol. 6, No. 2, pp. 442-449, 1991.

[5] A. Refenes and M. Azema-Barac, "Neural Network Applications in Financial Asset Management" in *Neural Computing & Applications*, Vol. 2, No. 1, pp. 13-39, Springer International, 1994.

[6] D. Rumelhart, G. Hinton, and R. Williams, "Learning Internal Representations by Error Propagation," in *D. Rumelhart, J. McClelland and the PDP Research Group eds., Parallel Distributed Processing*, Vol. 1, MIT Press, 1986.

[7] V. Vemuri and R. Rogers eds., Artificial Neural Networks: Forecasting Time Series, IEEE Computer Society Press, 1994.

[8] A. Weigend and N. Gershenfeld eds., Time Series Prediction: Forecasting the Future and Understanding the Past, Addison-Wesley, 1993.

Predicting Failures of Telecommunication Paths:

Limits on Learning Machine Accuracy Imposed by Data Quality

Corinna Cortes, L. D. Jackel, and Wan-Ping Chiang

AT&T Bell Laboratories, Holmdel, NJ 07733
corinna@neural.att.com, ldj@neural.att.com, and wpc1@corona.att.com

Abstract

Random errors and insufficiencies in databases limit the performance of any classifier trained from and applied to the database. In this paper we propose a method to estimate the limiting performance of classifiers imposed by the database. We demonstrate this technique on the task of predicting failure in telecommunication paths.

1 Introduction

Data collection for a classification or regression task is prone to random errors, e.g. inaccuracies in the measurements of the input or mis-labeling of the output. Missing or insufficient data are other sources that may complicate a learning task and hinder accurate performance of the trained machine. These insufficiencies of the data limit the performance of any learning machine or other statistical tool constructed from and applied to the data collection — no matter how complex the machine or how much data is used to train it.

In this paper we propose a method for estimating the limiting performance of learning machines imposed by the quality of the database used for the task. The method involves a series of learning experiments. The extracted result is, however, independent of the choice of learning machine used for these experiments since the estimated limiting performance expresses a characteristic of the data. The only requirements on the learning machines are that their capacity (VC-dimension) can be varied and can be made large, and that the learning machines with increasing capacity become capable of implementing *any* function.

We have applied the technique to data collected for the purpose of predicting failures in telecommunication channels of the AT&T network. We extracted information from one of AT&T's large databases that continuously logs performance parameters of the network. The character and amount of data comes to more material than humans can survey. The processing of the extracted information is therefore automated by learning machines. The currently best learning machine for this task, a decision tree constructed with the C4.5 package [1], achives 71% correct predictions.

We conjecture that the quality of the data imposes a limiting error rate on any learning machine of $\sim$ 25%, so that even with an unlimited amount of data, and an arbitrarily complex learning machine, the performance for this task will not exceed $\sim$ 75% correct. This conjecture is supported by experiments.

The relatively high noise-level of the data, which carries over to a poor performance of the trained classifier, is typical for many applications: the data collection was not designed for the task at hand and proved inadequate for constructing high performance classifiers.

Constructing classifiers of higher accuracy requires more information about the task at hand, or cleaner data. The given classification problem was hence repeated with more homogeneous data, and the accuracy improved to 82% correct.

Both a learning machine constructed from the original data and a system constructed from the more homogeneous data are implemented in the AT&T maintenance system and daily used to predict upcoming failures.

2 Basic Concepts of Machine Learning

We can picture a learning machine as a device that takes an unknown input vector and produces an output value. More formally, it performs some mapping from an input space to an output space. The particular mapping it implements depends of the setting of the internal parameters of the learning machine. These parameters are adjusted during a learning phase so that the labels produced on the training set match, as well as possible, the labels provided. The number of patterns that the machine can match is loosely called the "capacity" of the machine. Generally, the capacity of a machine increases with the number of free parameters. After training is complete, the generalization ability of of the machine is estimated by its performance on a test set which the machine has never seen before.

The test and training error depend on both the the number of training examples l, the capacity h of the machine, and, of course, how well suited the machine is to implement the task at hand. Let us first discuss the typical behavior of the test and training error for a noise corrupted task as we vary h but keep the amount l of training data fixed. This scenario can, e.g., be obtained by increasing the number of hidden units in a neural network or increasing the number of codebook vectors in a Learning Vector Quantization algorithm [7]. Figure 1a) shows typical training and test error as a function of the capacity of the learning machine. For $h \ll l$ we have many fewer free parameters than training examples and the machine is over constrained. It does not have enough complexity to model the regularities of the training data, so both the training and test error are large (underfitting). As we increase h the machine can begin to fit the general trends in the data which carries over to the test set, so both error measures decline. Because the performance of the machine is optimized on only part of the full pattern space the test error will always be larger than the training error. As we continue to increase the capacity of the learning machine the error on the training set continues to decline, and eventually it reaches zero as we get enough free parameters to completely model the training set. The behavior of the error on the test set is different. Initially it decreases, but at some capacity, $h^\star$, it starts to rise. The rise occurs because the now ample resources of the training machine are applied to learning vagaries of the training set, which are not reproduced in the test set (overfitting). Notice how in Figure 1a) the optimal test error is achieved at a capacity $h^\star$ that is smaller than the capacity for which zero error is achieved on the training set. The learning machine with capacity $h^\star$ will typically commit errors on misclassified or outlying patterns of the training set.

We can alternatively discuss the error on the test and training set as a function of the training set size l for fixed capacity h of the learning machine. Typical behavior is sketched in Figure 1b). For small l we have enough free parameters to completely model the training set, so the training error is zero. Excess capacity is used by the learning machine to model details in the training set, leading to a large test error. As we increase the training set size l we train on more and more patterns so the test error declines. For some critical size of the training set, l_c, the machine can no longer model all the training patterns and the training error starts to rise. As we further increase l the irregularities of the individual training patterns smooth out and the parameters of the learning machine is more and more used to model the true underlying function. The test error declines, and asymptotically the training and test error reach the same error value E_∞. This error

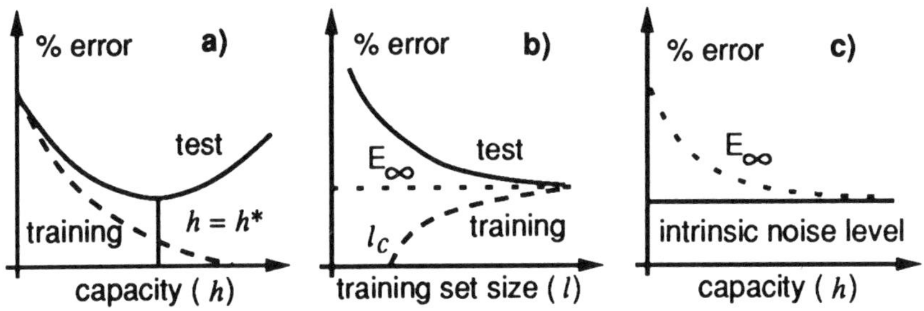

Figure 1: Errors as function of capacity and training set size. Figure 1a) shows characteristic plots of training and test error as a function of the learning machine capacity for fixed training set size. The test error reaches a minimum at $h = h^\star$ while the training error decreases as h increases. Figure 1b) shows the training and test errors at fixed h for varying l. The dotted line marks the asymptotic error E_∞ for infinite l. Figure 1c) shows the asymptotic error as a function of h. This error is limited from below by the intrinsic noise in the data.

value is the limiting performance of the given learning machine to the task. In practice we never have the infinite amount of training data needed to achieve E_∞. However, recent theoretical calculations [9, 2, 3, 8, 6] and experimental results [4] have shown that we can estimate E_∞ by averaging the training and test errors for $l > l_c$. This means we can predict the optimal performance of a given machine.

For a given type of learning machine the value of the asymptotic error E_∞ of the machine depends on the quality of the data and the set of functions it can implement. The set of available functions increases with the capacity of the machine: low capacity machines will typically exhibit a high asymptotic error due to a big difference between the true noise-free function of the patterns and the function implemented by the learning machine, but as we increase h this difference decreases. If the learning machine with increasing h becomes a *universal* machine capable of modeling *any* function the difference eventually reaches zero, so the asymptotic error E_∞ only measures the intrinsic noise level of the data. Once a capacity of the machine has been reached that matches the complexity of the true function no further improvement in E_∞ can be achieved. This is illustrated in Figure 1c). The intrinsic noise level of the data or the limiting performance of any learning machine may hence be estimated as the asymptotic value of E_∞ as obtained for asymptotically universal learning machines with increasing capacity applied to the task. This technique will be illustrated in the following section.

3 Experimental Results

In this section we estimate the limiting performance imposed by the data of any learning machine applied to the particular prediction task.

3.1 Task Description

To ensure the highest possible quality of service, the performance parameters of the AT&T network are constantly monitored. Due to the high complexity of the network this performance surveillance is mainly corrective: when certain measures exceed preset thresholds action is taken to maintain reliable, high quality service. These reorganizations can lead to short, minor impairments of the quality of the communication path. In contrast, the work reported here is preventive: our objective is to make use of the performance parameters to form predictions that are sufficiently accurate that preemptive repairs of the channels can be made during periods of low traffic.

In our study we have examined the characteristics of long-distance, 45 Mbits/s communication paths in the domestic AT&T network. The paths are specified from one city to another and may include different kinds of physical links to complete the paths. A path from New York City to Los Angeles might include both optical fiber and coaxial cable. To maintain high-quality service, particular links in a path may be switched out and replaced by other, redundant links.

There are two primary ways in which performance degradation is manifested in the path. First is the simple bit-error rate, the fraction of transmitted bits that are not correctly received at the termination of the path. Barring catastrophic failure (like a cable being cut), this error rate can be measured by examining the error-checking bits that are transmitted along with the data. The second instance of degradation, "framing error", is the failure of synchronization between the transmitter and receiver in a path. A framing error implies a high count of errored bits.

In order to better characterize the distribution of bit errors, several measures are historically used to quantify the path performance in a 15 minutes interval. These measures are:

Low-Rate The number of seconds with exactly 1 error.

Medium-Rate The number of seconds with more than one but less than 45 errors.

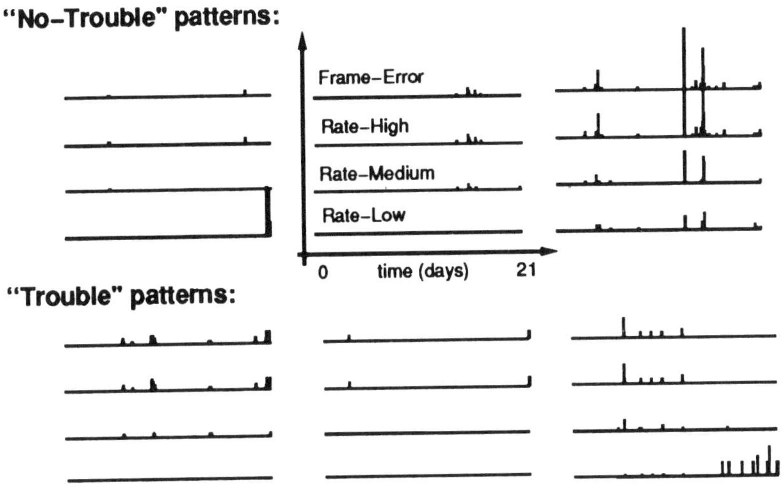

Figure 2: Errors as function of time. The 3 top patterns are members of the "No-Trouble" class. The 3 bottom ones are members of the "Trouble" class. Errors are here plotted as mean values over hours.

High-Rate The number of seconds with 45 or more errors, corresponding to a bit error rate of at least 10^{-6}.

Frame-Error The number of seconds with a framing error. A second with a frame-error is always accompanied by a second of High-Rate error.

Although the number of seconds with the errors described above in principle could be as high as 900, any value greater than 255 is automatically clipped back to 255 so that each error measure value can be stored in 8 bits.

Daily data that include these measures are continuously logged in an AT&T database that we call Perf(ormance)Mon(itor). Since a channel is error free most of the time, an entry in the database is only made if its error measures for a 15 minute period exceed fixed low thresholds, e.g. 4 Low-Rate seconds, 1 Medium- or High-Rate second, or 1 Frame-Error. In our research we "mined" PerfMon to formulate a prediction strategy. We extracted examples of path histories 28 days long where the path at day 21 had at least 1 entry in the PerfMon database. We labeled the examples according to the error-measures over the next 7 days. If the channel exhibited a 15-minute period with at least 5 High-Rate seconds we labeled it as belonging to the class "Trouble". Otherwise we labeled it as member of "No-Trouble".

The length of the history- and future-windows are set somewhat arbitrarily. The history has to be long enough to capture the state of the path but short enough that our learning machine will run in a reasonable time. Also the longer the history the more likely the physical implementation of the path was modified so the error measures correspond to different media. Such error histories could in principle be eliminated from the extracted examples using the record of the repairs and changes of the network. The complexity of this database, however, hinders this filtering of examples. The future-window of 7 days was set as a design criterion by the network system engineers.

Examples of histories drawn from PerfMon are shown in Figure 2. Each group of traces in the figure includes plots of the 4 error measures previously described. The 3 groups at the top are examples that resulted in No-Trouble while the examples at the bottom resulted in Trouble. Notice how bursty and irregular the errors are, and how the overall level of Frame- and High-Rate errors for the Trouble class seems only slightly higher than for the No-Trouble class, indicating the difficulty of the classification task as defined from the database PerfMon. PerfMon constitutes, however, the only stored information about the state of a given channel in its entirety and thus all the knowledge on which one can base channel end-to-end predictions: it is impossible to install extra monitoring equipment to provide other than the 4 mentioned end-to-end error measures.

The above criteria for constructing examples and labels for 3 month of PerfMon data resulted in 16325 examples from about 900 different paths with 33.2% of the examples in the class Trouble. This means, that always guessing the label of the largest class, No-Trouble, would produce an error rate of about 33%.

3.2 Estimating Limiting Performance

The 16325 path examples were randomly divided into a training set of 14512 examples and a test set of 1813 examples. Care was taken to ensure that a path only contributes to one of the sets so the two sets were independent, and that the two sets had similar statistical properties.

Our input data has a time-resolution of 15 minutes. For the results reported here the 4 error measures of the patterns were subsampled to mean values over days yielding an input dimensionality of 4×21.

We performed two sets of independent experiments. In one experiment we used fully connected neural networks with one layer of hidden units. In the other we used LVQ learning machines with an increasing number of codebook vectors. Both choices of machine have two advantages: the capacity of the machine can easily be increased by adding more hidden units, and by increasing the number of hidden units or number of codebook vectors we can eventually model *any* mapping [5]. We first discuss the results with neural networks.

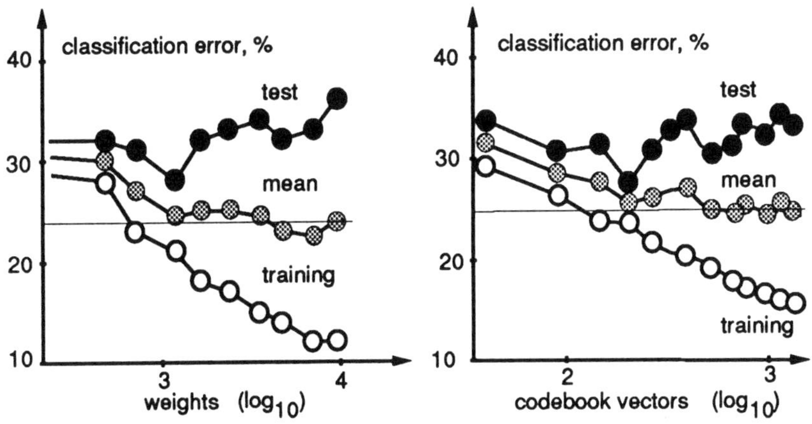

Figure 3: a) Measured classification errors for neural networks with increasing number of weights (capacity). The mean value between the test and training error estimates the performance of the given classifier trained with unlimited data. b) Measured classification errors for LVQ classifiers with increasing number of codebook vectors.

Baseline performance was obtained from a threshold classifier by averaging all the input signals and thresholding the result. The training data was used to adjust the single threshold parameter. With this classifier we obtained 32% error on the training set and 33% error on the test set. The small difference between the two error measures indicate statistically induced differences in the difficulty of the training and test sets. An analysis of the errors committed revealed that the performance of this classifier is almost identical to always guessing the label of the largest class "No-Trouble": close to 100% of the errors are false negative.

A linear classifier with about 200 weights (the network has two output units) obtained 28% error on the training set and 32% error on the test set.

Further experiments exploited neural nets with one layer of respectively 3, 5, 7, 10, 15, 20, 30, and 40 hidden units. All our results are summarized in Figure 3a). This figure illustrates several points mentioned in the text above. As the complexity of the network increases, the training error decreases because the networks get more free parameters to memorize the data. Compare to Figure 1a). The test error also decreases at first, going through a minimum of 29% at the network with 5 hidden units. This network apparently has a capacity that best matches the amount and character of the available training data. For higher capacity the networks overfit the data at the expense of increased error on the test set.

Figure 3a) should also be compared to Figure 1c). In Figure 3a) we plotted approximate values of E_∞ for the various networks — the minimal error of the network to the given task. The values of E_∞ are estimated as the mean of the training and test errors. The value of E_∞ appears to flatten out around the network with 30 units, asymptotically reaching a value of 24% error.

An asymptotic E_∞-value of 25% was obtained from LVQ-experiments with increasing number of codebook vectors. These results are summarized in Figure 3b). We therefore conjecture that the intrinsic noise level of the task is about 25%, and this number is the limiting error rate imposed by the quality of the data on any learning machine applied to the task.

Constructing classifiers with higher accuracy requires more information about the problem at hand, or cleaner data. The classification task was hence repeated for *links* of T3 paths of only one technology (high bandwidth fiber optics). With these data an accuracy of 82% correct classifications was obtained with decesion trees as implemented in the C4.5 software package.

Both a system for predicting failures in end-to-end paths and a system for predicting failures in fiber optics links are implemented in the AT&T maintenace platform and daily used to guide technichians to upcoming failures.

4 Conclusion

In this paper we have proposed a method for estimating the limits on performance imposed by the quality of the database on which a task is defined. The method involves a series of learning experiments. The extracted result is, however, independent of the choice of learning machine used for these experiments since the estimated limiting performance expresses a characteristic of the data. The only requirements on the learning machines are that their capacity can be varied and be made large, and that the machines with increasing capacity becomes capable of implementing *any* function. In this paper we have demonstrated the robustness of our method to the choice of classifiers: the result obtained with neural networks is in statistical agreement with the result obtained for LVQ classifiers.

Using the proposed method we have investigated how well prediction of upcoming trouble in a telecommunication path can be performed based on information extracted from a given database. The analysis has revealed a very high intrinsic noise level of the extracted information and demonstrated the inadequacy of the data to construct high performance classifiers. This study is typical for many applications where the data collection was not necessarily designed for the problem at hand.

Acknowledgments

We gratefully acknowledge Vladimir Vapnik who brought this application to the attention of the Holmdel authors. One of the authors (CC) would also like to thank Walter Dziama, Charlene Paul, Susan Blackwood, Eric Noel, and Harris Drucker for lengthy explanations and helpful discussions of the AT&T transport system.

References

[1] J. R. Quinlan. *C4.5: Programs for Machine Learning.* Morgan Kaufman, San Mateo, 1993.

[2] S. Bös, W. Kinzel, and M. Opper. The generalization ability of perceptrons with continuous output. *Physical Review E*, 47:1384–1391, 1993.

[3] Corinna Cortes. *Prediction of Generalization Ability in Learning Machines.* PhD thesis, University of Rochester, NY, 1993.

[4] Corinna Cortes, L. D. Jackel, Sara A. Solla, V. Vapnik, and John S. Denker. Learning curves: Asymptotic value and rate of convergence. In *Advances in Neural Information Processing Systems*, volume 6. Morgan Kaufman, 1994.

[5] G. Cybenko, K. Hornik, M. Stinchomb, and H. White. Multilayer feedforward neural networks are universal approximators. *Neural Networks*, 2:359–366, 1989.

[6] T. L. Fine. Statistical generalization and learning. Technical Report EE577, Cornell University, 1993.

[7] Teuvo Kohonen, György Barna, and Ronald Chrisley. Statistical pattern recognition with neural networks: Benchmarking studies. In *Proc. IEEE Int. Conf. on Neural Networks, IJCNN-88*, volume 1, pages I–61—I–68, 1988.

[8] N. Murata, S. Yoshizawa, and S. Amari. Learning curves, model selection, and complexity of neural networks. In *Advances in Neural Information Processing Systems*, volume 5, pages 607–614. Morgan Kaufman, 1992.

[9] H. S. Seung, H. Sompolinsky, and N. Tishby. Statistical mechanics of learning from examples. *Physical Review A*, 45:6056–6091, 1992.

A Neural System to Detect Faulty Components on Complex Boards in Digital Switches

M. Collobert*, D. Collobert

France Telecom CNET,
LAB/RIO/TNT,
2, avenue Pierre Marzin,
22307 Lannion, FRANCE.
collober@lannion.cnet.fr
** Décision International, 31400 Toulouse FRANCE*
collobem@lannion.cnet.fr

Abstract

We present a prototype, based on neural networks, of a computer-aided fault detection system for highly complex digital electronic board, which can be found in modern digital switches. This system can be configured for a specific board, simply using data from simulations done at the design stage.

1 Introduction

This work was realized within the framework of an industrial collaboration between the "Centre National d'Études des Télécommunications" (C.N.E.T.), Alcatel and Décision International.

Its purpose was the design of a system devoted to the localization of faulty components on a digital electronic board known to be deficient. The design led to the realization of an industrial prototype. The prototype was adapted to the CPU board of a modern digital switch manufactured by Alcatel, the OCB283. This board is one of the most complex found in this system. It is composed of the CPU, two circuits of similar complexity, 150 various components and 1400 connections.

Two approaches to fault identification were in competition: the present one and another which was based on expert systems.

2 Description of the Prototype

The system is based on the topological maps of Kohonen[2]. Its main advantages lies in its ability to work with VLSI process data[4] and in its capacity to be configured for a specific board by learning the results of the simulation of a number of arbitrary faults.

2.1 Learning

A set of arbitrary faults are generated by holding a number of connections to 0 or 1 during the simulation of the board, at the design stage. For each fault, we obtain an image of the board evolution. Figure 1 shows a partial view of such an image.

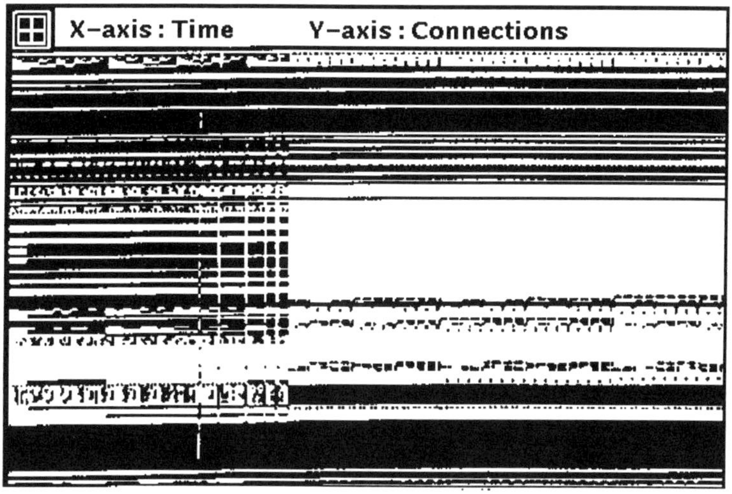

Figure 1: Fault Image : connections states during processor's cycles

Each line of the image represents the value of a connection during simulation (0 or 1). The connections are automatically selected to best discriminate the faults.

To each line we apply a function and obtain a unique real value for it. This function is automatically selected, from a panel, so as to maximize the discrimination of the different simulated faults. For example, this function could be the total number of "1" in the line. The resulting vector (one value for each line) is used as training pattern for the topological map. Figure 2 summarizes this preprocessing.

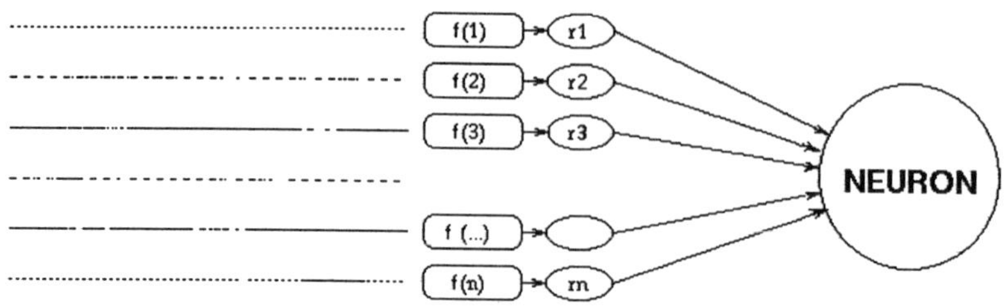

Figure 2: Data Reduction

The advantage of this preprocessing is to reduce the amount of data. Typically, an image fault have: 1000 connections * 1000 cycles = 1M pixels. As a result, a vector with approximately one hundred components is sufficient to characterize a fault.

2.2 Fault Detection

The data used as input for fault detection is obtained through the use of a digital analyzer. The number of lines is limited by the number of probe channels of this analyzer. The same processing of the resulting image is applied, and the resulting vector is used as input for the topological map.

2.3 The Topological Map

The topological map model was used as the base of the system for the following reasons:

- The statistical distribution of the faults and of their consequences (anomalous behavior frequency of a connection) are taken into account.

- The topological relations between faults are conserved: the map classifies on neighboring neurons the faults due to logically nearby components.

- The characteristic dimensions of the problem are compatible with the requirements of a topological map (taking into account images preprocessing).

The result of the learning phase is not only to change the network weights but also to associate to each emerging fault class one or more neurons.

The detection of a fault is realized by entering the processed data from the digital analyzer into the map. The neuron corresponding best to the input is selected. If it corresponds to a learned class, then this class of fault is given as output. Otherwise, the neighboring neurons corresponding to learned classes are given: these faults should be logically close to the real one.

As this kind of neural network is well known[5] [3], it will not be further described. To avoid folding, according to the general fractal dimension of VLSI Process[1], a two dimensional map was selected. Experiments shows that 10 neurons per fault's class is a good ratio.

2.4 Global fonctionnement

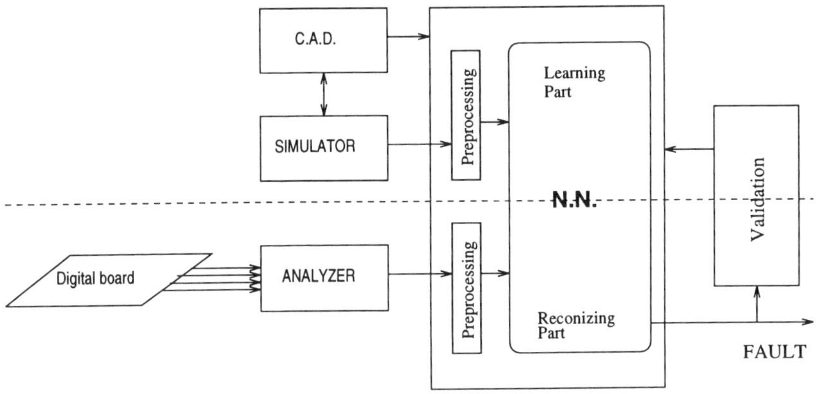

Figure 3: Global view of processings

3 Results

The system was tested in industrial environment on real faulty boards with the following learning parameters:

- 91 kinds of faults were simulated.
- The simulation was done for 1024 processor's cycles.
- The simulation's step was 0.1ns.
- The number of selected connections was 32.

The tested system was fast and efficient. On a workstation, data gathering and preprocessing took about 10s, and analysis by the network about 1s. For 80% of the tested boards, the system correctly recognized the fault or at least gave as result a direct consequence of the fault. This is sufficient for someone with basic knowledge of the board to find the fault. This was judged to be very good by the repairing team, considering the complexity of the board and competive techniques. Moreover, this result can be improved. During the test, the probe pins were installed manually and the number of tested connections was limited to 32. But in an industrialized version, hundreds of connections could be tested. This would lead to better discrimination of faults.

4 Conclusion

Compared to the approach based on expert systems, this system was found to be far better. In addition to the intrinsic difficulty to construct a good expert system, the problem we had to solve imposes a very difficult constraint. New versions of a board are designed continually, so the cost and time needed to acquire and analyze the human expertise for each version gives a net disadvantage to the expert system approach.

What the neural network approach needs is CPU time for the simulation of the faults (about one week on a standard workstation). This can be done directly and automatically from the files generated by CAD. The system can also be continuously improved by learning new faults observed on the functioning boards. Moreover, once the system has learned the faults on a workstation, it can be ported to a standard PC.

References

[1] G. R. H. Speckman and W. Rosenstiel, "Relations between generalized fractal dimension and self-organizing maps," in *Les reseaux neuromimetiques et leurs applications*, (Marseille), p. 155, 1994.

[2] T. Kohonen, *Self-Organization and Associative Memory.* Berlin: Springer Verlag, 1984.

[3] T. Kohonen, "The self-organizing map," *Proceedings of the IEEE*, vol. 78, no. 9, pp. 1464–1480, 1989.

[4] K. Marks and K. Goser, "Analysis of VLSI process data based on self-organizing feature maps," in *NeuroNimes*, pp. 337–349, 1988.

[5] H. Ritter and T. Kohonen, "Self-organizing semantic maps," *Biological Cybernetics*, vol. 61, pp. 241–254, 1989.

A Hybrid Expert System for Error Message Classification

H.C. Lau, K.Y. Szeto, K.Y.M. Wong and D.Y. Yeung*

Department of Physics,
** Department of Computer Science,*
The Hong Kong University of Science and Technology,
Clear Water Bay, Kowloon, Hong Kong.
phhclau@usthk.ust.hk, phkywong@usthk.ust.hk

Abstract

A hybrid intelligent classifier is built for pattern classification. It consists of a classification and regression tree (CART), a genetic algorithm (GA) and a neural network (NN). CART extracts features of the patterns by setting up decision rules. Rule improvement by GA is explored. The rules act as a pre-processing layer of NN, a multi-class neural classifier, through which the most probable class is determined. A realistic test of classifying error messages generated from a telephone exchange shows that the CART-NN hybrid system has comparable performance with Bayesian neural network.

1 Introduction

In a telephone network, error messages are continuously generated. Usually the number of combinations of error messages is enormous, making it a difficult diagnosis problem. Thus it is desirable to have an efficient and reliable expert system developed on the classification of error messages for intelligent network management.

Neural networks and related intelligent systems are found useful in analyzing complex technical devices [1]. Here we introduce three approaches which have complementary advantages, and their hybridization yields results improved over the individual systems.

2 A Hybrid Network Architecture

To formulate the classification problem, we consider a pattern coded in binary vector form, $\mathbf{x} = (x_1, \ldots, x_N)$. Each attribute x_i can take either 0 or 1, it is an indication of a special feature. The purpose of a classifier is to map the pattern vector $\mathbf{x}$ to a class variable c, where c takes discrete values in $\{1, \ldots, C\}$.

The hybrid classifier is composed of a rule-based hidden layer and a linear neural network [2], as shown in Figure 1. Each node in the input layer corresponds to an input attribute x_i, while that in output layer corresponds to an output class c. The hidden layer contains R nodes representing R classification rule vectors $\{\mathbf{r}^1, \ldots, \mathbf{r}^R\}$. They are fully connected to the input attributes. Each rule $\mathbf{r}^j = (r_1, \ldots, r_N)$ has the same dimension as the input vector, but the component r_i can take 0, 1 or 'don't care'.

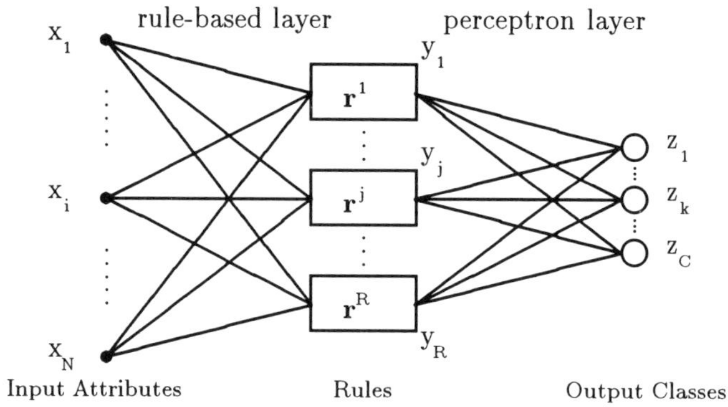

Figure 1: Hybrid Classifier Network Architecture

The output of a hidden node, y_j, is related to a matching function Φ between attributes in $\mathbf{x}$ and $\mathbf{r}^j$. It is defined as:

$$y_j = \Phi\left(1 - \frac{H(\mathbf{x}, \mathbf{r}^j)}{D(\mathbf{r}^j)}\right) S_j \qquad 1 \leq j \leq R, \tag{1}$$

where $H(\mathbf{x}, \mathbf{r}^j)$ is the hamming distance between $\mathbf{x}$ and $\mathbf{r}^j$, $D(\mathbf{r}^j)$ is the effective dimension of the rule $\mathbf{r}^j$ and the output of Φ is normalized in $[0, 1]$. In evaluating H and D, the 'don't care' attributes in $\mathbf{r}^j$ and the corresponding attributes in $\mathbf{x}$ will be ignored. S_j is the strength of rule $\mathbf{r}^j$, showing how important the rule is. As a result, each node will fire an output between S_j and 0, with perfectly matched and perfectly unmatched being the two extremes. y_{R+1} is set to 1 as a bias term for the perceptron.

The rule-based pre-processor aims at selecting the features of the input pattern. With $R < N$, it also serves the purpose of dimensionality reduction to the network. The set of y's are then fed to a perceptron, with weights w_{kj} and output node activation z_k, given by:

$$z_k = \sum_{j=1}^{R+1} w_{kj} y_j \qquad 1 \leq k \leq c. \tag{2}$$

It is an estimate of the corresponding class probability. Therefore

$$c = \arg \max_{1 \leq k \leq c} \{z_k\} \tag{3}$$

offers the first choice of our class prediction. The second and third alternatives, and so on, can be determined in a similar manner by finding the second and third largest argument in $\{z_k\}$.

3 Rule Extraction: CART

We now turn to the problem of finding the rules in the hidden layer. The rules are extracted from a classification and regression tree (CART) [3]. When a set of training patterns is presented, the tree is grown by recursively finding splitting conditions until all terminal nodes have pure class membership.

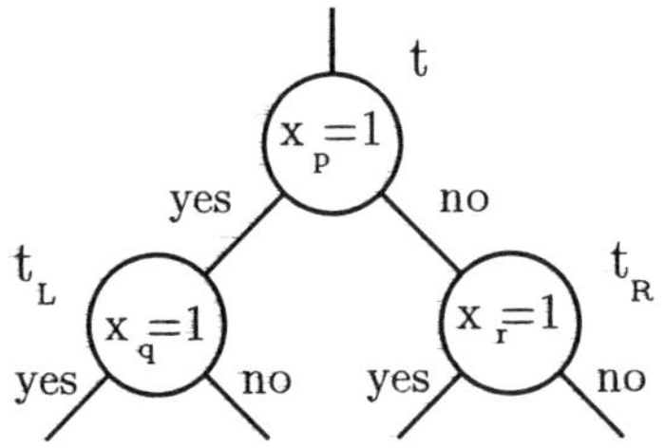

Figure 2: Decision Node of a Classification Tree

Consider a branch node t and its left and right child t_L and t_R respectively, as in Figure 2. Denote $p(i|t)$ as the conditional probability that the pattern belongs to class i, given that it lands in node t. Define the node impurity function by the Gini criterion [3]:

$$i(t) = \sum_j \sum_{i \neq j} p(i|t)p(j|t). \tag{4}$$

An efficient discrimination is achieved when the attribute is selected that provides the greatest reduction in impurity among the patterns. In other words, x_p is chosen to maximize

$$\Delta i(x_p; t) = i(t) - i(t_L)p_L(t) - i(t_R)p_R(t) \tag{5}$$

where $p_L(t)$ and $p_R(t)$ are the conditional probability that the pattern lands in node t_L and t_R respectively given that it lands in node t.

After the tree is grown, it is pruned by minimizing the error complexity using a pruning factor which represents the cost per node. The number of rules, R, is thus kept below the dimension of the pattern vector, N.

CART can be used independently to classify patterns, but here we use it to produce classification rules for subsequent neural network processing. The rule vectors $\mathbf{r}^j$ are generated by exhausting all routes of the hierarchical tree, setting up appropriate attributes in traversing the decision nodes of each route, and inserting 'don't cares' when those attributes are not examined in the splitting criteria. These rules constitute the rule-based layer in Figure 1, and for this purpose, the class outcome of each rule is not important at this stage.

Alternatively, rule improvement by the genetic algorithm can be explored before building the rule-based layer, as described in the following section.

4 Rule Evolution: Genetic Algorithm

The rules generated by CART, each with its corresponding class output, act as the initial population in GA. We investigate several simple variations on the application of genetic operators, reporting here the results of adaptive crossover [4, 5].

We randomly generate a fixed number N_a of additional rules per class according to the bit frequencies of the training set. All the rules belonging to class c are initially given the same strength $P(c)$, the probability of occurrence of class c based on the statistics of the training set. The j-th rule with class output c will have its strength S_j modified according to

$$S_j(t+1) = S_j(t) \pm p_{c,j} \tag{6}$$

The strength is increased(plus sign) if the rule gives the correct class, otherwise it is decreased(minus sign). The probability $p_{c,j}$ is a measure of how close the rule matches the given training example and is computed by a Boltzmann distribution at temperature $T = 1/\beta$,

$$p_{c,j} = \frac{e^{-\beta d_{c,j}}}{\sum_{(c',j')} e^{-\beta d_{c',j'}}} \tag{7}$$

where

$$d_{c,j} = v_{c,j} log(\frac{v_{c,j}}{P(c)}) + P(c) log(\frac{P(c)}{v_{c,j}}) \tag{8}$$

is the symmetric Kullback-Leibler's measure of cross entropy between the distance $v_{c,j}$ and the probability $P(c)$ of occurence of class c. Here $v_{c,j} = H(\mathbf{x}, \mathbf{r}^j)/D(\mathbf{r}^j)$ is the distance between the rule $\mathbf{r}^j$ and the input $\mathbf{x}$. The population of rules $(R + C * N_a)$ is maintained at constant as we replace the rule with the lowest strength after each generation. At the end of the evolution, we remove N_a rules of the weakest strength in each class, so that the final number of rules is R, although some of them are different from the initial set. Fine tuning of GA is achieved by adjusting the temperature parameter.

5 Neural Network Learning

In this section, we describe the supervised learning in the perceptron layer, based on a modified training procedure, to include multi-class outputs [6]. Suppose we are given a training set, which consists of pairs of pattern vector $\mathbf{x}$ and its corresponding class ξ. Initially, the weights of the perceptron are set with random values between 0 and 1. Then a training pair $\{\mathbf{x}, \xi\}$ is selected randomly and fed to the network, through which the most probable class will be predicted.

For correct classification, it is desirable that the predicted class c should be equal to the actual class ξ with high certainty. Otherwise, update of weights takes place when the desired node output z_ξ cannot exceed other outputs by a learning threshold τ. Then the weights are modified according to a multi-class proportional increment training procedure:

$$\Delta w_{kj} = 0, \qquad \text{if } z_k + \tau < z_\xi \tag{9}$$

$$\Delta w_{kj} = -\eta y_j, \qquad \text{if } z_k + \tau \geq z_\xi \tag{10}$$

$$\Delta w_{\xi j} = \mu \eta y_j \tag{11}$$

where μ is the number of k's that accounts to the modification of (10), and η is the learning rate. The learning threshold is introduced to increase the stability of the network.

The above learning procedure is repeated until a satisfactory percentage of training patterns are classified correctly. This algorithm has the advantages of fast convergence and fault tolerance.

6 Diagnosis of Error Messages

In order to compare with existing expert systems on realistic data, we use error messages generated from a telephone exchange computer, indicating which circuit card is malfunctioning [1]. The training set consists of 442 samples and the test set of 112 samples. Each sample is in the format of a bit string consisting of an error vector of 122 bits (N) and 36 possible error classes (C). A simple measure of the performance of the system is the generalization rate.

Using CART, a decision tree is constructed and pruned. The individual tree gives a generalization of 68.8% when unpruned and a best generalization of 69.6% when pruned by a factor of 0.9.

A multi-class perceptron yields a generalization rate of 61.6% with learning rate $\eta = 1$ and learning threshold $\tau = 1000$, the result goes up to 83.0% when the first three choices are included. Next we incorporate the perceptron with a pre-processing layer of CART rules, with pruning factor 1.3, which individually gives 63.4% only on generalization. However, it reduces the input dimension from 122 to 72. With no GA taking part, all rules are given with equal strength, $S_j = 1$, and a linear function Φ is selected for the matching criterion (1). The resulting CART-NN hybrid network is found to have a performance boost up to 74.1% on first prediction and 88.4% up to the third alternatives. It is obviously better than the individual results of CART and NN.

Using GA, with initial rules generated by CART, around 70% on the first prediction is achieved with NN. We observe that the number of additional rules, N_a, per class during evolution has to be small in order for the perceptron to achieve good performance. Fine tuning of the temperature parameter and systematic elimination of weak rules at the end of evolution are necessary for GA to be compatible with NN.

A comparison of the classification results among different methods is summarized in Table 1. The hybridization of CART and NN yields a slightly better performance than the Bayesian neural network [1].

Table 1: Classification Results of Various Techniques. The figures indicate the successful rate of classification when the specified choices are included.

Choice	RBF	Backprop	Multi-Layer Bayesian	NN	CART-NN Hybrid
Training set results (in %)					
1st	61.3	91.2	80.8	92.8	82.6
$1^{st} - 2^{nd}$	75.3	94.6	92.3	98.4	94.8
$1^{st} - 3^{rd}$	82.3	95.1	96.1	99.5	98.2
Rest	17.7	4.9	3.9	0.5	1.8
Testing set results (in %)					
1st	43.8	67.0	72.3	61.6	74.1
$1^{st} - 2^{nd}$	57.2	75.9	82.1	75.9	83.0
$1^{st} - 3^{rd}$	62.6	76.8	87.4	83.9	88.4
Rest	37.4	23.2	12.6	16.1	11.6

7 Conclusions

Perceptron classifier performs well when the data are separable. However, in most real life situations, patterns are found to have irregular distributions. Therefore it may require a combination of different classification techniques to achieve better performance. In this specific application, there is a strong correlation inherent in the data (e.g. concurrent error bits caused by simultaneous break down of components). This probably accounts for the success of rule-based approaches such as CART and the Bayesian neural network.

NN alone does not perform well when input dimension is large and few training examples are available. Therefore, we use CART as a pre-processing layer of NN for dimensionality reduction and feature extraction. Other conventional ways of pre-processing such as radial basis functions and principal component analysis were found to be inferior to CART. On the other hand, CART provides understandable rules and easily interpreted concepts regarding the nature of the patterns. However, it may generate irrevocable errors at an early stage of the hierarchy, and improvement of rules is hard to implement. GA has the potential for improvement. Finally, NN provides the necessary flexibility and fault tolerance to the generated rules. In conclusion, the combination of these methods gives not only a good success rate of classification, but also some rules of thumb which are generally inaccessible in a neural network.

Acknowledgements

We thank Dr. A. Holst for providing us the data on fault diagnosis of the telephone exchange computer, and Mr. D. Chan for technical support. This project is supported by the Hong Kong Telecom Institute of Information Technology, HKUST.

References

[1] A. Holst and A. Lansner, "Diagnosis of Technical Equipment using a Bayesian Neural Network," in *Proceedings of the International Workshop on Applications of Neural Networks to Telecommunications,* ed. by J. Alspector, R. Goodman, T.X. Brown, pp. 147–153, 1993.

[2] R.M. Goodman, C.M. Higgins, J.W. Miller and P. Smyth, "Rule-Based Neural Networks for Classification and Probability Estimation," *Neural Computation,* vol. 4, no. 6, pp. 781–804, 1992.

[3] L. Breiman, J.H. Friedman, R.A. Olshen and C.J. Stone, *Classification and Regression Trees,* Pacific Grove, CA: Wadsworth, 1984.

[4] J.D. Schaffer and A. Morishima, "An Adaptive Crossover Distribution Mechanism for Genetic Algorithm from Genetic Algorithms and their applications," in *Proceedings of the Second International Conference on Genetic Algorithms,* pp. 36, 1987.

[5] S.J. Louis and G.J.E. Rawlins, "Syntactic Analysis of Convergence in Genetic Algorithms," *Foundations of Genetic Algorithms 2* ed. by L.D. Whitley, San Mateo, CA: Morgan Kaufmann, pp. 141, 1993.

[6] J. Sklansky and G.N. Wassel, *Pattern Classifiers and Trainable Machines,* New York: Springer-Verlag, 1981.

A Higher Order Bayesian Neural Network for Classification and Diagnosis

Anders Holst and Anders Lansner

SANS—Studies of Artificial Neural Systems
Department of Numerical Analysis and Computing Science
Royal Institute of Technology, S–100 44 Stockholm, Sweden
aho@sans.kth.se, ala@sans.kth.se

Abstract

Here we will consider classification in domains where the attributes are causally connected to each other, which are often the case in *e. g.* a technical system. By first using the data to estimate the causal structure of the domain, and then to estimate the conditional probabilities of the classes given that causal structure, we have attained good classification results with a Bayesian neural network on a realistic task of classifying faults in a telephone exchange computer.

1 The Classification Task

In the type of classification task we are interested in here, we are given a set of attributes (*e. g.* symptoms), and we want to decide which of a number of classes that set belongs to (*e. g.* the cause of the symptoms). The goal is to train a neural network to perform the classification, given a set of previous example patterns, each consisting of a set of attributes and their corresponding class. The kind of domain we will focus on, is a technical system, which we want to diagnose in case of malfunctioning.

The approach to classification will be to estimate $P(q \mid \boldsymbol{x})$, the probability of the class q, given a vector $\boldsymbol{x}$ of measurements on the system. If we can do this, and then select the class with the highest probability, we have an optimal Bayesian classifier [1]. The learning task is thus to estimate the above probability distribution from the given database of patterns.

We will in the following concentrate on how to estimate $P(\boldsymbol{x})$, since $P(\boldsymbol{x} \mid q)$ can be achieved in a similar way, and then Bayes rule can be applied to get the desired result:

$$P(q \mid \boldsymbol{x}) = P(q)\frac{P(\boldsymbol{x} \mid q)}{P(\boldsymbol{x})} \tag{1}$$

Unfortunately, $P(\boldsymbol{x})$ can not be estimated from the database directly, since there are too many possible outcomes of $\boldsymbol{x}$ (exponential in the number of attributes), and there is usually a very limited set of training examples. We have to express it in probabilities of variables of much lower order.

If the attributes are all completely independent, the calculation of $P(\boldsymbol{x})$ is straightforward:

$$P(\boldsymbol{x}) = P(x_1) \cdot P(x_2) \cdots P(x_n) \tag{2}$$

However, usually this is of course not the case. Many components of the system might depend on each other, and thus measurement on different components can be correlated.

One method to solve this, is to "merge" correlated attributes into groups, or joint variables y_j, such that attributes from different groups are independent. Then the total probability is instead calculated from these (independent) joint variables:

$$P(\boldsymbol{x}) = P(y_1) \cdot P(y_2) \cdots P(y_m) \tag{3}$$

The larger a group of attributes gets, the more possible outcomes will y_j have, and the more uncertain will the estimation of probabilities be. Therefore, when many attributes are correlated, another approach has to be taken.

In many domains it is meaningful to talk about *causal structure* or *dependency graph*, a graph that describes how the attributes are causally connected [8, 7]. Even if all attributes are indirectly affected by most other attributes, the direct dependencies are probably of lower order. If we know the causal structure, it is possible to utilize it to calculate $P(\boldsymbol{x})$.

Assuming that there are no loops in the causal structure, the attributes can be sorted into the chain rule for joint probabilities, such that they are conditioned only on variables they depend on. By writing the conditional probabilities as fractions, and multiplying both numerator and denominator with some factors, it can be rewritten as a non directed expression. This means that to use it there is no need to know the causal direction, but only the statistical dependence between attributes. This rewriting process is illustrated in figure 1.

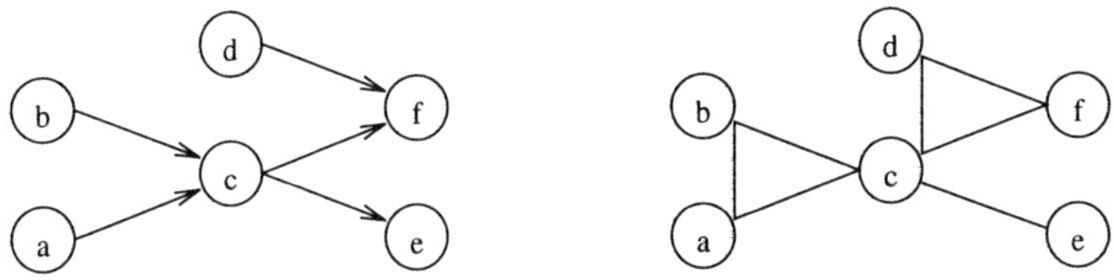

Figure 1: Starting from a dependency graph that is a pure tree, it is possible to rewrite the probability distribution over the graph, to a similar expression, using a non-directed generalized graph. First the attributes are sorted into the chain rule for probabilities:

$P(\boldsymbol{x}) = P(a)P(b \mid a)P(c \mid ab)P(d \mid abc)P(e \mid abcd)P(f \mid abcde)$

Keep only direct dependencies according to the graph:

$P(\boldsymbol{x}) = P(a)P(b)P(c \mid ab)P(d)P(e \mid c)P(f \mid cd)$

Rewrite the conditional probabilities as fractions:

$P(\boldsymbol{x}) = P(a)P(b)\frac{P(abc)}{P(ab)}P(d)\frac{P(ce)}{P(c)}\frac{P(cdf)}{P(cd)}$

Note independencies and rewrite:

$P(\boldsymbol{x}) = P(a)P(b)P(c)P(d)P(e)P(f)\left(\frac{P(abc)}{P(a)P(b)P(c)}\right)\left(\frac{P(ce)}{P(c)P(e)}\right)\left(\frac{P(cdf)}{P(c)P(d)P(f)}\right)$

The result is of the form (4), *i. e.* all first order factors and then additional factors to compensate for the dependencies among these.

With this method (see [2] for details), the form of the resulting expression will be:

$$P(\boldsymbol{x}) = P(x_1) \cdot P(x_2) \cdots P(x_n) \cdot \Psi(y_1) \cdot \Psi(y_2) \cdots \Psi(y_m) \qquad (4)$$

where y_j corresponds to "edges" in the graph, *i. e.* groups of attributes with direct dependencies between them. $\Psi(y_j)$ are *compensation factors*, fractions of probabilities to compensate (2) for dependencies between attributes in y_j.

If we can find the causal structure, we can thus calculate $P(\boldsymbol{x})$ with the help of this. The main idea here is to first use the database to estimate the causal structure in the domain. Then the database is used again to estimate the conditional probabilities of the classes, given that causal structure. The algorithm is well suited for implementation in a Bayesian neural network.

2 The Bayesian Neural Network

The Bayesian neural network model [4, 3] is designed to calculate the probabilities of some output attributes (classes) given some input attributes. If the attributes are independent it is possible to use a one-layer network [6]. In the network there is one unit for each outcome of each input attribute

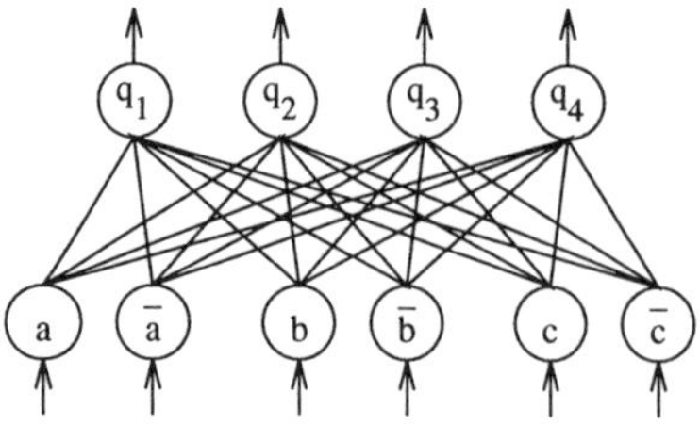

Figure 2: A one-layer Bayesian neural network with three binary input attributes and four output classes. See the text for details.

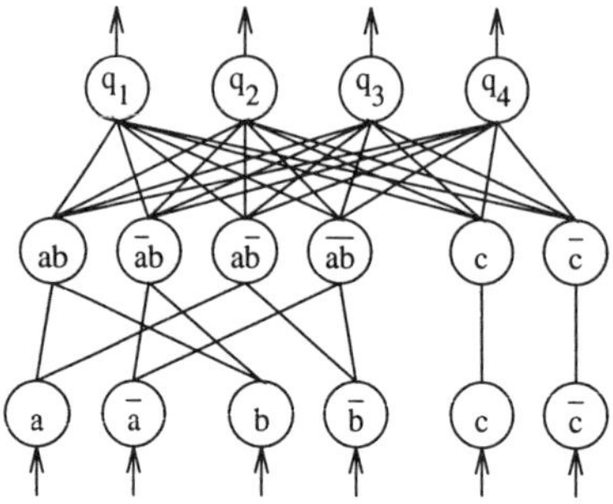

Figure 3: A multi-layer Bayesian neural network with a hidden layer consisting of one first-order column c and one second-order complex column ab.

x_i, as well as one unit for each class q (see figure 2). If Bayes rule (1) is combined with the independence assumption (2) (and a corresponding assumption conditionally given q) we get:

$$P(q \mid \boldsymbol{x}) = P(q)\frac{P(x_1 \mid q)}{P(x_1)}\frac{P(x_2 \mid q)}{P(x_2)}\cdots\frac{P(x_n \mid q)}{P(x_n)} \tag{5}$$

When taking the logarithm of (5) the right hand side becomes a sum. Each term $\log(P(x_i \mid q)/P(x_i))$ in this sum corresponds to the weight from the corresponding input unit x_i to the class unit q. Each class unit sums its inputs and exponentiates the result, yielding the desired probability.

If there are dependencies between attributes, we have to introduce a hidden layer in the network. This hidden layer consists in the Bayesian neural network of *complex columns*, *i. e.* groups with units for every possible outcome of some joint variables y_j. If we base our calculation on (3) these variables will represent groups of attributes that are dependent (see figure 3). Let us call the corresponding columns *partitioning* complex columns. Analogous with the above case, the weights are set to the logarithms of the factors $P(y_i \mid q)/P(y_i)$.

The alternative is to use (4), in which case the hidden layer will consist of first order columns for all input attributes (corresponding to the factors $P(x_i)$) and higher order columns for all "edges" in the dependency graph (corresponding to the compensation factors $\Psi(y_j)$). These latter columns are called *overlapping* complex columns (see figure 4). The weight from a

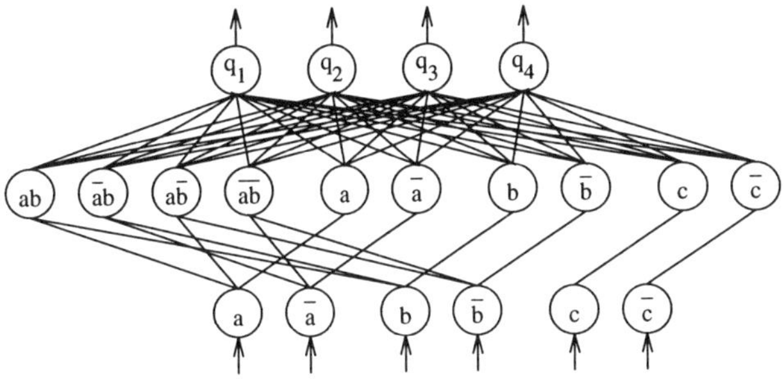

Figure 4: A multi-layer Bayesian neural network with a hidden layer consisting of units for the input attributes, plus an overlapping column for a dependency between a and b.

unit in an overlapping column is in analogy with the previous cases calculated as $\log(\Psi(y_i \mid q)/\Psi(y_i))$.

To construct the hidden layer, we have to find which attributes are correlated. The approach taken here, is to first search for low order correlations, and then search for higher order correlations only where we have already found lower order correlations. The reason for this is that an exhaustive search for correlations is infeasible due to its exponential complexity.

The measure used to check if a set of attributes are correlated, comes from information theory. The mutual information between two variables X and Y is defined as:

$$I(X,Y) = \sum_{x_i, y_j} P(x_i, y_j) \log \left(\frac{P(x_i, y_j)}{P(x_i)P(y_j)} \right) \tag{6}$$

If the mutual information between two attributes (simple or complex) are above some threshold, they are considered dependent, and a complex column is created for them. This is done in a few passes creating successively higher order columns.

After creating the hidden layer, the weights between the hidden units and the output units are determined in one pass by estimating the required joint probabilities $P(y_i)$ and $P(y_i \mid q)$ from the database and calculating the weights from them according to the above prescriptions.

3 The Diagnosis Application

This Bayesian model has been tested on a realistic task: diagnosis of telephone exchange computers. In this case, one malfunctioning circuit card out of 36 is to be identified from an error vector of 122 bits. The training set consists of 442 real examples and the test set of 112 examples.

Classification of these data is done with the one-layer Bayesian neural network as well as both versions of the multi-layer Bayesian neural network. The first multi-layer network ("Multi-layer 1") has 303 hidden units

Table 1: Classification Results. The different neural networks are tested with both the training and the test set. The percent of correct answers among the first, second and third alternatives from the networks are shown.

Tested on training data	Back-prop	One-layer	Multi-layer 1	Multi-layer 2
First	91.2%	75.1%	83.5%	86.7%
Second	3.4%	15.4%	10.6%	8.1%
Third	0.5%	4.3%	2.7%	2.3%
Rest	5.0%	5.2%	3.2%	2.9%

Generalization	Back-prop	One-layer	Multi-layer 1	Multi-layer 2
First	67.0%	66.1%	70.5%	75.0%
Second	8.9%	12.5%	10.7%	5.4%
Third	0.9%	8.0%	5.4%	8.0%
Rest	23.2%	13.4%	13.4%	11.6%

organized into 73 partitioning complex columns of up to fourth order. The second multi-layer network ("Multi-layer 2") has 831 hidden units, and uses 123 overlapping columns, also of up to fourth order. The results are compared to those from another investigation of the same database with a back propagation network. Also, in connection with the telephone exchange computer system is a conventional diagnosis program which manages to classify about half of the examples correctly.

Table 1 summarizes the results with the neural networks. The fraction of correct answers among the first, second, and third alternatives from the networks are shown. Compared to the one-layer and multi-layer Bayesian networks, back propagation seems a little better when tested on the training data, but for generalization the Bayesian networks achieve better results. If the three first alternatives are considered together, the Bayesian networks come better off on the training set as well. The second version of the multi-layer Bayesian networks (which uses overlapping complex columns) gives a result on the training data close to those of back propagation, and the best results for generalization.

It is also interesting to compare the causal structure found and compensated for by the first and second version of the multi-layer Bayesian neural networks. In figures 5 and 6 are the dependencies detected in the telephone exchange computer database shown by arcs between units. The overlapping complex columns compensate for much more dependencies than the partitioning complex columns, which explains the better results on both training

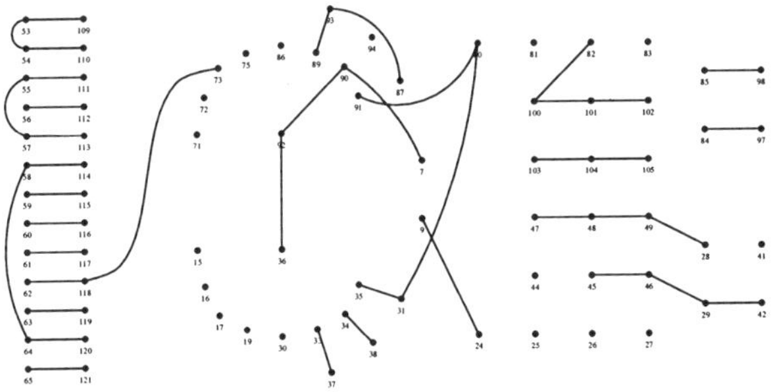

Figure 5: The dependencies accounted for by the first multi-layer Bayesian neural network. The arcs connect groups of units that are gathered together in partitioning complex columns.

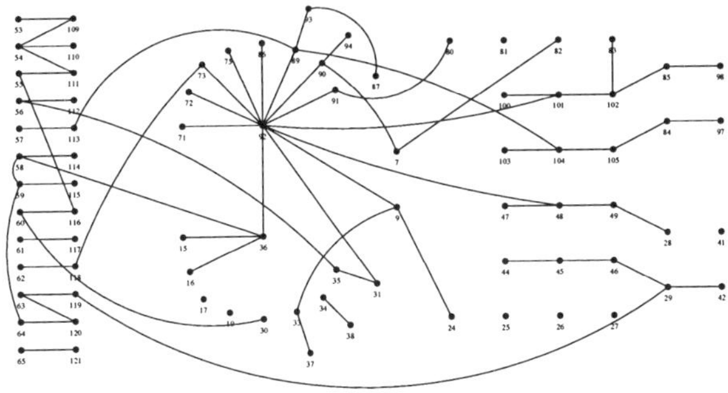

Figure 6: The dependencies accounted for by the second multi-layer Bayesian neural network. The arcs represent second or higher order correlations between units, and correspond to overlapping complex columns.

and test data. The possibility to depict the found dependency graph may also in itself provide a valuable tool in many domains.

4 Discussion and Conclusions

Complicated technical systems which are hard to analyze, is a good example of an area in which neural networks can be useful. The Bayesian neural network is especially suited for use in this kind of domains, since they often involve attributes connected in a causal structure. This study has shown good results when applying Bayesian neural networks to diagnosis of a telephone exchange computer. The performance is comparable to that of a back propagation neural network, but the Bayesian network tends to be slightly better at generalization, especially when overlapping complex columns are used. Training of the Bayesian neural network is also much faster, since this is done mainly in one pass, preceded by a few passes for construction of complex columns.

The Bayesian model has a large potential for further development. For example, we have here only treated binary input features. It is however also possible to handle continuous attributes by introducing a kind of soft interval coding [5]. Further, there are several ways to improve the selection of which primary events to merge into complex columns. But already this network model has proven itself a powerful and general tool for classification and diagnosis.

Acknowledgments

This work was supported by Ellemtel Telecommunication Systems Laboratories (Ellemtel Utvecklings AB) and the Swedish Research Council for Engineering Sciences (TFR) under grant TFR 93-672. We thank M. Boda and H. Brandt for useful comments and discussion.

References

[1] R. O. Duda and P. E. Hart, *Pattern Classification and Scene Analysis*. Wiley, New York, 1973.

[2] A. Holst and A. Lansner, "A Bayesian neural network with extensions," Tech. Rep. TRITA-NA-P9325, Dept. of Numerical Analysis and Computing Science, Royal Institute of Technology, Stockholm, Sweden, 1993.

[3] I. Kononenko, "Bayesian neural networks," *Biol. Cybernetics*, vol. 61, pp. 361–370, 1989.

[4] A. Lansner and Ö. Ekeberg, "A one-layer feedback, artificial neural network with a Bayesian learning rule," *Int. J. Neural Systems*, vol. 1, no. 1, pp. 77–87, 1989.

[5] A. Lansner and A. Holst, "A higher order Bayesian neural network with spiking units," 1994. (Submitted).

[6] M. L. Minsky and S. A. Papert, *Perceptrons*. MIT Press, 1988. Expanded Edition.

[7] J. Pearl, *Probabilistic Reasoning in Intelligent Systems: Networks of Plausible Inference*. Morgan Kaufmann, San Mateo, 1988.

[8] D. Wedelin, *Efficient Algorithms for Probabilistic Interference, Combinatorial Optimization and the Discovery of Causal Structure from Data*. PhD thesis, Dept. of Computer Sciences, Chalmers University of Technology, Göteborg, Sweden, 1993.

A Strong Combination of Neural Networks and Deep Reasoning in Fault Diagnosis

Tadashi Sone

NTT Network Service Systems Labs.
9-11,Midori-Cho 3-Chome,Musashino-Shi,
Tokyo 180 Japan
`sone@luna.nslab.ntt.jp`

Abstract

The performance requirements of complex faults diagnosis in switching systems are increasing. To solve this subject, both the neural network and deep reasoning are indispensable technique. Especially, the combination method of these algorithms is an important fact to stress. This paper defines parts of these algorithms and describes a strong combination method between these algorithms. The diagnostic rate was improved by a strong combination with the individual procedure of these algorithms.

1 Introduction

Switching systems have become very reliable and the failure rate has decreased to where it is difficult to acquire experimental and heuristic knowledge on maintenance. However, complex faults remain, when an infrequent but complex failure occurs, inexperienced maintenance engineers cannot efficiently troubleshoot the fault. Examples of complex failure are intermittent faults, mismatch faults between a software module and a hardware module, and boundary faults of hardware modules.

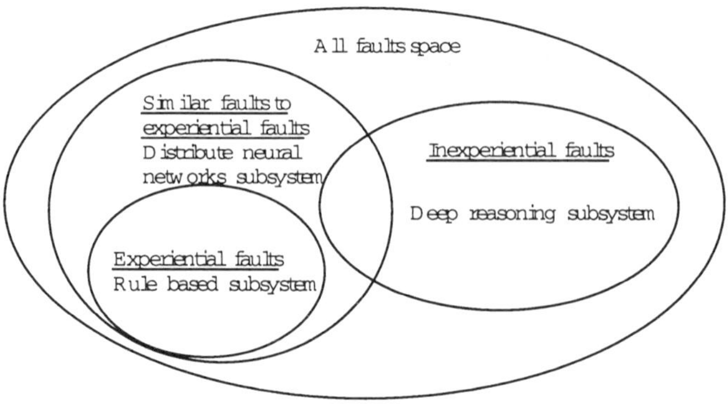

Figure 1: The faults through three types of organization

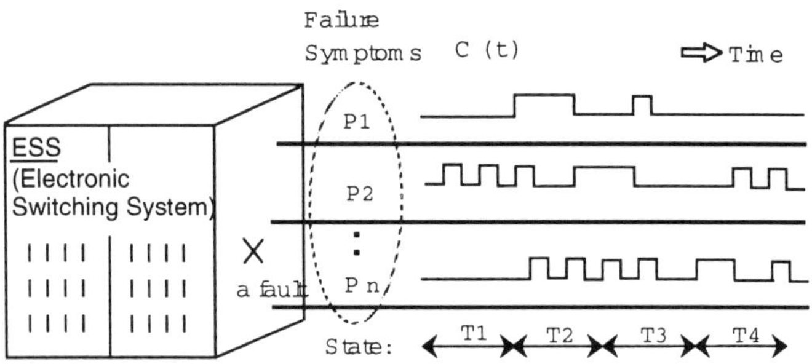

Figure 2: Time series of sensor signals

2 Characteristics of Switching systems failure

It seems reasonable to consider the faults through three types of organization and can be best summarized in the Figure 1.

In the first place, it is easy to identify the faults location from the error message from Switching systems. A rule based subsystem is a useful method for this faults type.[1] In the second place, it can not clear up the fault location by the error message. We classify the failure patterns from more detailed failure symptoms from Switching systems like a time series of sensor signals (Figure 2). To identify the failure pattern, we have developed a fault identification system that applies a distributed neural networks architecture (see Figure 3).[2][3][4] As mentioned above, both fault types mean experiential faults. In the third place, the characteristic of this fault type means outside human experience by now. Therefore, we need the deep reasoning technology based on Switching systems architecture or physical design knowledges(Figure 4).[5][6]

So far, Through three faults type of organization, diagnosis systems should be differentiated into a Rule based subsystem, a Distributed neural network subsystem and a Deep reasoning subsystem as shown in Figure 5.

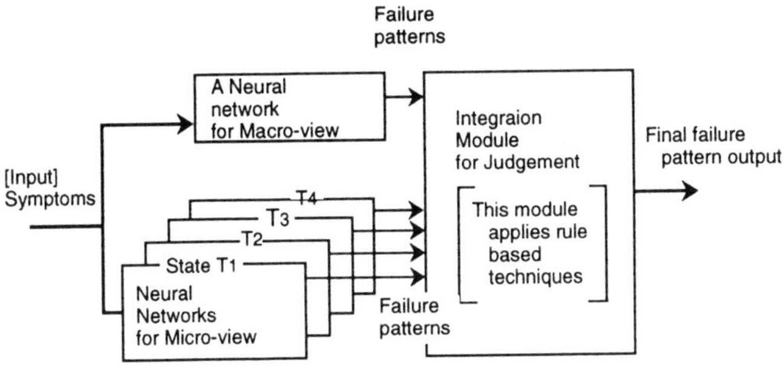

Figure 3: Distributed neural network

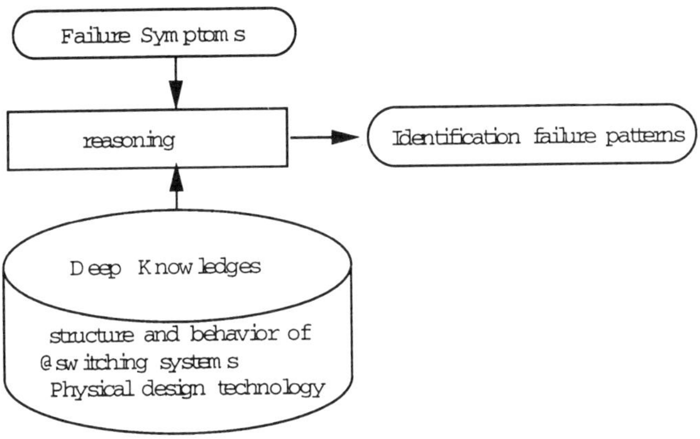

Figure 4: Deep reasoning subsystem

3 Fault diagnostic system architectures investigate

In practical use of a fault diagnostic system, both the Neural Network (NN) and Deep Reasoning (DP) are indispensable technique. This is an important fact to stress. To get the perfect rule-base is an ideal, but I think it is impossible for a fault diagnostic system. Using the Hybrid AI (The NN and DR) , what matters is rather that how strong combination of the NN and DR we will make.

At the present time, we use the RB subsystem and we develop the Neural Network subsystem to improve the recognition rate of the RB for assumed faults. Also, the term "inexperiential faults" can be defined as more difficult faults which are very difficult to suppose the failure symptoms by experiment failure date base. To solve this inexperiental faults, we depend human expert (Figure 6-a). And to improve human skill, they learn the architecture and function of switching system, and recognize the failure symptoms of inexperiental faults using dummy fault generator. The traditional various configurations for both the NN and DR are shown in Figure 6 -b,-c-d. These combination of the NN and DR are not tight. Therefore, when both the NN and DR got confused at recognition fault patterns, the Distribute module (Figure 6-c) or the Judgement module (Figure 6-d) cannot efficiently troubleshoot the faults.

So, we arrive at the conclusion that the NN and DR should be combined tightly with each judgement informations. The effectiveness of a strong combination is shown in Figure 7.

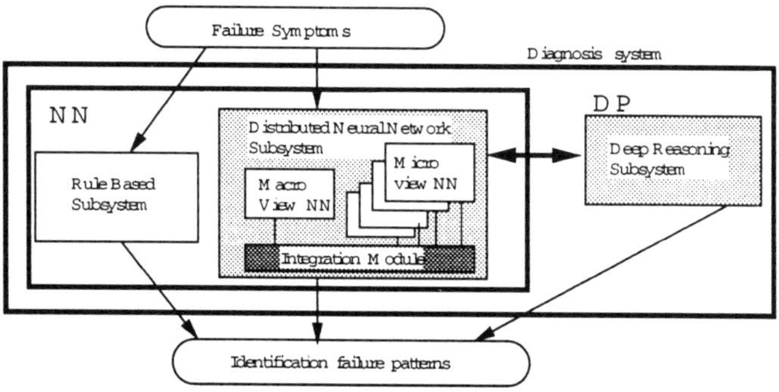

Figure 5: Basic configuration of Hybrid AI

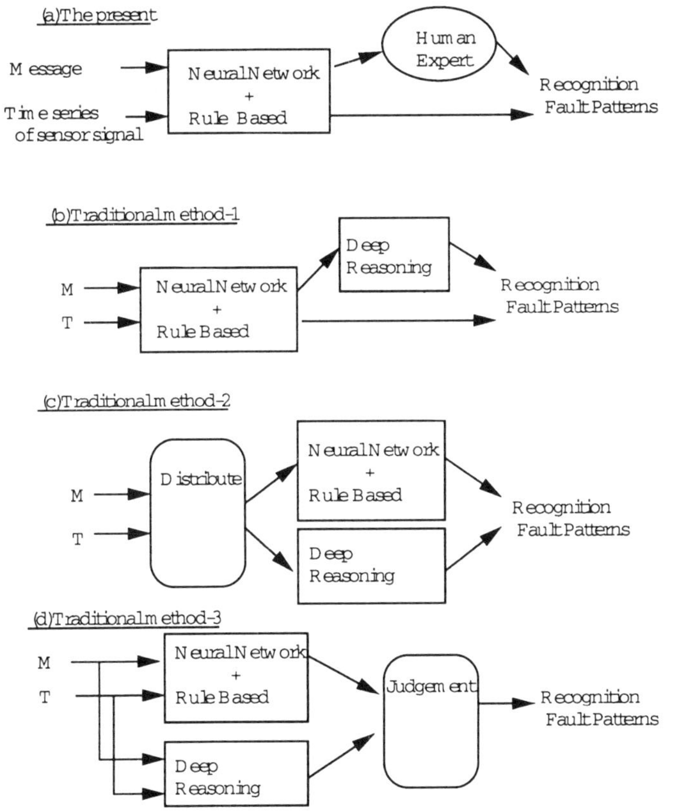

Figure 6: Faults Diagnostic system architectures

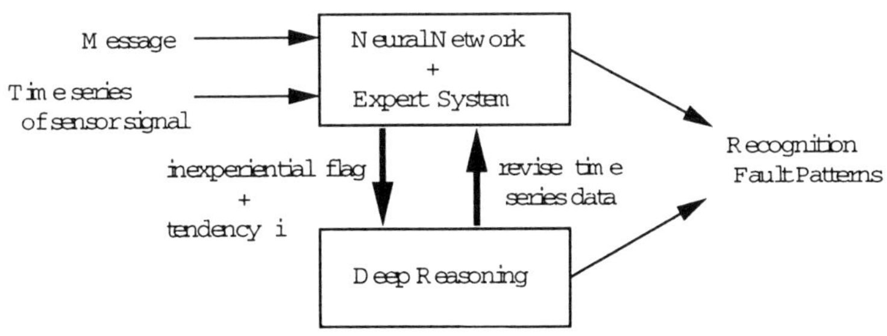

Figure 7: Proposed architectures

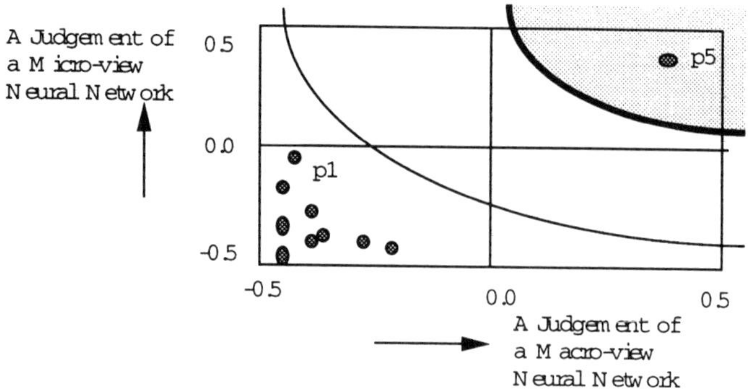

Figure 8: Experiential failure patterns

4 Examine a Strong Combination

To be concrete a strong combination, two informations between the NN and DP are defined. The NN gives the DP about a inexperiential flag and that's tendency. And the DP gives the NN about correcting time series date. These informations are just one example of a strong combination in my proposed system.

A inexperiential flag are experiential or not on the failure symptoms from switching systems. Also, this NN is distributed neural networks, and consists of several micro-viewing neural networks and one macro-viewing neural network. The vertical axis of Figure 8 indicates the judgements of a micro-viewing neural network. Also, the horizontal axis indicates the judgements of macro-viewing neural network. The highest point shows determined and the lowest point show the undecided. The diagram for distinct experientail faults would look something like Figure 8. In this case, it is clear that a fault pattern is p 5. The other side, we can represented inexperiential faults diagrammatically as shown Figure 9 (a pattern x1 tendency) and Figure 10 (a pattern x2 tendency) by experiment. The deep reasoning subsystem is carried out when we observe the trend like X1 and X2. If X1 should occur, the DP would hypothesize that fault locations are closely related to fault pattern p 2, p 5, p 7, p 4. The DP examine the state transitions in switching system based on Deep knowledges, and revises the time series of sensor signals. And the NN restart using the DP' correct date.

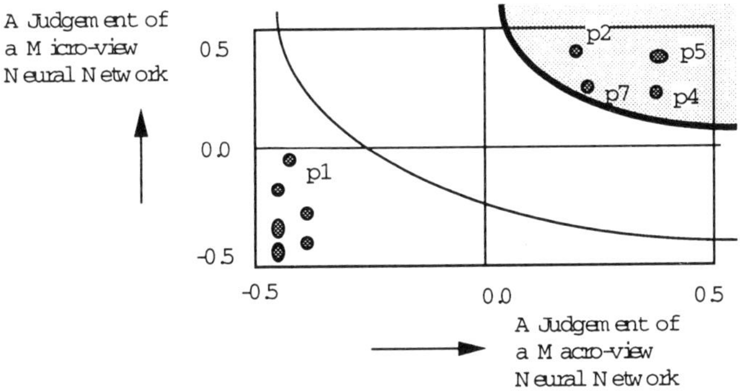

Figure 9: Inexperiential failure patterns X1

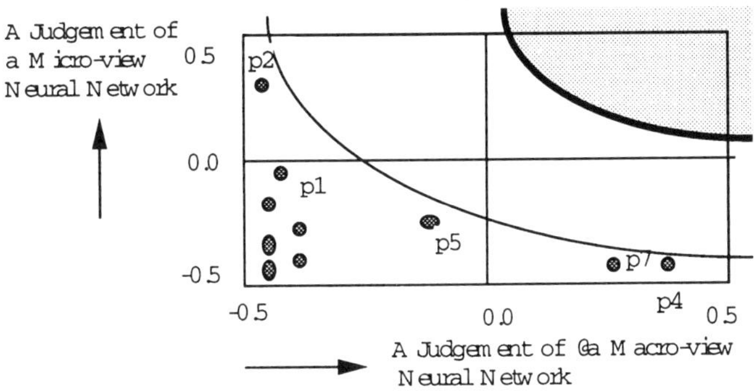

Figure 10: Inexperiential failure patterns X2

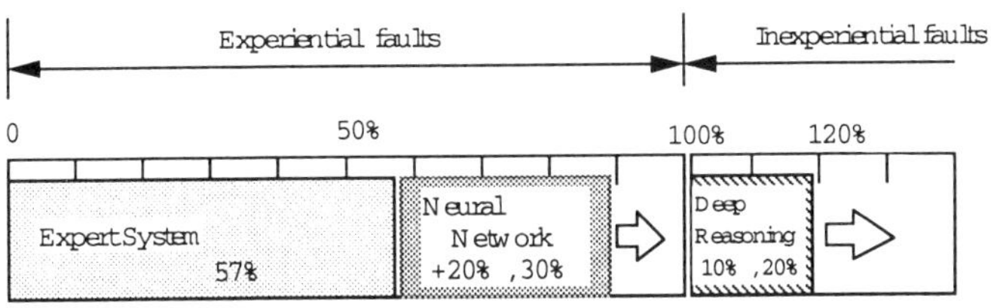

Figure 11: A example of results in complex faults

5 Conclusion

Therefore, I propose a strong combination of the NN and DP in Diagnosis at the switching systems. And to be concrete, I examine the information between the NN and DP. In this case, I select the inexperiential tendency from the characteristic of distributed neural networks for the DP. And revising time series symptoms based on the switching architecture for the NN are selected.

It follows from what has been said that the NN is effective in the experiential faults and the DP is effective in the inexperiential faults. The diagnostic rate was improved compared with the individual action of each method. For a example, the effectiveness is shown in Figure 11.

References

[1] A.D.Chattell, J.B.Brook, "A Neural Network Pre-processor for a Fault Diagnosis Expert System," in *Proceedings IWANNT.* 1993,p. 297-305.

[2] T.Sone, "Using distributed Neural Networks to Identify Faults in Switching System" in *Proceedings IWANNT.* 1993,p. 288-296.

[3] T.Nogami, Y.Yokoi, M.Kasai, "Failure Diagnosis System on Pneumatic Control Valves by Neural Network," in *Proceedings ICNN.* 1993, p. 1876–1881.

[4] M.Boda, H.Brandt, "Application of Neural Networks in Fault Diagnose," in *Proceedings ISS,* vol. 1, October 1992, p. 254-257.

[5] R.Davis, "Diagnostic Reasoning Based on Structure and Behavior," *Artificial Intelligence,* vol. 24, 1984, p. 347–410.

[6] J.Kleer, B.C.Williams, "Diagnosis with behavioral Modes," in *Proceedings IJCAI,* 1989, p. 104–109.

Neural Network Detection of Fraudulent Calling Card Patterns

Jerome T. Connor*, Laurence R. Brothers, and Joshua Alspector

Bellcore
445 South Street, Morristown, New Jersey 07960-6438 USA
**present address: London Business School*
London, NW1 4SA, United Kingdom
jconnor@lbs.lon.ac.uk
quasar@bellcore.com
josh@bellcore.com

Abstract

We apply neural networks to automatic calling card fraud detection. Inputs for the neural network based on geographical information obtained from the originating, terminating and billing phone numbers is used. This geographic information is expressed in terms of distance and entropy measures. The resulting neural network is shown to be a good indicator of calling card fraud.

1 Purpose

Telecommunications fraud is a multi-billion dollar problem around the world. In addition to the substantial loss of revenue due to uncollectible billing, there are many "soft" or unquantifiable costs, including network congestion, loss of network integrity, reduction of customer satisfaction, and others.

In North America, a major form of toll fraud is "calling card fraud", in which customer accounts and PINs (personal identification numbers) are stolen or otherwise misused to enable local and international calls to be made from any telephone on the continental SS7 network. Some carriers permit such calls to be made from overseas locations as well.

If fraudulent usage of a customer account cannot be detected through on-line active surveillance of account activity, it may go undetected for one or even two billing cycles before the customer realizes that fraudulent activity is taking place. It is of highest importance to detect fraudulent usage as soon as possible in order to reduce the active lifetime of a stolen account and PIN number. All major carriers, therefore, maintain work-centers whose sole purpose is fraud detection. They employ systems for fraud detection whose characteristics vary widely; almost all of these systems, however, rely on simple aggregation of call-attempts ("peg count") for their primary means of fraud-detection. These simple time- and usage-based thresholds can often be defeated by clever thieves who realize what fraud-detection techniques are being used against them (see Yuhas, 1993).

In this paper we discuss a neural-net based method for fraud- detection which demonstrates several improvements over simple thresholding:

- Our method can differentiate fraudulent usage from legitimate high usage.
- Our method prioritizes potential cases of fraud, allowing fraud investigators to optimize their time, and thus allowing carriers to spend less money on fraud-detection.
- Our method is adaptive and can respond to changes in the methods used by criminals. It should be much less apparent to criminals what criteria are being applied to detect fraudulent usage than with simple thresholds.

We examine calling card usage for trends related to fraud. Fraud is shown to follow trends with respect to geographic measures. The geographic information can be obtained from the NPA-NXX numbers associated with the billing, originating and terminating phone numbers. This information is available in present fraud detection systems and practical for use in a real module.

Section 4 investigates the entropy of calling card patterns. The entropy, defined later, roughly corresponds to the disorder in the calling card pattern. Calls that are scattered about have a higher entropy than those that are narrowly focused. We introduce three types of entropy measures of calling card usage. *International entropy* is the disorder in the international terminating location. *Domestic entropy* is the disorder in the domestic terminating location. *Originating entropy* is the disorder in the originating location. We demonstrate that entropy measures are better indicators of fraud than counts of calling card usage.

Lastly in section 5, we present a prioritizer of fraud candidates. This prioritizer is a neural network which ranks a list of calling card numbers according to which one is most likely to be fraud. The neural network takes as inputs quantities introduced in previous sections that are correlated with fraud.

2 Data

Information from three datasets is used. The Anti-Fraud Module (AFM) database produced labels (i.e. fraudulent or legitimate), LIDB produced detailed calling card records and the geographic database produced a more convenient representation of phone numbers. Only calling card users which are labeled by the AFM database for June 5, 6, 7 and 8, 1993 and August 14, 15, 16, 17, 18 and 19, 1993 are used. These dates correspond to LIDB records that are available to us.

A geographic database is used to transform terminating, originating and billing area codes and exchanges, denoted NPA-NXX, into horizontal and vertical coordinates. A unique geographic coordinate is not available for each telephone number, but the resolution of the geographic position is still adequate as shown in Fig. 1.

Figure 1: Geographic locations of NPA-NXX numbers.

There exist problems in high density urban areas. For instance, the 212 area code corresponding to Manhattan has only 2 different geographic locations associated with it, upper and lower. Subsequent work could make great improvements by obtaining finer resolution geographic information in urban areas.

3 Distance Information

The implicit assumption of the following study is that the calling card distribution depends on the distance of the call. For example, it may be that fraudulent calls tend to be longer distance than legitimate calls. The following data analysis verifies this.

The distance of the calls is obtained by transforming the terminating and originating phone numbers into their geographic coordinates by way of the geographic database. These geographic coordinates can then be transformed into the distance of the calls. It should be noted that a large part, about one-third of the data, does not have complete records of the terminating and originating phone numbers and therefore lacks any distance information.

The majority of the calls are local, this corresponds to the large peak in the left most bin of the histogram.We desire a measure of how unlikely is a particular calling card pattern. If the probability of a call of distance d_i is used, the measure would be multiplicative

$$P(d_1, d_2, \ldots, d_N) = \prod_{i=1}^{N} p(d_i). \tag{1}$$

The probability can be turned into an additive distance penalty measure, DM, by taking the negative log of (1)

$$DM(d_1, d_2, \ldots, d_N) = -\sum_{i=1}^{N} \log p(d_i). \tag{2}$$

This corresponds to a individual penalty for each call of $-\log p(d)$. The individual probabilities, $p(d)$, are estimated from the histogram. In cases where the estimated probability of a call is 0, the -log of the probability would be infinite. We replace the infinite measures by the largest non-infinite penalty measure.

4 Entropy of Credit Card Fraud

The general principle behind these metrics is that individuals tend to have specific locations of areas from which, and to which, their calls are generally made. The intrusion of additional fraudulent usage into a legitimate pattern tends to add "disorder" to the call distribution. If, as often happens, stolen ac count numbers and PINs are distributed to multiple criminals, the total disorder can be expected to rise even higher.

"Entropy" is defined by

$$E = -\sum_{i=1}^{N} p_i \log p_i \quad (3)$$

where p_i denotes the probability of an observation being a member of class i. If the observations have a probability equal to 1 of being a member of one class, then the entropy is equal to its minimum value of 1. If the probabilities of each class are equal, the entropy reaches its maximum value. Thus, entropy can be viewed as a measure of disorder. We will use it in several ways to measure the disorder of a calling cards usage. This will later be used to detect fraud with a neural network.

4.1 International Entropy

Ten classes of telephone calls are defined, corresponding to geographic regions of international calls: Australia, Latin America, Africa, etc. This information is easily obtained from the first digit in a country code of the terminating phone number. All domestic calls fall into category 1, the first digit of a call to the U.S. from overseas.

The entropy of calling card calls was calculated for the week of June 5,6,7 and 8, 1993. The fraudulent call tends to have noticeably higher entropy that the legitimate calls. The legitimate calls tend, with a few exceptions, to have very low entropy.

The entropy measure tends to be better at identifying fraud, but in cases with zero entropy, the distance measure can be used to identify fraud. This last case often corresponds to many long distance domestic calls and no international calls.

A large concentration of legitimate calls is observed near the minimum of the distance penalty and the entropy measure. This leads to the conclusion that if there is both high international entropy and a corresponding high distance penalty, it is very likely to be a fraudulent call.

4.2 Originating Entropy

We investigated the entropy of the terminating number in the previous section, here we look into an independent information source, the originating number. The entropy of the originating phone number is calculated by defining classes for each phone number. The class chosen corresponded to the RBOC from which the call was made. This definition of class differs from the international entropy, investigated in section 4.1, because all originating numbers are domestic, hence the originating numbers would all be in the same class of section 4.1.

As in section 4.1, the fraudulent calling card users have a higher entropy than does the legitimate calling card users. The discrimination is not as good as the international entropy, but since the information is independent of the international entropy it is useful.

4.3 Domestic Entropy

In section 4.1, we investigated the international entropy of the terminating number. Here we look at the domestic entropy of the terminating number. The domestic entropy of the terminating number is calculated with the classes defined in section 4.2. This information is lost in the international entropy measure, because the domestic classes are lumped into one international class.

As with the international and originating entropy, the fraudulent calling card users have a higher entropy 4.1 than does the legitimate calling card users. The discrimination of the domestic entropy of the terminating line number is the least useful of the three entropies.

5 Neural Network Prioritization of Credit Card Fraud

We now show that a neural network can prioritize a list of suspicious users. This should lead to a more efficient use of a fraud investigator's time as well as the ability to catch "more important" cases of fraud more quickly.

The measures described above vary in utility and applicability. In isolation, any given measure is quite likely to be incorrect about the fraudulence or legitimacy of a given pattern of calls. However, in combination, the measures become much more accurate. When the above measures are combined with conventional metrics such as checks for simultaneity of account usage, we expect to greatly surpass the accuracy and efficiency of threshold-based systems.

A neural network provides a convenient, trainable, adaptive platform for the automatic combination of these various fraud detection measures, metrics and algorithms.

Preliminary analysis (not described in detail in this paper), bears out this prediction of higher fraud detection accuracy – it appears that an increase in accuracy of 50% or more is not unreasonable to expect.

5.1 Model Definition

A neural network is defined

$$\hat{y} = f(\sum_{i=1}^{H} W_i f(\sum_{j=1}^{P} w_{ij} x_k)) \tag{4}$$

where $\hat{y}$ is the neural network predicted class, $\mathbf{x}$ contains information used to make predictions, $\mathbf{W}$ and $\mathbf{w}$ are parameters of the neural network that are estimated from the data, H is the number of hidden neurons, P is the number of inputs, and f is a nonlinear sigmoidal function going smoothly from 0 to 1.

For classifying suspicious calling card users, the following information is used:

- x_1 - Distance Penalty of Calling Card calls
- x_2 - Calling Card counts
- x_3 - International Entropy of termination numbers
- x_4 - Domestic Entropy of originating numbers
- x_5 - Domestic Entropy of terminating numbers
- y - 1 if the User is Fraudulent, 0 otherwise.

5.2 Results

Data from the dates June 5, 6, 7, and 8, 1993 were used to obtain 432 input-output pairs. The data was randomly split into three different sets: 200 for an estimation set, 82 for a cross-validation set, and 150 for a test set.

In Fig. 2, the results on the independent test set of the best neural network model, chosen by its performance on the cross-validation set, is shown. The outer sigmoid of (4) was removed to better observe the spread of neural network predictions. This corresponds to a confidence,

$$c = \sum_{i=1}^{H} W_i f(\sum_{j=1}^{P} w_i j x_k) \tag{5}$$

which is a monotonic transform of the neural network prediction.

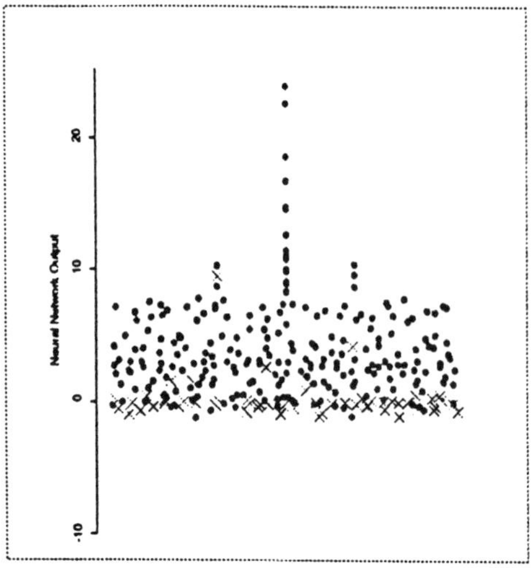

Figure 2: Neural Network Penalties of Fraudulent (dots) and Legitimate (x's) calling card users.

6 Conclusions

We have shown that fraud can be correlated with geographic quantities such as distances and the entropy of calling card patterns. A neural network, using geographic information, was then shown to be a good reprioritizer of suspicious lists for investigation of fraud. The neural network produces a list with the most likely example of fraud at the top, the most likely legitimate users at the bottom.

These priorities can be used not only to order investigations for the benefit of human fraud investigators, but also to automatically and pre-emptively deactivate the accounts of high priority cases, and to filter very low priority cases out of consideration entirely. Thus, not only will the time and resources of a fraud-detection work-center be better employed, but the overall accuracy of the efforts of the work-center will improve, detecting more fraud, with more precision, sooner than with conventional threshold-counting systems.

References

[1] B. P. Yuhas, Toll-Fraud Detection, *Proceedings of the International Workshop on Applications of Neural Networks to Telecommunications*, ed. J. Alspector, R. Goodman, and T.X. Brown, Lawrence Erlbaum Associates, 1993, pp. 239–244.

Subject Index